SCIENTIFIC BASIS OF FERTILITY REGULATION

Cellular and molecular events in spermiogenesis

SCIENTIFIC BASIS OF FERTILITY REGULATION

Cellular and molecular events in spermiogenesis

Proceedings of a Symposium held at Oaxtepec, Mexico, 11–13 March 1987

EDITED BY

D. W. HAMILTON

Department of Cell Biology and Neuroanatomy, University of Minnesota, Minneapolis, USA.

G. M. H. WAITES

Special Programme of Research, Development and Research Training in Human Reproduction, World Health Organization, Geneva, Switzerland.

The right of the University of Cambridge to print and sell all manner of books was granted by Henry VIII in 1534. The University has printed and published continuously since 1584.

Published on behalf of
the World Health Organization

CAMBRIDGE UNIVERSITY PRESS

CAMBRIDGE

NEW YORK · PORT CHESTER

MELBOURNE · SYDNEY

Published by the Press Syndicate of the University of Cambridge
The Pitt Building, Trumpington Street, Cambridge CB2 1RP
40 West 20th Street, New York, NY10011, USA
10 Stamford Road, Oakleigh, Melbourne 3166, Australia

First published 1990

Printed in Great Britain by The Bath Press, Avon

British Library cataloguing in publication data
Cellular and molecular events in spermiogenesis:
proceedings of a symposium held at Oaxtepec, Mexico,
11–13 March 1987.
1. Mammals. Spermatogenesis
I. Hamilton, D. W. II. Waites, G. M. H. III. World
Health Organization IV. Series
599.03′2

Library of Congress cataloging-in-publication data
Cellular and molecular events in spermiogenesis:
scientific basis of fertility regulation: proceedings of a symposium held at Oaxtepec,
Mexico, 11–13 March 1987 / edited by D.W. Hamilton, G.M.H. Waites.
p. cm.
Proceedings of a symposium sponsored by the WHO Special Programme
of Research, Development and Research Training in Human
Reproduction and Family Health International.
"Published on behalf of the World Health Organization."
ISBN 0-521-37265-8
1. Spermiogenesis in animals—Congresses. I. Hamilton, David,
1935 Nov. 29– II. Waites, G. M. H. III. Special Programme of
Research, Development, and Research Training in Human Reproduction
(World Health Organization) IV. Family Health International
(Organization) V. World Health Organization.
[DNLM: 1. Spermatogenesis—congresses. 2. Spermatogenesis—drug
effects—congresses. WJ 834 C393 1987]
QL966.C45 1989
691.1′88—dc20 89–22248 CIP

ISBN 0 521 37265 8

MU

Contents

Preface

The WHO Special Programme of Research, Development and Research Training in Human Reproduction has been pleased to co-sponsor this Symposium with Family Health International. It is the first in a series of Basic Science Symposia to be supported jointly with other agencies.* The topics selected for these Symposia will represent areas of basic biomedical science perceived to have relevance for the development of new applications in the regulation of fertility and in the understanding of infertility.

Despite considerable research over the past several decades, there presently is no systemic method of male contraception that is safe, highly effective and reversible. Over 30 years ago, tumor-inhibiting substances, such as tretamine, were studied for their antispermatogenic properties. Since that time, numerous classes of compounds ranging from sulfonic esters and sulfamates to chlorinated hydrocarbons and gossypol have been examined. Only a very few have reached the stage of clinical testing. While hormonal regulation of male fertility with the use of steroids or LHRH analogues, or both, has achieved some degree of success in limited clinical trials, its widespread applicability still depends upon considerable further development.

Clearly, a totally novel approach to finding ways to control male fertility is needed. Thus, it was considered important to bring together experts from various fields outside that of reproductive biology to discuss the latest findings in cellular and molecular biology in relation to the regulatory mechanisms of mammalian spermiogenesis. It is expected that this interdisciplinary approach will provide new insights and may lay the foundation for the development of a modern male contraceptive for use throughout the world.

Many of the cellular and molecular events that occur during spermiogenesis are unique and yet are essential for the production of fertile sperm. From the precision with which the cellular changes occur in spermiogenesis it is clear that there are control mechanisms of which we have no knowledge at present. The specific properties of spermatids could offer

* The second symposium, which took place 22–24 September 1988, was entitled "Potentials of Molecular Biology in Fertility Regulation: Growth Regulatory Factors" and was co-sponsored with the U.S. Center for Population Research of the National Institute of Child Health and Human Development.

vulnerable points for targeted intervention without generalized effects on the early stages of spermatogenesis, including the genome of the developing germ cells. A drug intervening with one or other of the unique processes in spermatids would have several advantages, offering a specific action which should be safe, rapid in onset and reversible.

The WHO Special Programme and FHI want to acknowledge the support of the Population Council and, in particular, the hospitality of the Mexican hosts, at the Instituto Nacional de la Nutricion and the Department of Social Security which provided the conference facilities.

I Introduction and overview

Spermiogenesis: An introduction to the Symposium

DAVID W. HAMILTON

Department of Cell Biology and Neuroanatomy, University of Minnesota, Minneapolis, MN 55455, USA.

Introduction

Spermiogenesis* is the final phase of differentiation of male germ cells, just prior to their release from the seminiferous epithelium. During spermiogenesis the round, relatively undifferentiated haploid germ cells, the spermatids, undergo enormously complex morphological, biochemical and physiological changes that result in the formation of highly asymmetrical, flagellated spermatozoa. Many of the changes that occur during this time are unique events that do not happen in any other cell type in the body and yet are essential for the production of fertile sperm. Spermiogenesis, therefore, offers the potential for targeted intervention in fertility control in a manner that may not have generalized effects in the body and which may not have deleterious effects on the genome.

Spermiogenesis in mammals is poorly understood, even though the basic morphological events have been known for many years (Leblond & Clermont, 1952; Clermont & Leblond, 1955); it is only through the recent development of experimental techniques in cellular and molecular biology that it has been possible to study spermiogenesis in sufficient detail necessary to begin to understand it. The relative paucity of knowledge about spermiogenesis is in stark contrast to the explosive increase in knowledge in biology at the cellular and molecular levels in other mammalian cells and organs that has occurred over the past decade.

Among the many transforming events occuring during spermiogenesis that are unique to the differentiating haploid cell, two events – formation of the acrosome and formation of the sperm tail – have been chosen as the major focuses for this Symposium. In order to provide a general background for the topics that comprise the Symposium an overview of spermiogenesis is appropriate, not only to highlight important events in the process, but also to point out areas in which focused research at a basic cellular and molecular level might provide leads for development of contraceptive agents.

* The term spermateliosis has also been used as a designation for spermiogenesis (Bishop & Walton, 1966), particularly by European workers (e.g., Ploen, 1971). We have adopted the term spermiogenesis for the purposes of this Symposium.

Morphology of spermiogenesis

Spermiogenesis is a relatively long process in mammals occupying for example ~ 13.5 days in mice (Monesi *et al.*, 1978), and ~ 22 days in humans (Heller & Clermont, 1964). Although it is a continuous process, for descriptive and analytical purposes it can be divided conveniently into a number of discrete morphological steps (Fig. 1) based on structural modifications that occur in the developing acrosome (Leblond & Clermont, 1952; Clermont & Leblond, 1955). The number of discrete steps varies from species to species (see the chapter by Clermont *et al.* in this volume), but in all species they can be grouped into four major stages: Golgi, cap, acrosome and maturation.

Golgi phase

The earliest spermatids (Fig. 2) are round and somewhat smaller than the secondary spermatocytes from which they develop. Their nuclei are also round and occupy about half the volume of the cell. The cytoplasm, although undistinguished at the light microscopical level, contains abundant organelles when viewed in the electron microscope that give indications of development of the polarity that is characteristic of fully formed sperm.

The most obvious morphological changes that occur during this phase involve the Golgi apparatus–lysosomal complex, the differentiation of the flagellum and disposition of mitochondria within the cell.

The Golgi apparatus–lysosomal complex in spermatids is structurally much more complex than is typical of cells in general (Figs. 4 and 5). Early in this phase a number of lysosomal vesicles found near the *cis* face of the Golgi apparatus begin to form granules. These vesicles coalesce to form the acrosomal vesicle and the granules also coalesce to form the acrosomal granule. At an indeterminate time the membrane of the vesicle attaches to the nuclear envelope (Figs. 4 and 5). The mechanisms that determine which lysosomal vesicles will become the acrosome, that regulate the synthesis of the acrosomal granule, the factors that initiate the attachment of the acrosomal vesicle to the nuclear envelope and the actual identification of the 'glue' that keeps the vesicle and nucleus together are completely unknown, but are clearly of great importance in spermatogenesis, for without an acrosome, a mammalian sperm is not able to fertilize.

At about the time acrosomal vesicle development begins, the centrioles, which generally can be found near the Golgi apparatus (although not in the intimate relationship often seen in other cells), translocate to the pole of the cell opposite to the developing acrosome, thus establishing the longitudinal polarity of the cell. The distal centriole elaborates the flagellum, and the proximal centriole, which also elaborates a short-lived extension of microtubules (e.g., Fawcett, 1972), attaches to the nuclear envelope in the implantation fossa and eventually will form the connecting piece in the

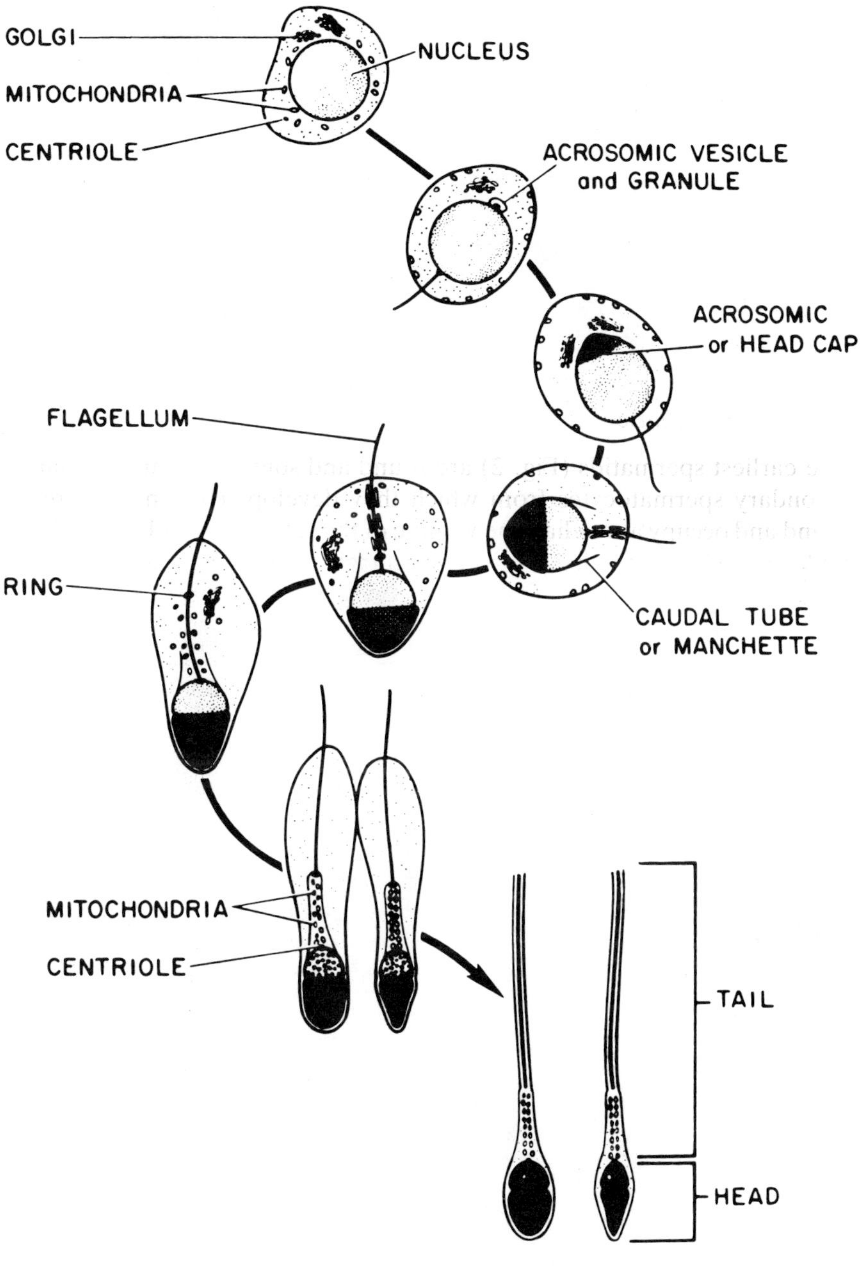

Fig. 1 Diagrammatic representation of spermiogenesis in the human (modified from Clermont & Leblond, 1955).

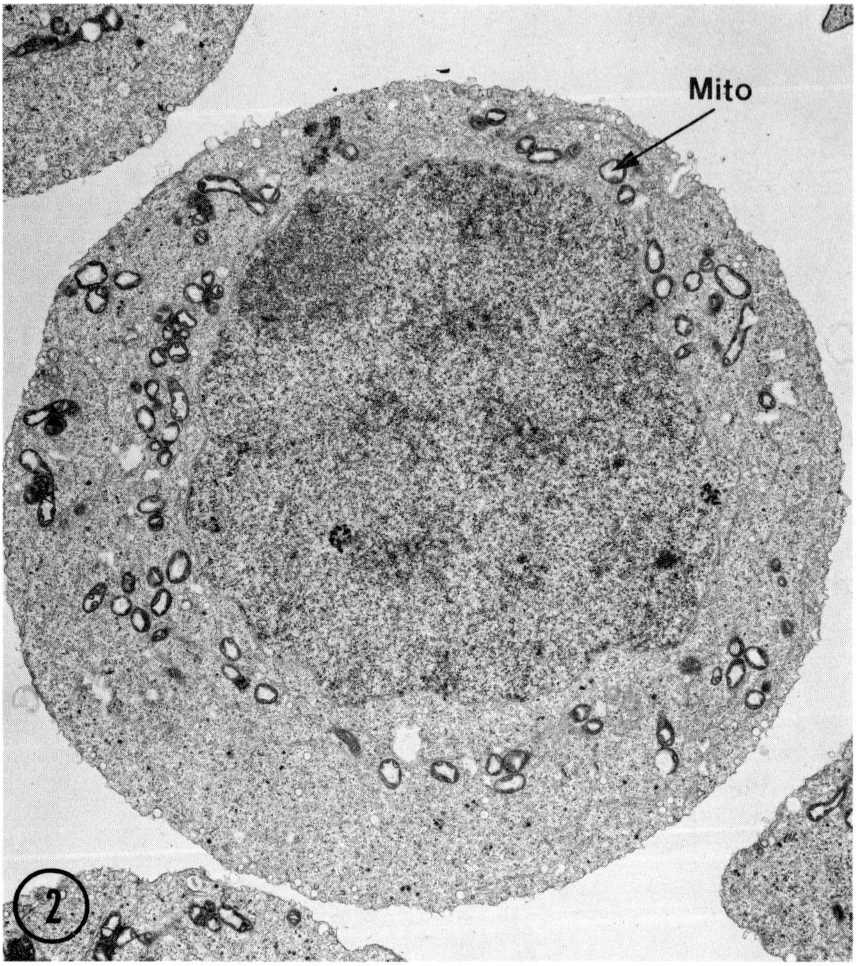

Figs. 2–6 are illustrations of cells isolated from rat testis by unit gravity sedimentation through a 2–4% bovine serum albumin gradient using the technique of Wolgemuth *et al.* (1985). This technique, which is reasonably quick and simple, yields highly purified populations of spermatids (85–95% pure) that can be used for structural, biochemical or physiological experiments.

Fig. 2 Round spermatid. Mitochondria (Mito) are grouped around the nucleus. The cytoplasm contains numerous free ribosomes, scattered membrane elements and a few microtubules. × 7 400.

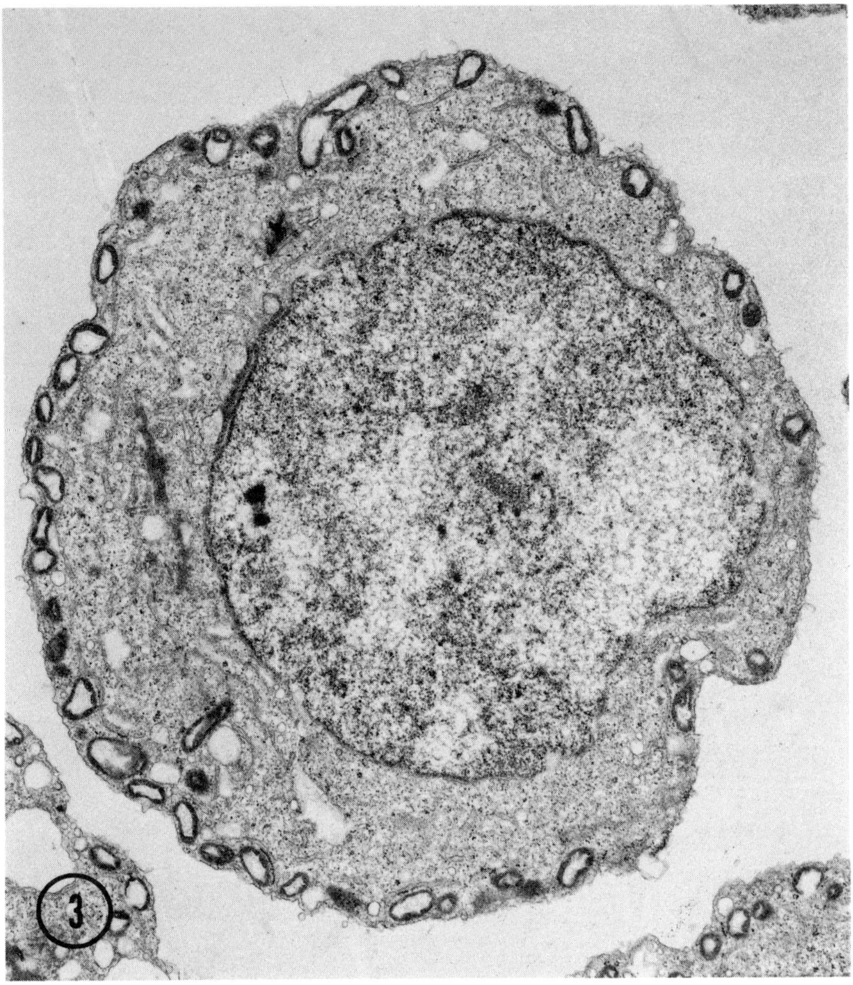

Fig. 3 Round spermatid in which the mitochondria have moved to the periphery of the cell, to lie just deep to the plasma membrane. × 9 500.

neck region of the sperm (Fawcett & Phillips, 1969; Irons, 1983). Here, again, the stimuli and mechanisms for centriolar translocation, attachment to the nucleus and formation of the neck region are unknown, although it is known that defects in the process can lead to infertility in humans (Perotti *et al.*, 1981; Holstein *et al.*, 1986).

A third obvious morphological event involves mitochondria. At the start of spermiogenesis mitochondria are grouped around the nucleus (Fig. 2). During the Golgi phase these disperse and come to lie at the periphery of

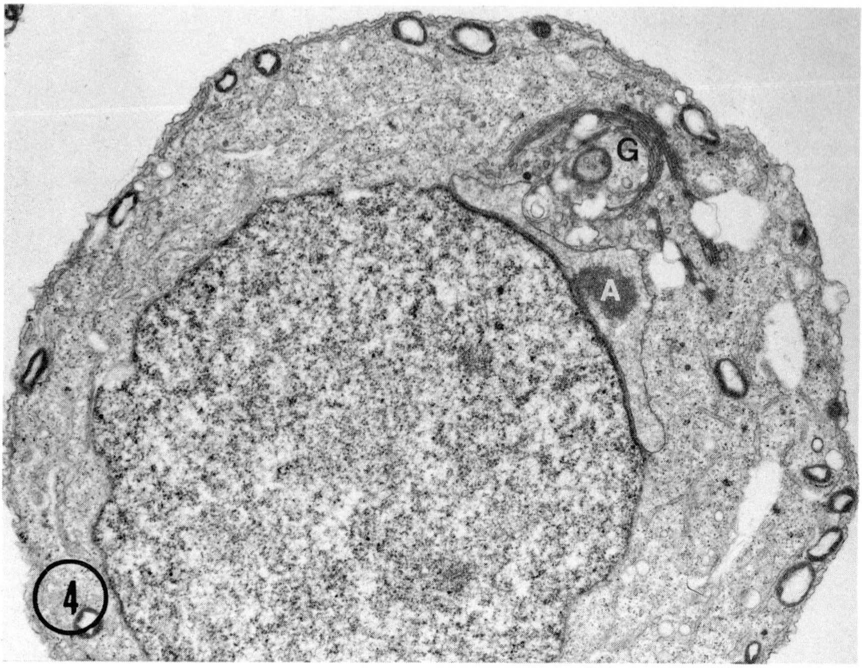

Fig. 4 Late Golgi phase spermatid. The acrosomal granule (A) and developing acrosome are obvious at this stage, and the complexity of the Golgi apparatus (G) is also apparent. × 7 900.

the cell, just deep to the plasma membrane (Figs. 3–5). In many other cell types mitochondrial distribution in the cytoplasm seems to be random. Why they are non-randomly distributed in early spermatids is not known and why they migrate to the periphery, only to migrate again to take positions around the flagellum (see below), is equally mysterious.

Cap phase

Shortly after the acrosomal vesicle attaches to the nuclear envelope the attachment site begins to extend distally towards the caudal end of the cell, resembling a partially inflated balloon into which a blunt object has been pushed, until it covers almost half the nuclear surface. The factors that govern the extension of the acrosomal vesicle are unknown.

Acrosome phase

During this phase the acrosome completes its morphological differentiation, mitochondria migrate from the cell periphery and come to lie next to the

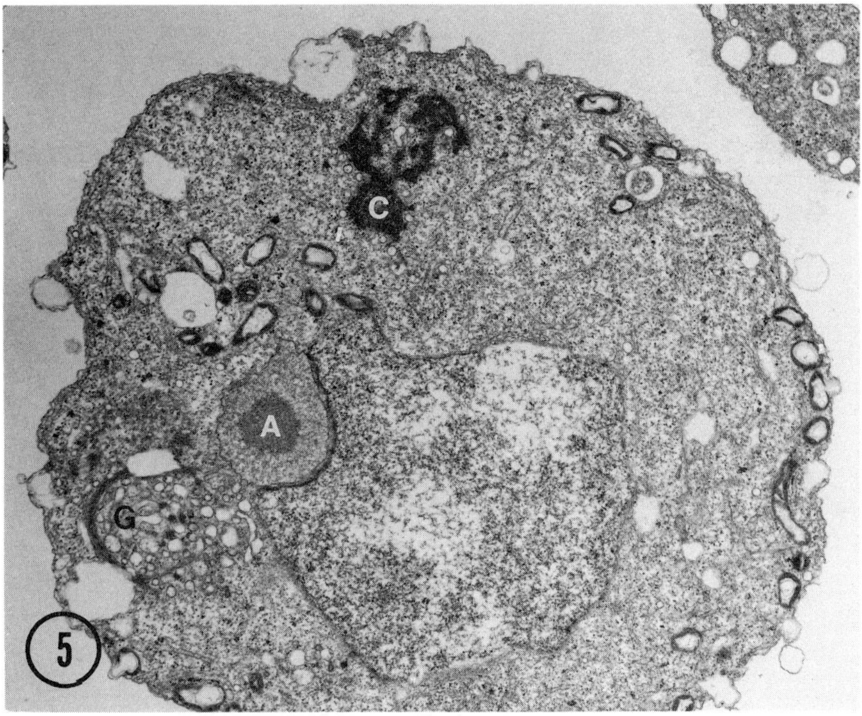

Fig. 5 Spermatid at approximately the same stage of development as that shown in Fig. 4, but here the chromatoid body (C) is present. × 7 400.

developing flagellum, and the nucleus becomes condensed. It is during this phase, also, that the spermatid begins to assume the elongated shape characteristic of a mature sperm (Fig. 6).

Although the above are major events during this stage of differentiation, they are coupled to a number of other events that are also important in forming sperm. Thus, associated with the movement of the mitochondria towards the developing axoneme is development of the annulus (Phillips, 1974), which will eventually delimit the lower border of the mitochondrial sheath. Also, at this time there is elaboration of outer dense fibers associated with the axoneme (Irons & Clermont, 1982a), and the fibrous sheath (Irons & Clermont, 1982b), which develops distal to the annulus.

An event that has been little considered in spermiogenesis is the structural bonding that occurs among mitochondria to form the mitochondrial sheath, although it has been known that mitochondria are arranged very precisely in a helix around the axoneme, and the number of mitochondria in the sheath appears to be genetically determined (Woolley & Beatty, 1967). The whole topic of regulation of assembly of the mitochondrial sheath is poorly

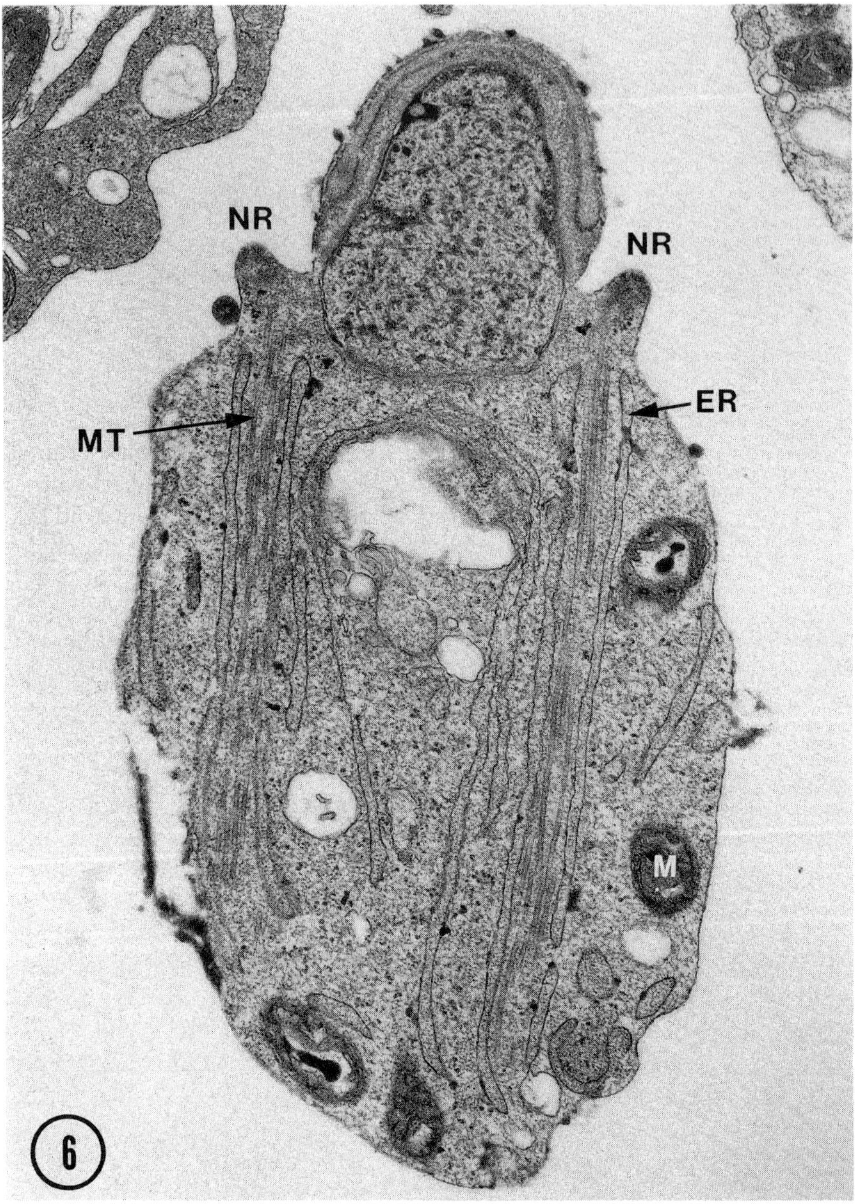

Fig. 6 Elongating spermatid in the acrosome phase. The condensing nucleus (at the top of the figure) and acrosome are clearly aligned along the long axis of the cell. Microtubules (MT) comprising the manchette arise from densities on either side of the base of the nucleus. These densities have been designated nuclear ring (NR), since if the two dimensions of

understood and yet must be of great importance since mitochondria generate energy for the flagellar beat.

In addition to the processes that lead to permanent structures in sperm, this phase of spermiogenesis is also the time at which both the manchette and the chromatoid body appear, and then quickly disappear. The manchette is an ordered array of microtubules that surround the posterior region of the developing sperm head (Fig. 6). It arises from a ring of dense material just distal to the acrosome and extends caudally along the first part of the developing sperm tail (Fig. 6), and frequently elements of smooth endoplasmic reticulum are closely associated with the manchette. The function of this organelle is unknown: it has been implicated in shaping of the sperm head (Rattner & Brinkley, 1972), although Fawcett *et al.* (1971) concluded from a comparative study that this probably is not its function.

The chromatoid body is another transient organelle that may arise in spermatocytes (Eddy, 1970; Fawcett *et al.*, 1970). It disappears in spermatids after having migrated along the length of the midpiece in concert with the annulus (Fawcett *et al.*, 1970). Again, the function of the chromatoid body is not known, but evidence indicates both that it is derived from the nucleus (Comings & Okada, 1972) and that it is not derived from the nucleus (Fawcett *et al.*, 1970), and it has been reported both to contain (Sud, 1961) and not to contain (Eddy, 1970) ribonucleic acid.

By the end of the acrosome phase the major morphogenetic events in formation of sperm have occurred and seemingly all that is left to do is 'fine tuning' to produce the final shape of the cell during the maturation phase.

Maturation phase

This phase of spermiogenesis involves primarily 'house-cleaning' events. Thus, the disposition of excess cytoplasm by Sertoli cells results in formation of residual bodies, and the final elongated shape of the spermatid is achieved. For all intents and purposes the spermatid at this stage looks like a fully-formed sperm. However, as is well-known (e.g., Bedford, 1975) testicular sperm are not fertile and other maturational processes must occur in the epididymis to produce a cell able to fertilize an egg.

Caption for Fig. 6 (cont.)

the micrograph are extended into three dimensions the nuclear ring would encircle the base of the nucleus. In this isolated cell preparation the nuclear ring appears to be a membrane modification not closely related to the nucleus. The density resembles that found in other microtubule organizing centers. The manchette is surrounded on both its inner and outer surfaces by highly organized elements of smooth endoplasmic reticulum (ER). M, mitochondria. × 28 500.

Other aspects of spermiogenesis

The morphological observations given above comprise only a part of spermiogenesis. There are discrete events at the biochemical and physiological levels that also may be unique to spermatids and thus may be amenable to control. For instance, synthesis of testis-specific, stage-specific proteins that associate with chromatin (Hecht's Chapter, this volume), or other stage-specific proteins (Kramer & Erickson, 1982; O'Brien, 1987), such as testis-specific/spermatid-specific isoforms of enzymes, would be very attractive targets for research on control of their synthesis. Equally, understanding control of stage-specific lipid synthesis (Beckman *et al.*, 1978) or of membrane events such as fucosylation (Grootegoed *et al.*, 1982) could yield strategies for intervention in fertility. In large part, topics similar to these are the focus of this Symposium, and will not be expanded on here.

Summary and relevance to the Symposium

While the review given above is not exhaustive, it is clear that the precision with which the events of spermiogenesis occur is awe-inspiring and suggests that there are exquisite control mechanisms at the cellular level of which we at present have no knowledge. Since the majority of morphogenetic changes that occur during spermiogenesis are found only in spermatids, it seems likely that deeper understanding of the basic biology of these cells will yield important information that can be used in developing strategies for fertility regulation in the male.

The purposes of this Symposium were twofold:

> to acquaint reproductive biologists with the current status of research in cellular and molecular biology of the Golgi apparatus–lysosomal complex and of centrioles and flagellae, and
>
> to acquaint non-reproductive biologists with some of the interesting and important problems that could be addressed using spermatids as model cells.

In the chapters that follow various aspects of acrosome/Golgi function, flagellar structure/tubulin assembly and molecular biology of spermatids are presented. At the end of the sessions discussions were held to attempt to identify research strategies that might yield information useful in the development of male antifertility drugs.

References

Beckman, J.K., Gray, M.E. & Coniglio, J.G. (1978). The lipid composition of isolated rat spermatids and spermatocytes. **Biochimica Biophysica Acta** 530:367–374.

Bedford, J.M. (1975). Maturation, transport, and fate of spermatozoa in the epididymis. (in) **Handbook of Physiology, Section 7: Endocrinology, Volume V: Male Reproductive System**. (D.W. Hamilton & R.O. Greep, eds.):303–318.

Bishop, M.W.H. & Walton, A. (1966). Spermatogenesis and the structure of the mammalian spermatozoa. (in) **Marshall's Physiology of Reproduction** (A.S. Parkes, ed.), Vol I, part 2:1–129.

Clermont, Y. & Leblond, C.P. (1955). Spermiogenesis of man, monkey, ram and other mammals as shown by the "periodic acid-Schiff" technique. **American Journal of Anatomy** 96:229–254.

Comings, D.E. & Okada, T.A. (1972). The chromatoid body in mouse spermatogenesis: Evidence that it may be formed by the extension of nucleolar components. **Journal of Ultrastructure Research** 39:15–23.

Eddy, E.M. (1970). Cytochemical observations on the chromatoid body of male germ cells. **Biology of Reproduction** 2:114–128.

Fawcett, D.W. (1972). Observations of cell differentiation and organelle continuity in spermatogenesis. (in) **International Symposium on the Genetics of the Spermatozoon** (R.A. Beatty & S. Gluecksonn-Waelsh, eds.) pp. 37– 68.

Fawcett, D.W. & Phillips, D.M. (1969). The fine structure and development of the neck region of the mammalian spermatozoon. **Anatomical Record** 165:153–183.

Fawcett, D.W., Eddy, E.M. & Phillips, D.M. (1970). Observations on the fine structure and relationships of the chromatoid body in mammalian spermatogenesis. **Biology of Reproduction** 2:129–153.

Fawcett, D.W., Anderson, W.A. & Phillips, D.M. (1971). Morphogenetic factors influencing the shape of the sperm head. **Developmental Biology** 26:220–251.

Grootegoed, J.A., Krueger-Sewnarain, B.C., Jutte, N.H.P.M., Rommerts, F.F.G. & van der Molen, H.J. (1982). Fucosylation of glycoproteins in rat spermatocytes and spermatids. **Gamete Research** 5:303–315.

Heller, C.G. & Clermont, Y. (1964). Kinetics of the germinal epithelium in man. **Recent Progress in Hormone Research** 20:545–575.

Holstein, A.F., Schill, W.B. & Bruecker, H. (1986). Dissociated centriole development as a cause of spermatid malformation in man. **Journal of Reproduction and Fertility** 78:719–725.

Irons, M.J. (1983). Synthesis and assembly of connecting-piece proteins as revealed by radioautography. **Journal of Ultrastructure Research** 82:27–34.

Irons, M.J. & Clermont, Y. (1982a). Formation of the outer dense fibers during spermiogenesis in the rat. **Anatomical Record** 202:463–471.

Irons, M.J. & Clermont, Y. (1982b). Kinetics of fibrous sheath formation in the rat spermatid. **American Journal of Anatomy** 165:121–130.

Kramer, J.M. & Erickson, R.P. (1982). Analysis of stage-specific protein synthesis during spermatogenesis of the mouse by two-dimensional gel electrophoresis. **Journal of Reproduction and Fertility** 64:139–144.

Leblond, C.P. & Clermont, Y. (1952). Spermiogenesis of rat, mouse, hamster and guinea pig as revealed by the periodic acid-fuchsin sulfurous acid technique. **American Journal of Anatomy** 90:167–215.

Monesi, V., Geremia, R., D'Agostino, A. & Boitani, C. (1978). Biochemistry of male germ cell differentiation in mammals: RNA synthesis in meiotic and post-meiotic cells. (in) **Current Topics in Developmental Biology** 12:11–36.

O'Brien, D. (1987). Stage-specific protein synthesis by isolated spermatogenic cells throughout meiosis and early spermiogenesis in the mouse. **Biology of Reproduction** 37:147–157.

Perotti, M.E., Giarola, A. & Gioria, M. (1981). Ultrastructural study of the decapitated sperm defect in an infertile man. **Journal of Reproduction and Fertility** 63:543–549.

Phillips, D.M. (1974). **Spermiogenesis**. Academic Press Inc., New York and London. pp.+68.

Ploen, L. (1971). A scheme of rabbit spermateliosis based upon electron microscopical observations. **Zeitschrift fuer Zellforschung und Mikroskopische Anatomie** 115:553–564.

Rattner, J.B. & Brinkley, B.R. (1972). Ultrastructure of mammalian spermiogenesis. III. The organization and morphogenesis of the manchette during rodent spermiogenesis. **Journal of Ultrastructure Research** 41:209–218.

Sud, B.P. (1961). Morphological and histochemical studies of the chromatoid body and related elements in the spermatogenesis of the rat. **Quarterly Journal of Microscopical Science** 102:495–505.

Wolgemuth, D.J., Gizang-Ginsburg, E., Engelmyer, E., Gavin, B.J. & Ponzetto, C. (1985). Separation of mouse testis cells on a Celsep^{tm} apparatus and their usefulness as a source of high molecular weight DNA and RNA. **Gamete Research** 12:1–10.

Woolley, D.M. & Beatty, R.A. (1967). Inheritance of mid-piece length in mouse spermatozoa. **Nature** 215:94–95.

II The Golgi apparatus, lysosomes and acrosome formation

Structural and cytochemical characteristics of the Golgi apparatus during the formation of the acrosomic system in rat spermatids

YVES CLERMONT*, LOUIS HERMO*, A. RAMBOURG** AND G. THORNE-TJOMSLAND*

*Department of Anatomy, McGill University, 3640 University Street, Montreal, Canada H3A 2B2 and **Département de Biologie, CEA, Saclay, France.

Abstract

The acrosomic system, containing lysosomal enzymes, is formed by the Golgi apparatus of rat spermatids during steps 1–7 of spermiogenesis. The spheroidal or hemispherical Golgi apparatus shows a cortex composed of stacks of saccules bridged by membranous tubules and a medulla containing many vesicles of various sizes. In steps 1–2 spermatids, several proacrosomic granules form and coalesce to give a single acrosomic granule which is attached to the nucleus at step 3. In the following steps, 4–7, the acrosomic granule grows by addition of material elaborated by the associated Golgi apparatus and gives rise to the acrosome and headcap (acrosomic system). During step 8, the Golgi apparatus separates from the acrosomic system, which continues to change in shape as the nucleus, to which it is applied, elongates, condenses and becomes falciform.

The contribution of the components of the Golgi apparatus of steps 1–7 spermatids to the formation of the acrosomic system was investigated with the electron microscope in sections of glutaraldehyde-fixed testes post-fixed with potassium ferrocyanide-reduced osmium and in sections treated to demonstrate phosphatase activities (NADPase, TPPase and CMPase). While the *cis*-(osmiophilic)-element and the subjacent first saccule of the Golgi stacks were negative for the phosphatases tested, the next 5 or 6 saccules were NADPase-positive. Underlying these saccules, one flattened saccule was strongly reactive for TPPase and reacted slightly for CMPase. The last two elements on the *trans*-face of the stacks, elements that showed a slightly distended saccular portion continuous with a network of anastomotic tubules and which presented a 'peeling off' configuration, were NADPase and TPPase-negative but were strongly positive for CMPase. Attached to these *trans* sacculotubular elements, CMPase-positive buds were present. In the medulla several NADPase-positive, a few TPPase-positive and many CMPase-positive vesicles were observed. Some of these vesicles were attached to the acrosomal membrane seemingly delivering their content into the acrosomic system, which itself showed weak NADPase and TPPase but strong CMPase activity. These observations indicated, firstly, that the

3 types of saccular components of the Golgi stacks, positive for NADPase, TPPase and CMPase activity respectively, directly delivered their enzymatic content into the growing acrosomic system by means of carrier vesicles budding at their edges; secondly the CMPase-positive *trans* sacculotubular elements are seemingly continuously produced, peel off the stacks and fragment into CMPase-positive vesicles which migrate toward and merge with the developing acrosomic system.

Key words

Acrosome, Golgi apparatus, lysosome, spermiogenesis, spermatids.

Introduction

The formation of the acrosome of mammalian spermatids in close association with the Golgi apparatus was established early by microscopists despite the long debate concerning the existence or non-existence of this cytoplasmic organelle (Gatenby & Woodger, 1921; Bowen, 1926). Using the periodic acid–Schiff method to demonstrate the presence of carbohydrates in tissue sections Leblond (1950) observed the presence of sugars in the developing acrosomic system of rat spermatids. The formation and evolution of this structure was then analyzed in detail with the light microscope in the spermatids of rodents and other mammalian species (Leblond & Clermont, 1952; Clermont & Leblond, 1955). Using paper chromatography of acrosomal hydrolysates it was determined that the sugars responsible for the periodic acid–Schiff reactivity of guinea pig acrosomic system were galactose, mannose, fucose and hexosamines (Clermont *et al.*, 1955).

Leblond and collaborators, using the radioautographic method on sections of tissues from animals injected with radioactive sugars later demonstrated that one of the main functions of the Golgi apparatus was to complete the glycosylation of oligosaccharide side chains of glycoproteins (reviews by Leblond & Bennett, 1977; Bennett & Leblond, 1977; Bennett, 1984). While some of these glycoproteins are packaged in secretory granules or lysosomes, others are incorporated into the plasma membrane. The carbohydrate-rich acrosomic system is obviously formed as a result of a similar biosynthetic process taking place in the Golgi apparatus of early spermatids, although it should be remembered that this structure is neither a secretory granule nor a secondary lysosome. It appears as a peculiar primary lysosome which becomes tightly bound to the surface of the nucleus of the spermatid to be utilized by the spermatozoon at the time of fertilization. Indeed the acrosomic system contains a number of hydrolytic enzymes includinq acid phosphatases (Allison & Hartree, 1970; Morton, 1976) which are recognized glycoproteins.

The obvious association of the Golgi apparatus with the developing acrosomic system has been analyzed repeatedly in numerous electron microscopic studies since the initial investigation of the cat spermatid by Burgos & Fawcett (1955). However, despite these investigations, the detailed tridimensional architecture of the spermatids' Golgi apparatus, the cytochemical characteristics of its components as well as the exact mode of formation of proacrosomic and acrosomic structures from Golgi elements remained unclear. Over the past few years we have attempted to clarify some of these issues (Hermo *et al.*, 1979, 1980; Clermont *et al.*, 1981; Tang *et al.*, 1982; Clermont & Tang, 1985). The purpose of the present chapter is to summarize as well as update with unpublished data, the structure of the Golgi apparatus' components, their cytochemical characteristics as well as their exact role in the formation of the acrosomic system of rat spermatids.

Formation of acrosomic system in rat spermatids

Spermiogenesis, which lasts approximately 20 days in the rat (Clermont *et al.*, 1959) has been divided into four main phases i.e. the Golgi, Cap, Acrosome and Maturation phases, on the basis of the main structural changes of the acrosomic system and nucleus of spermatids (Leblond & Clermont, 1952). Each phase has been subdivided into steps; in the rat a total of 19 steps has been described during spermiogenesis and this classification is still used in electron microscope studies of differentiating spermatids (see Fig. 1 in Clermont & Rambourg, 1978 or Fig. 1 in the first chapter in this volume).

The Golgi apparatus is actively involved in the formation of the acrosomic system during the Golgi phase (steps 1–3) and Cap phase (steps 4–7) i.e. over a period of 7.5 days. During the Acrosome phase (steps 8–14) and Maturation phase (steps 15–19), the acrosomic system undergoes structural transformation and condensation, but without the contribution of the Golgi apparatus which has separated from it (Fig. 1).

In step 1 of spermiogenesis several minute proacrosomic granules containing electron dense material appear in the Golgi apparatus. These coalesce and result in fewer and larger proacrosomic granules which characterize step 2 (Fig. 1). Each proacrosomic granule shows an electron dense inner zone surrounded by a less dense outer zone (Fig. 2). In step 3, the proacrosomic granules fuse to yield a single acrosomic granule which establishes contact with the nucleus. As in the case of proacrosomic granules, this acrosomic granule contains a dense inner zone and less dense outer zone. During step 4 the acrosomic granule continues to increase in size and becomes hemispherical (Fig. 6). Toward the end of step 4 and during the following steps 5, 6 and 7 this growing acrosomic granule gives rise to a cap-like structure at the surface of the nucleus (Figs. 1, 3 and 6). In step 5 the inner dense zone, previously seen in the acrosomic granule, establishes

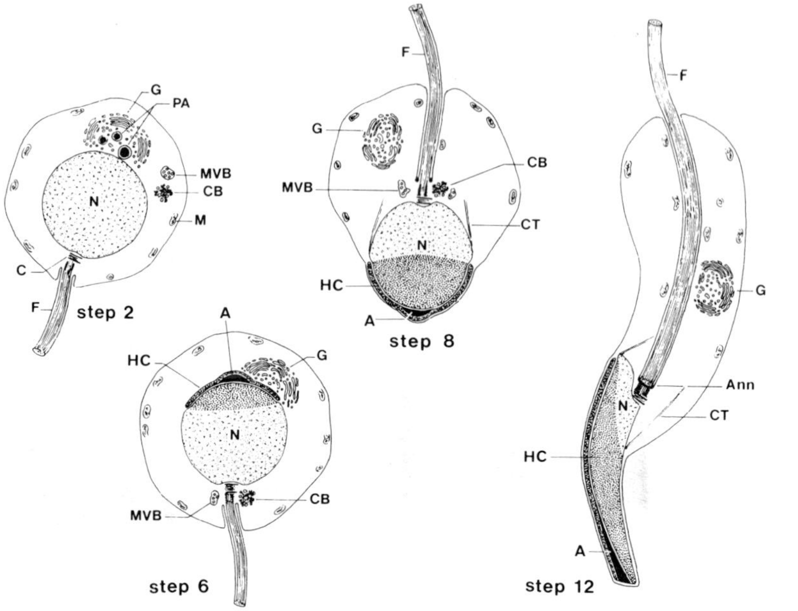

Fig. 1 Rat spermiogenesis has been subdivided into 19 steps on the basis of the changes taking place in the acrosomic system and nucleus of the spermatid (Leblond & Clermont 1952). The four diagrams represent steps 2, 6, 8 and 12 of spermiogenesis to illustrate the main changes taking place in the Golgi apparatus (G) and associated structures, i.e. proacrosomic granules, PA; acrosome system (acrosome, A; head cap, HC), multivesicular bodies (MVB) and chromatoid body (CB) during the evolution of the spermatid into a mature spermatozoon.

In step 2 proacrosomic granules (PA) form in the medulla of the juxta-nuclear hemispherical Golgi apparatus (G). Each proacrosomic granule shows a dense inner zone and a pale outer zone. These proacrosomic granules coalesce to form a single acrosomic granule which associates with the nucleus. The inner zone of the granule becomes associated with the inner acrosomal membrane to form the acrosome (A), while the outer zone due to the continuous addition of material from the Golgi apparatus grows at the surface of the nucleus to form the head cap (HC), as seen in step 6 spermatid. In step 8 of spermiogenesis the nucleus, partly covered by the acrosome and head cap, is now excentrically positioned within the cytoplasm and the plasma membrane is applied closely to the outer acrosomal membrane. The Golgi apparatus separates at this step from the acrosomic system, which stops growing, and occupies a position in the middle of the elongated cytoplasm next to the flagellum. Following step 8, the nucleus of the spermatid elongates to take a falciform shape as shown in step 12, and the acrosomic system, with its acrosome (A) seen along the dorsal edge of the nucleus and its head cap (HC) covering the lateral surfaces of the nucleus continues to change in shape.

The changes of the acrosomic system during the later steps of spermiogenesis, not illustrated here, are complex and have been described in detail in Lalli & Clermont (1981). Other labels: M, mitochondria; CT, caudal tube; Ann, annulus; F, flagellum; N, nucleus. Modified from Fig. 1 in Clermont & Tang (1981).

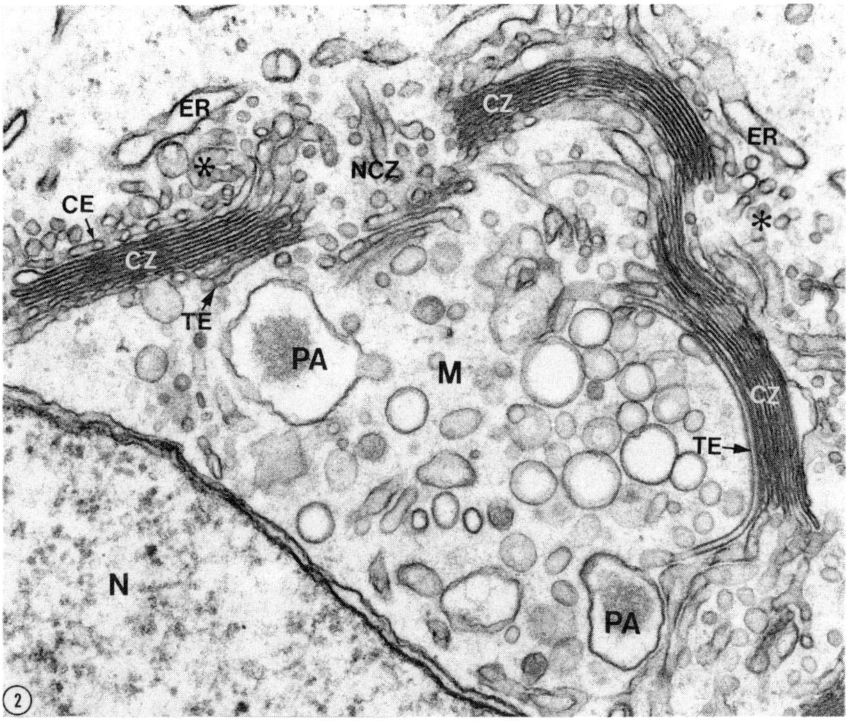

Fig. 2 EM photograph of step 2 spermatid showing two proacrosomic granules (PA) in the medulla (M) of the Golgi apparatus. The proacrosomic granules have a dense inner zone and a pale outer zone. The cortex of the Golgi apparatus has compact zones (CZ) made up of stacks of saccules and non-compact zones (NCZ) showing membranous tubules and vesicles. The *cis*-element (CE) on the *cis* face of the stacks of flattened saccules as well as the slightly thicker *trans* sacculotubular elements (TE) on the *trans* face of the stacks can be identified. The medulla (M) contains numerous vesicular or tubular profiles of various sizes. Between the ER cisternae (ER) seen at the periphery of the Golgi apparatus and the *cis*-element, small vesicles and membranous tubules are seen in what may be referred to as the peripheral Golgi region (asterisks). N, nucleus. × 62 500.

contact with the inner acrosomal membrane, i.e. the membrane which delimits the acrosomic apparatus facing the nuclear envelope, and takes a hemispherical shape without changing significantly in volume (Figs. 1, 3, 6 and 9). This inner zone may be considered to be the acrosome proper, i.e. a body (soma) at the top (akros) of the nucleus. In contrast, the outer zone of the acrosomic granule is the one which increases in volume and expands at the surface of the nucleus during steps 4–7 to form the head cap (Figs. 1, 3 and 6). The expression 'acrosomic system' is used to designate this derivative of the Golgi apparatus, since it is made up of an acrosome and head cap regions, which remain distinct entities in spermatids and spermatozoa (Lalli & Clermont, 1981).

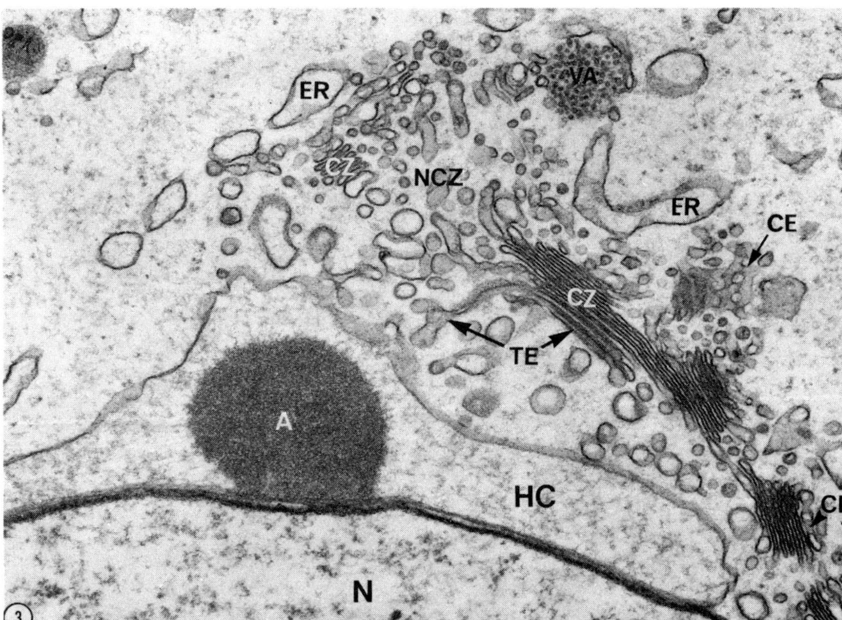

Fig. 3 Golgi apparatus and acrosomic system of a step 5 spermatid. The acrosomic system shows the electron dense acrosome (A) in contact with the inner acrosomal membrane which is applied to the surface of the nucleus (N). The head cap (HC), filled with wispy material extends at the surface of the nucleus. The acrosome, head cap and enclosing membrane form the acrosomic system. The following structures in the Golgi apparatus can be identified. Stacks of saccules forming the compact zones of the Golgi cortex (CZ) separated by non compact zones (NCZ); *cis*-element (CE); 'peeling off' *trans* sacculotubular elements (TE). A vesicular aggregate in the peripheral Golgi region is visible × 48 750.

In step 8 of spermiogenesis the nucleus migrates towards and becomes closely applied to the spermatid's plasma membrane (Fig. 1). As this takes place, the Golgi apparatus separates from the acrosomic structures and moves caudally to occupy a position in the mid-portion of the elongated cytoplasm next to the axoneme. Following this separation, although the Golgi apparatus remains actively involved in the biosynthesis of glycoproteins, it does not add carbohydrates to the acrosomic system (Clermont & Tang, 1985).

During the Acrosome and Maturation phases the acrosomic system undergoes marked structural transformations as the nucleus of the spermatid elongates, condenses and becomes falciform (Fig. 1). While the acrosome proper occupies the dorsal ridge of the nucleus, the head cap extends on its two lateral surfaces. Towards the end of spermiogenesis, this organelle becomes separated from the nuclear surface by a layer of electron dense material known as the perforatorium. A detailed description of the complex terminal transformations of the acrosomic system has been given by Lalli & Clermont (1981).

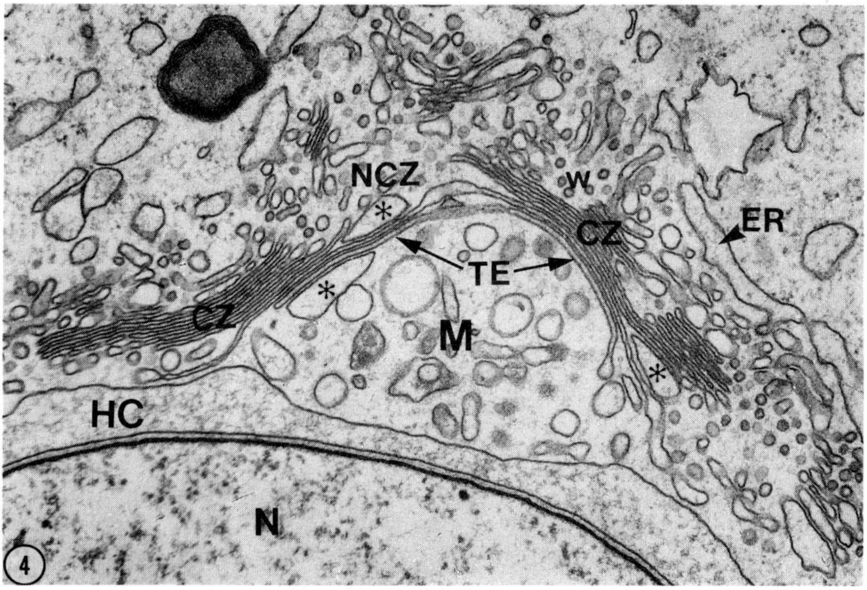

Fig. 4 Golgi apparatus of a step 6 spermatid. This photograph shows in addition to the various components of the compact (CZ) and non compact (NCZ) zones of the Golgi cortex, the characteristic association of cisternae of the endoplasmic reticulum (ER and asterisks) to the various saccules and *trans* sacculotubular elements (TE). N, nucleus; HC, head cap portion of the acrosomic system; M, medulla; W, well. × 37 500.

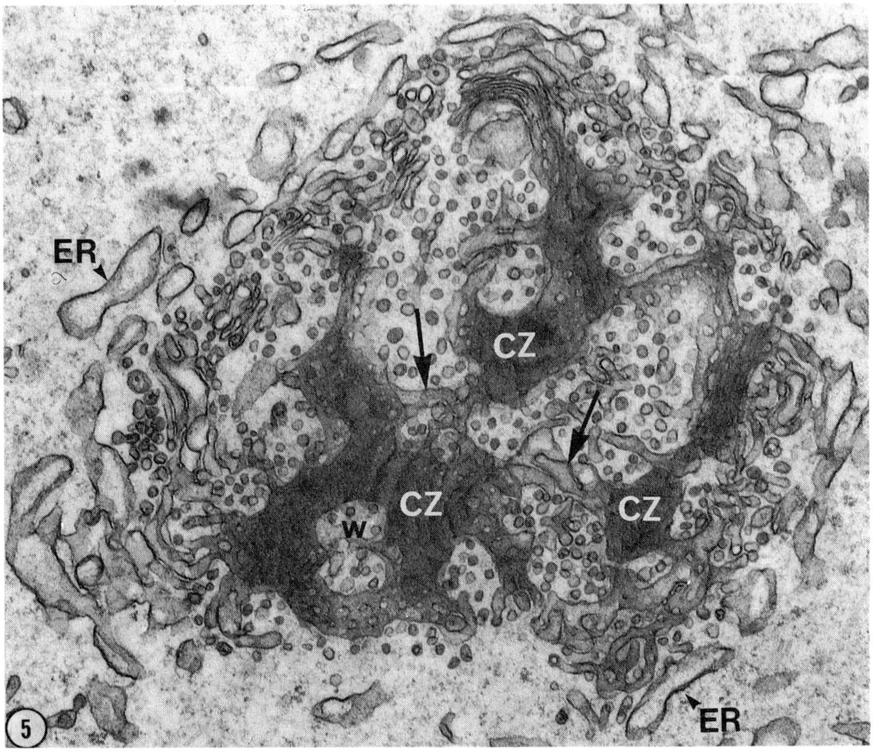

Fig. 5 A face view of the cortex of the Golgi apparatus of a step 2 spermatid. The compact zones (CZ) are stellate in shape and connected to each other by membranous tubules which occupy the non compact zones (arrows). The latter zones contain numerous small vesicles. ER, cisternae of endoplasmic reticulum. × 32 500.

Overall appearance of the Golgi apparatus of spermatids

Throughout the greater part of spermiogenesis, i.e. steps 1–16, the Golgi apparatus forms a compact organelle with two distinct regions referred to as cortex and medulla. The cortex is composed of stacks of saccules, bridging intersaccular regions made up of tubular structures and numerous small vesicles, while the medulla contains loosely arranged large vesicles, short membranous tubules and sacculotubular elements (Figs. 2, 3 and 4). The shape of the Golgi apparatus is generally spheroidal in step 1 spermatids. During the following 6 steps it becomes hemispherical in shape with the concavity of the hemisphere oriented toward the developing acrosomic system (Figs. 3, 4, 7 and 10). Thus the cortex assumes a dome-shaped appearance with the edges of the cupola approximating the acrosomic system and the medulla forming a cavity between the cortex and the

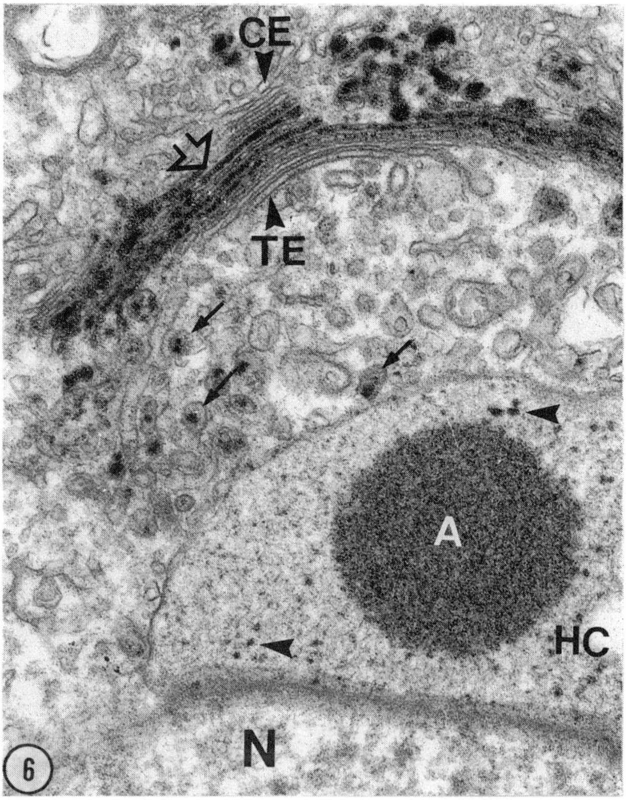

Fig. 6 Photograph showing the acrosomic system (acrosome, A; head cap, HC) and Golgi apparatus of step 4 spermatid stained to show NADPase activity. While saccules in the middle of the Golgi stack are reactive (large open arrows), the *cis* element (CE) and *trans* sacculotubular elements (TE) are not. Some vesicles in the medulla (small arrows) have some NADPase activity. Some reactivity is also observed in the head cap (arrowheads). × 38 000.

acrosomic elements. Following its separation from the acrosomic system, the Golgi apparatus resumes its spheroidal shape (Fig. 12). The organelle degenerates by vacuolation in step 16 spermatids, which is approximately 6 days before the completion of spermiogenesis.

Tridimensional structure of the cortex of the Golgi apparatus of steps 2–7 spermatids

The Golgi apparatus when viewed in sections cut perpendicular to its cortex, presents two distinct zones referred to as compact and non-compact zones (Figs. 2, 3, 4 and 7). The compact zones are composed mainly of superimposed flattened saccules and correspond to the stacks of saccules observed in the Golgi apparatus of most cell types. The non-compact zones, intercalated between the stacks of saccules, are made up of distended perforated saccules or membranous tubules which bridge the saccules of adjacent stacks, and of numerous small spherical vesicles, 60–80 nm in diameter (Figs. 2–4).

When the cortex is seen face view in tangential sections, the compact zones are stellate in shape and are separated from each other by the non-compact porous zones (Fig. 5). Clearly, the stacks of saccules of the compact zones are bridged by porous saccules, but more often by membranous tubules of the non-compact zones (Fig. 5). Numerous small vesicles are seen in the latter areas. Thus, the whole cortex is made up, with the exception of small vesicles, of continuous elements.

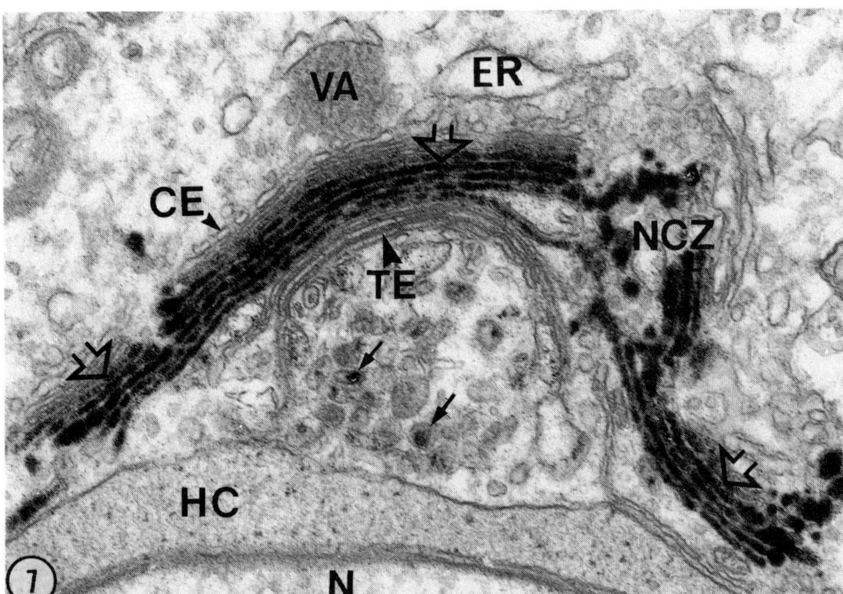

Fig. 7 Golgi apparatus of a step 6 spermatid treated to show NADPase activity. Saccules in the middle of the stack (open arrows) as well as tubules connected to them seen in the non compact zone (NCZ) are reactive. A few vesicles in the medulla contain reactive products (small arrows) and the head cap (HC) contains some discrete precipitate distributed throughout. A vesicular aggregate (VA) in the peripheral Golgi regions, the *cis*-element (CE) and subjacent saccule, as well as the *trans* sacculotubular elements (TE) are non reactive. ER, endoplasmic reticulum; N, nucleus. × 38 000.

Detailed structure of the Golgi cortex

The peripheral Golgi region

This region is the layer between the network of rough ER cisternae which is applied to the Golgi apparatus and the first, or *cis*, element of the stack of saccules. This layer contains small vesicles and membranous tubules. Some of the latter appear to be continuous with the *cis*-element of the stack or with the edges of the underlying saccules (Figs. 2 and 3). Vesicles in this region often form clusters (Fig. 3). While many vesicles at the periphery of these vesicular aggregates have a standard size of approximately 60–80 nm, the vesicles in the core of the aggregate are much smaller (30–40 nm) and their membranes often appear diffuse (Fig. 3). There is evidence that these vesicular aggregates are sites of disintegration of vesicles (Hermo *et al.*, 1980; Clermont, Y. & Hermo, L., unpublished observations). Characteristically, such vesicular aggregates are capped by ER cisternae

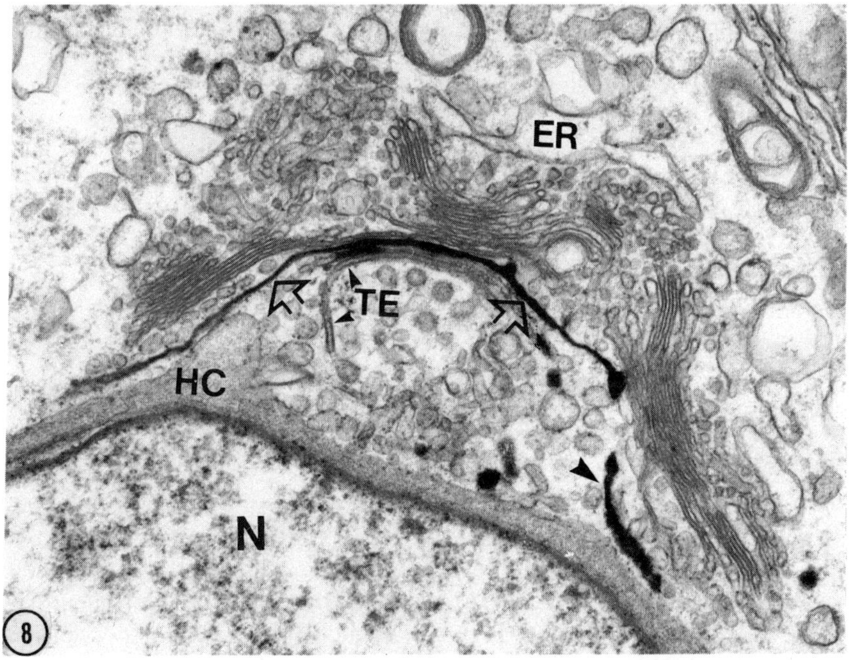

Fig. 8 Golgi apparatus of a step 6 spermatid stained to show TPPase activity. One saccule on the *trans* side of the stack is reactive (arrows), a portion of which (arrowhead) is seen next to the head cap (HC). The other Golgi elements and in particular the *trans* sacculotubular element (TE) are non reactive. N, nucleus; ER, endoplasmic reticulum. × 37 500.

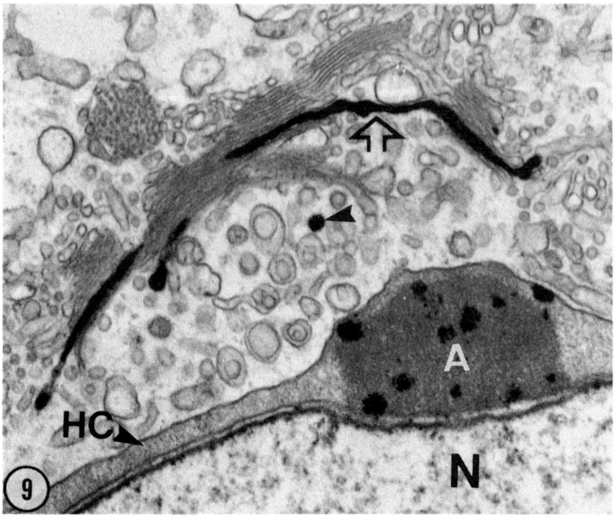

Fig. 9 Golgi apparatus and associated acrosomic system of a step 6 spermatid stained to showTPPase activity. One saccule in the Golgi cortex is reactive (arrow). Some spotty reaction is observed in the acrosome (A), while the head cap (HC) is non reactive. One vesicle in the medulla (arrowhead) is also reactive. N, nucleus. × 37 500.

(Figs. 3 and 7). ER cisternae frequently have fuzz coated protrusions along their membrane facing the Golgi apparatus (Figs. 4 and 13). These bud-like structures are usually interpreted as being the precursors of small vesicles transporting proteins from the endoplasmic reticulum to the Golgi elements. The actual existence of budding vesicles with a narrow stem are however rarely seen (Hermo *et al.*, 1980).

Structural and cytochemical characteristics of the Golgi elements of the compact zones

In side views of the compact zones, some 6–10 distinct stacked flattened elements, each delimited by a unit membrane, are present (Figs. 3 and 4). The first element on the *cis*-side of the stack appears as a layer of ovoid or elongated membranous profiles connected by electron dense bridges (Figs. 2, 3 and 7). In face view this element appears as a network of anastomotic tubules (Fig. 3). It corresponds structurally to the *cis*-osmiophilic element described in the Golgi apparatus of various cell types (Friend & Murray, 1965; Rambourg *et al.*, 1974). Subjacent to the *cis*-element, 3 to 5 flattened saccules delineated by paired parallel unit membranes have few pores or interruptions except for large perforations which are in register from one saccule to another (Figs. 4 and 7). These pan-shaped

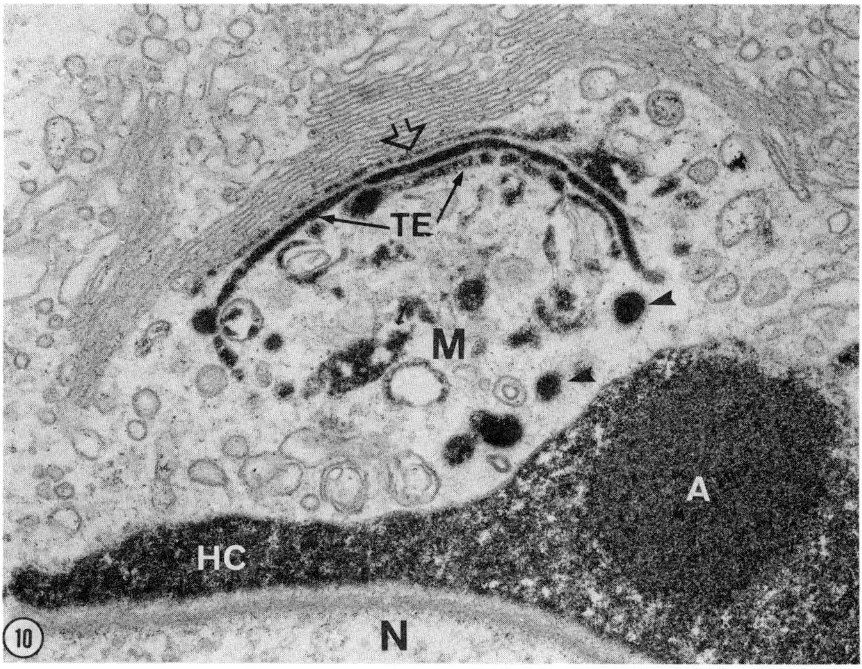

Fig. 10 Spermatid at step 5 of spermiogenesis treated to show CMPase activity in the Golgi apparatus and acrosomic system. The acrosome (A) and head cap (HC) have an intense and uniform reaction throughout. In the Golgi apparatus the slightly thickened *trans* sacculotubular elements (TE) have a strong reactivity for CMPase as well some of the vesicles (arrow-heads) in the medulla (M). One thinner saccule overlying the *trans* sacculotubular elements shows some reactivity (open arrow). N, nucleus. × 47 500.

cavities on the *cis*-side of the stack were called 'wells' by Hermo *et al.* (1980). The bottom of the wells are usually closed by a continuous saccule and contain numerous small spherical vesicles 60–80 nm in diameter (Figs. 4 and 7). In face views of the stacks the perforations forming the vesicle-containing wells can be identified readily and have a circular outline (Fig. 5). Initially described in the spermatid (Hermo *et al.*, 1980), such wells have been observed in the Golgi apparatus of several other cell types (Rambourg & Clermont, 1986; Rambourg *et al.*, 1987). Rambourg & Clermont (1986) working on the Golgi apparatus of ganglion cells presented evidence to indicate that such wells are seemingly the sites of formation of the small vesicles seen therein.

With the exception of the *cis* element and the underlying first saccule, all the saccules involved in the formation of the wells plus two or more saccules in the middle of the stack are NADPase positive (Figs 6 and 7; see also Clermont *et al.*, 1981).

Subjacent to these NADPase-positive saccules, i.e. on the *trans*-side of the stacks, there are some 3 or 4 saccules that have a 'peeling off' configuration (Figs. 2–4). One of these saccules subjacent to the NADPase-positive saccules is flattened and has only minute pores when seen in face view (Clermont *et al.*, 1981). This saccule presents a strong TPPase-reaction (Figs. 8 and 9) and a light CMPase-reactivity (Figs. 10 and 11). The remaining 2 or 3 'peeling off' saccules on the *trans*-side of the stacks, when seen in side view, have a lumen which is slightly distended that contains some finely granular material (Figs. 2–4). These distended saccules are always strongly reactive for CMPase (Figs. 10 and 11). These elements usually have a saccular portion continuous with an anastomotic tubular part and hence are referred to as trans sacculotubular elements (Rambourg *et al.*, 1979). Not infrequently one or more *trans*-sacculotubular elements may be seen completely separated from the overlying stacks and found free in the medulla of the Golgi apparatus where they appear to fragment. CMPase-positive *trans*-sacculotubular elements frequently have coated buds at their edges (Figs. 3, 4, 10, 11 and 13).

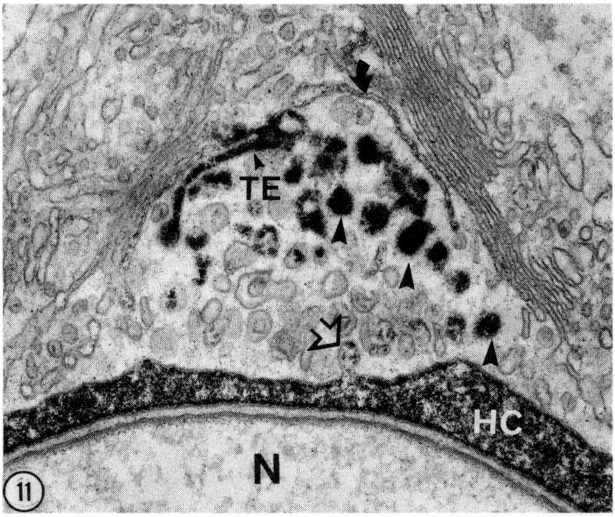

Fig. 11 Spermatid at step 6 treated to show CMPase activity. A *trans* sacculotubular element (TE), some vesicles in the medulla (arrowheads) as well as the head cap of the acrosomic system (HC) are strongly reactive. One CMPase-reactive vesicle is connected to the head cap (open arrow). Another 'peeling off' saccule on the *trans* face of the stack shows a weak reaction (curved arrow), while all other elements of the Golgi cortex are nonreactive. N, nucleus. × 47 500.

Structure of the non-compact zones of the Golgi cortex

All the saccular elements of the compact zone just described are continuous with slightly distended highly fenestrated saccules or membranous anastomotic tubules. The latter structures are loosely arranged and thus give the non-compact areas a porous and disorganized appearance (Figs. 2, 3, 4, 5 and 7). The following features characterize these zones. Firstly, such areas frequently protrude onto the *cis*-side of the Golgi cortex to form 'puff-like' projections or mounds (Figs. 7 and 8). Secondly, these non-compact zones contain an abundance of small 60–80 nm vesicles (Figs. 2, 3 and 5). Thirdly, the membranous tubules seen in these zones are reactive for phosphatases in a manner similar to that of the saccules to which they are connected (Figs. 7, 8 and 9). In contrast, none of the small 60–80 nm vesicles seen in these zones are reactive to the various phosphatases tested.

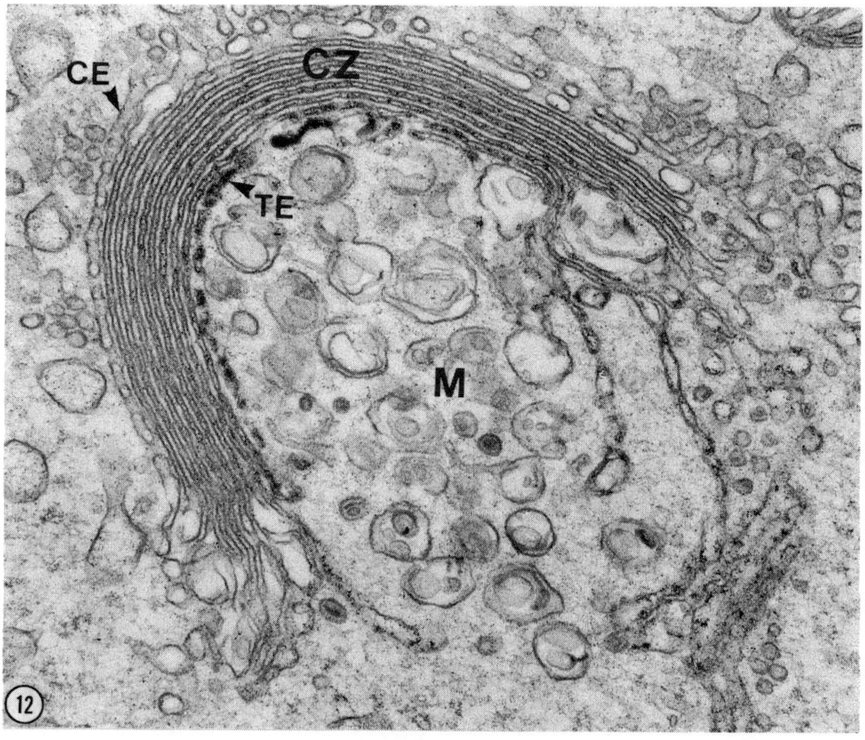

Fig. 12 Golgi apparatus of a late spermatid, detached from the acrosomic system, treated to show CMPase activity. In this Golgi apparatus which shows a large compact zone (CZ), CMPase reactivity is weak and limited to a spotty reaction on a *trans* element (TE); CE, *cis*-element; M, medulla containing non reactive vesicular profiles. × 62 500.

Fourthly, in face view of these zones, the vesicle-containing spaces seen between the membranous tubules tend to be circular and are reminiscent of the wells observed in the compact zones (Fig. 4), but in this case the gaps formed by the perforations may traverse the whole cortex of the Golgi apparatus. Fifthly, ER cisternae, connected to the peripheral network of

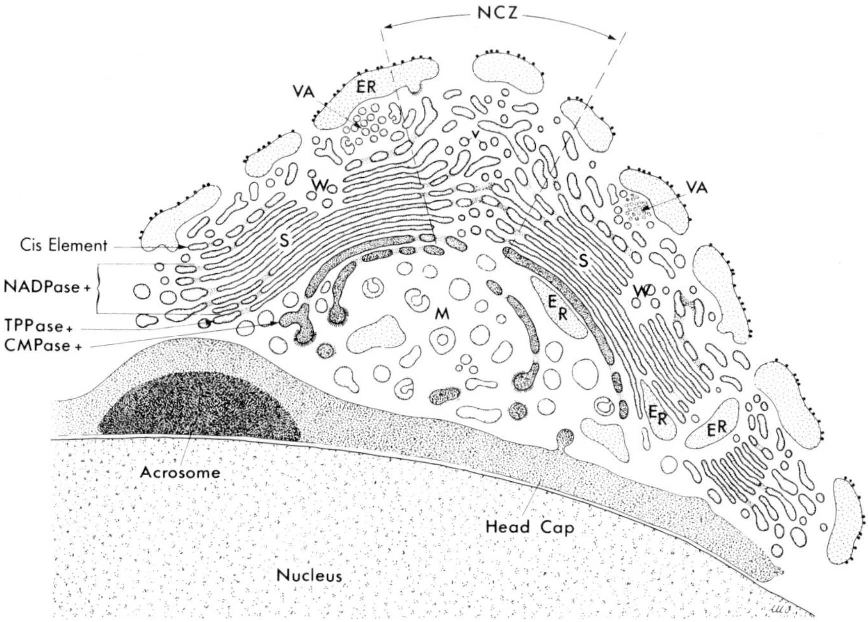

Fig. 13 Diagram showing the structure of the hemispherical Golgi apparatus of a step 6 spermatid overlying the acrosomic system consisting of the acrosome and head cap. The cortex of the Golgi apparatus is made up of compact zones composed of stacks of saccules (S) and of non compact zones (NCZ) containing membranous tubules that bridge the saccules of adjacent stacks. The stacks of saccules consist of the *cis*-(osmiophilic) element, the NADPase-positive saccules, some showing perforations in register called wells (W) and 'peeling-off' TPPase and CMPase-positive elements. The latter are considered as *trans* sacculotubular elements. The medulla (M) contains in addition to some *trans* sacculotubular elements detached from the stacks, numerous vesicles of various sizes, some coated, which have NADPase, TPPase or CMPase-activity. The acrosomic system also shows a CMPase-, NADPase-, and TPPase-positivity (see text). Rough ER cisternae are seen at the periphery of the Golgi apparatus (ER) but ER cisternae are also seen in the medulla, in close apposition to the trans sacculotubular elements or inserted between the saccules. Small vesicles are seen in the lumen of the wells, in the non compact zone and in the peripheral Golgi region where they form aggregates (VA).

ER cisternae, are occasionally seen traversing the Golgi cortex through the non-compact zones (Hermo *et al.*, 1979).

Detailed structure of the medulla of the Golgi apparatus

In steps 2–7 spermatids the medulla is the area present between the stacks of saccules and the developing acrosomic system (Figs. 2–4). At later steps of spermiogenesis it is the area enclosed by the C-shaped cortex (Fig. 12). In steps 3–7 the medulla contains vesicles of various sizes. Large vesicles, some C-shaped are abundant (Figs. 2–4). Some vesicles with a denser content have a bristle coat (Fig. 2). Some of the vesicles have NADPase (Figs. 6 and 7), TPPase (Figs. 8 and 9) and/or CMPase-activity (Figs. 10 and 11). This zone occasionally contains tubular or sacculotubular elements which have separated from the *trans* face of the Golgi cortex. Images of fusion of vesicles to the outer acrosomal membrane are occasionally seen (Fig. 11). The acrosomic system itself shows a strong CMPase-reactivity (Figs. 10 and 11), a spotty NADPase-reactivity (Figs. 6 and 7) and weak TPPase-reactivity limited to the acrosome (Fig. 9).

The medulla was found to be a compartment of its own since it appeared from steps 1 to 16 of spermiogenesis as a station in which labeled glycoproteins temporarily accumulated before being released to the acrosomic system, multivesicular bodies or cytoplasmic vesicles en route towards the plasma membrane (Clermont & Tang, 1985).

Relation of ER cisternae to the various components of the Golgi apparatus

The ER cisternae are not only seen at the periphery of the Golgi apparatus where they form a tight network (Figs. 2 and 7). They are also present in the cortex at the level of the non compact zones or in the spaces seen between the edges of the dome shaped cortex and the acrosomic system. Thus cisternae of the endoplasmic reticulum connected with the rest of the ER cisternae are present in the medulla. In addition such cisternae are not infrequently seen apposed to, without being continuous with, the *trans* sacculotubular elements (Fig. 4). Occasionally, ER cisternae are deeply inserted between the saccules of the Golgi cortex (Figs. 4 and 13; see full description in Hermo *et al.*, 1979). The significance of this close and unique relation of ER cisternae to the saccules of the Golgi stacks, has yet to be clarified.

Evolution of the Golgi apparatus during spermiogenesis

The overall architecture of the Golgi apparatus remains basically the same from steps 1 to 16 of spermiogenesis, although the following modifications are noted. The compact zones (stacks of saccules) tend to be small in step 1, these zones enlarge in steps 2–7 (Figs. 3–11) and become dominant in the following steps 8–16 (Fig. 12). Correspondingly the inter-saccular, non compact zones are numerous and prominent in early steps of spermiogenesis and progressively decrease in size as spermiogenesis progresses. The wells seen in the compact zones appear to be more conspicuous in steps 3–7 than at other steps of spermiogenesis.

The slightly distended sacculotubular elements seen on the *trans* face of the Golgi stacks are conspicuous from steps 1 to 7 of spermiogenesis, but are not distinguishable thereafter. Correspondingly, while these *trans* sacculotubular elements have a strong CMPase reactivity during steps 1–7, the CMPase reactivity is considerably reduced in the Golgi apparatus following its separation from the acrosomic system (Fig. 10). This CMPase reactivity, although present in *trans* elements of the Golgi stacks, is in fact limited to a weak and spotty reaction and few, if any, CMPase-positive vesicles are seen in the medulla (Fig. 12). On the other hand, NADPase and TPPase reactivity remain intense in the various elements of the stacks in the Golgi apparatus of late spermatids (Clermont *et al.*, 1981; Tang *et al.*, 1982). It might be recalled that the rate of uptake of labeled sugars by the Golgi apparatus of spermatids and its transfer from the cortex to the medulla remain unchanged following the separation of the organelle from the acrosomic system (Clermont & Tang, 1985).

Discussion

The juxtaposition of the Golgi apparatus to the developing acrosomic system of spermatids has traditionally been taken as evidence by light microscopists (Bowen, 1926; Leblond & Clermont, 1952; others) for a direct contribution of the Golgi apparatus to the formation of the acrosomic system of spermatozoa. This view has been reinforced by early electron microscope investigations of spermiogenesis of various mammals (Burgos & Fawcett, 1955; Clermont, 1956; Fawcett & Hollenberg, 1963; Sandoz, 1970; Susi *et al.*, 1971; others). The role of the Golgi apparatus in the formation of the acrosomic system was more directly demonstrated, however by EM radioautography following injection of radioactive tracers. Labeled sugars initially incorporated in the Golgi apparatus were clearly delivered, presumably bound to proteins, to the growing acrosomic system (Sandoz, 1972; Sandoz & Roland, 1976; Tang *et al.*, 1982; Clermont & Tang, 1985). However, the limited resolution of the radioautographic method did not permit the determination of the exact contribution of the Golgi apparatus components to the developing acrosomic system. Various EM studies led to the suggestion that proacrosomic or acrosomic granules of early spermatids as well as the head cap portion of the acrosomic system of later spermatids are derived from the addition of material elaborated in the stacks of saccules and carried by vesicles derived from the *trans* saccules (Fawcett & Hollenberg, 1963; Susi *et al.*, 1971). Earlier cytochemical studies on the Golgi apparatus of rat spermatids (Clermont *et al.*, 1981; Tang *et al.*, 1982) as well as the data presented here on the CMPase-reactivity of various Golgi components, i.e. *trans*-sacculotubular elements, vesicles in the medulla, as well as the CMPase-reactivity of the proacrosomic and acrosomic system, support this notion. However, it is possible that in addition to the *trans*-sacculotubular elements, other elements of the stack could equally contribute material to the growing acrosomic system directly *via* carrier vesicles forming at their edges. This is suggested by the fact that NADPase-positive or TPPase-positive vesicles are seen in the medulla of the Golgi apparatus and that NADPase- and TPPase-reactivity are also detected in the growing acrosomic system itself. This raises the possibility that the NADPase-reactive saccules, seen in the middle of the stacks or even the TPPase-reactive saccule on the *trans*-side of the stacks contribute material directly to the acrosomic system without passing through the NADPase- and TPPase-negative *trans*-sacculotubular elements (Fig. 13).

It should be noted that the mechanism described here in the formation of the acrosomic system, which obviously requires the formation of carrier vesicles from Golgi saccules, is not characteristic of the formation of secretory granules. Indeed the procollagen containing secretory granules in fibroblasts (Marchi & Leblond, 1983), secretory granules in atrial cardiocytes (Rambourg *et al.*, 1984), mucous granules in Brunner's gland cells (Rambourg *et al.*, in the press), and zymogen granules in pancreatic acinar cells and several other exocrine secretory cells (Rambourg, A., Clermont, Y. & Hermo, L., unpublished observations) are all formed directly within the saccules of Golgi stacks. It is therefore possible that the process described above in the formation of the acrosomic system is more representative of the formation of primary lysosomes or of lysosome-like organelles.

The formation of proacrosomic granules during steps 1 and 2 appears to occur through a process which is similar to the one taking place during the following steps 3–7, i.e. mainly *via* the formation of carrier vesicles from the *trans*-sacculotubular elements. It is noticeable however that the proacrosomic and acrosomic granules contain a dense inner zone and a less dense outer zone. The material of the inner dense zone becomes associated with the inner acrosomal membrane (in steps 4 or 5) and does not change appreciably in volume thereafter. In contrast, the less dense material of the outer zone of the acrosomic granule accumulates rapidly within the acrosomic system to form the head cap during step 4 to step 7 of spermiogenesis. It thus appears that the Golgi apparatus of steps 1–2 spermatids elaborates substances destined to the acrosomic system which have different physico-chemical or chemical properties than that elaborated by the Golgi apparatus of steps 3–7 spermatids. Similar observations have already been made in an EM investigation of guinea pig spermatids by Fawcett & Hollenberg (1963) who concluded that 'the outer and inner zones of the acrosome appear to result from two temporarily distinct phases of Golgi activity'. Such a biphasic activity of the Golgi apparatus in the elaboration of the acrosomic system may well correspond to the elaboration of different glycoproteins that would segregate in different parts of the acrosomic system.

The contribution of CMPase-positive *trans*-sacculotubular elements to the formation of the phosphatase-positive acrosomic system appears evident. Their 'peeling off' configuration or detachment from the stacks of saccules and their fragmentation into coated vesicles or tubules destined to fuse with the growing acrosomic system indicate that there must be a continuous formation of new saccules in the stacks. However the exact mode of formation of new saccules in the stacks is still unclear. The formation of new Golgi saccules is usually assumed to take place on the *cis*-face of the Golgi stacks as a result of the fusion of vesicles originating from the proximal ER cisternae. Such saccules would then migrate from the *cis* to the *trans* face where they would desquamate (reviewed by Farquhar & Palade, 1981). This hypothetical scheme inspired earlier discussions of the contribution of the Golgi apparatus to the formation of the acrosomic system (Susi *et al.*, 1971; Mollenhauer *et al.*, 1976). This scheme is still however the object of discussion (reviewed by Goldfischer, 1982; Bennett, 1984) and the exact mode of formation of new saccules or the pathway(s) of transport of proteins from the ER cisternae to the secretory granules or lysosomes remains to be clarified.

Materials and methods

Testes of adult rats were fixed by perfusion through the abdominal aorta with 2.5% glutaraldehyde buffered with sodium cacodylate (0.1 M) containing 0.05% $CaCl_2$ (pH 7.4). After perfusion (5–15 min), the testes were trimmed and placed in the same fixative for 1 h at 4°C. The tissue blocks were then washed in 0.1 M sodium cacodylate buffer, postfixed in potassium ferrocyanide reduced osmium tetroxide for 2 hr at ~ -4°C, dehydrated and embedded in Epon.

For cytochemical studies, the tissue was sectioned at 75 μm with a Sorval TC-2 tissue chopper. For the localization of cytidine monophosphatase (CMPase) the sections were incubated for 1 h at 37°C in a medium at pH 5.0 as described by Novikoff (1963); cerium chloride was used as a capture agent. Thiamine pyrophosphatase (TPPase) was demonstrated by incubating sections for 2 h at 37°C in a medium at pH 7.2 proposed by Novikoff & Goldfischer (1961). Nicotinamide adenine dinucleotide phosphatase (NADPase) was detected by incubating sections for 4 h at 37°C in a medium at pH 5.0 according to the method of Smith (1980). Controls for these cytochemical reactions consisted in incubating sections in a medium which did not contain the substrate. After incubation the sections were washed in the buffer of the incubating medium and then rinsed in sodium cacodylate buffer containing 4% sucrose and 0.05% $CaCl_2$ at pH 7.2. They were then postfixed in ferrocyanide-reduced osmium as above and embedded in Epon. Thin sections of all Epon blocks were stained with lead citrate and examined with a Philips 400 electron microscope.

Stereopairs of photographs were frequently examined to obtain three dimensional images of the Golgi apparatus' components. To obtain such stereopairs, sections were placed on the goniometer stage of the electron microscope and two pictures of the same field were taken before and after tilting the specimen. For sections of 0.5 μm or less the sections were tilted at $-18°$ and $+18°$ from the original (0°) position of the stage. For sections of 0.5 μm or more the tilt was $-7°$ and $+7°$ from the 0° position: A tridimensional image of the structure was obtained by inspecting properly adjusted pairs of such photographs with a stereoscopic binocular lens.

Acknowledgments

The work done in the Department of Anatomy, McGill University, was supported by a grant from the Medical Research Council of Canada. The assistance of Drs M. Lalli and X. Tang during the course of this work is gratefully acknowledged. The drawings were prepared by Mrs Margo Oeltzschner.

References

Allison, A.C. & Hartree, E.E. (1970). Lysosomal enzymes in the acrosome and their possible role in fertilization. **Journal of Reproduction and Fertility** 21:501–515.

Bennett, G. (1984). Role of the Golgi complex in the secretory process. In: Cantin M., ed. **Cell Biology of the Secretory Process**. Basel, S. Karger, pp.102–147.

Bennett, G. & Leblond, C.P. (1977). Biosynthesis of the glycoproteins present in plasma membrane, lysosomes and secretory material as visualized by radioautography. **Histochemistry Journal** 9:393–417.

Bowen, R.I. (1926). The Golgi apparatus. Its structure and functional significance. **The Anatomical Record** 32:151–193.

Burgos, M.H. & Fawcett, D.W. (1955). Studies on the fine structure of the mammalian testis. I. Differentiation of the spermatids in the cat *(Felis domestica)*. **The Journal of Biophysical and Biochemical Cytology** 1:287–300.

Clermont, Y. (1956). The Golgi zone of the rat spermatid and its role in the formation of cytoplasmic vesicles. **The Journal of Biophysical and Biochemical Cytology**, Supplement 2:119–122.

Clermont, Y. & Leblond, C.P. (1955). Spermiogenesis of man, monkey, ram and other mammals as shown by the periodic acid–Schiff technique. **The American Journal of Anatomy** 96:229–254.

Clermont, Y. & Rambourg, A. (1978). Evolution of the endoplasmic reticulum during rat spermiogenesis. **The American Journal of Anatomy** 151:191–212.

Clermont, Y. & Tang, X.M. (1985). Glycoprotein synthesis in the Golgi apparatus of spermatids during spermiogenesis of the rat. **The Anatomical Record** 213:33–43.

Clermont, Y., Glegg, R.E. & Leblond, C.P. (1955). Presence of carbohydrates in the acrosome of the Guinea pig spermatozoon. **Experimental Cell Research** 8:453–458.

Clermont, Y., Lalli, M. & Rambourg, A. (1981). Ultrastructural localization of nicotinamide adenine dinucleotide phosphatase (NADPase), thiamine pyrophosphatase (TPPase), and cytidine monophosphatase (CMPase) in the Golgi apparatus of early spermatids of the rat. **The Anatomical Record** 201:613–622.

Clermont, Y., Leblond, C.P. & Messier, B. (1959). Durée du cycle de l'épithélium seminal du rat. **Archives d'anatomie microscopique et de morphologie expérimentale** 48*bis*:37–56.

Farquhar, M.G. & Palade, G. (1981). The Golgi apparatus (complex) – (1954–1981) from artefact to center stage. **The Journal of Cell Biology** 91:971–1030.

Fawcett, D.W. & Hollenberg, R.D. (1963). Changes in the acrosome of guinea pig spermatozoa during passage through the epididymis. **Zeitschrift für Zellforschung** 60:276–292.

Friend, D.S. & Murray, M. (1965). Osmium impregnation of the Golgi apparatus. **The American Journal of Anatomy** 117:135–150.

Gatenby, J.B. & Woodger, J.H. (1921). The cytoplasmic inclusions in germ cells. IX. On the origin of the Golgi apparatus, on the middle piece of the ripe sperm of Cavia, and the development of the acrosome. **Quarterly Journal of Microscopical Sciences** 65:265–291.

Goldfischer, S. (1982). The internal reticular apparatus of Camillo Golgi: A complex, heterogeneous organelle, enriched in acid, neutral, and alkaline phosphatase and involved in glycosylation, secretion, membrane flow, lysosome formation and intracellular digestion. **The Journal of Histochemistry and Cytochemistry** 30:717–733.

Hermo, L., Clermont, Y. & Rambourg, A. (1979). Endoplasmic reticulum-Golgi apparatus relationships in the rat spermatid. **The Anatomical Record** 193:243–256.

Hermo, L., Rambourg, A. & Clermont, Y. (1980). Three-dimensional architecture of the cortical region of the Golgi apparatus in rat spermatids. **The American Journal of Anatomy** 157:357–373.

Lalli, M. & Clermont, Y. (1981). Structural changes of the head components of the rat spermatid during late spermiogenesis. **The American Journal of Anatomy** 160:419–434.

Leblond, C.P. (1950). Distribution of periodic acid reactive carbohydrates in adult rat. **The American Journal of Anatomy** 86:1–49.

Leblond, C.P. & Bennett, G. (1977). Role of the Golgi apparatus in terminal glycosylation. In: Brinkley, B. & Porter, K., eds. **International Cell Biology** 1976–1977. New York, Rockefeller University Press, pp.326–336.

Leblond, C.P. & Clermont, Y. (1952). Spermiogenesis of the rat, mouse, hamster and guinea pig as revealed by the 'periodic acid-fuchsin sulfurous acid technique'. **The American Journal of Anatomy** 90:167–216.

Marchi, F. & Leblond, C.P. (1983). Collagen biogenesis and assembly into fibrils, as shown by ultrastructure and ^{3}H-proline radioautographic studies on the fibroblasts of the rat foot pad. **The American Journal of Anatomy** 168:167–197.

Mollenhauer, H.H., Hass, B.S. & Morré, D.J. (1976). Membrane transformations in Golgi apparatus of rat spermatids. A role for thick cisternae and two classes of coated vesicles in acrosome formation. **Journal de Microscopie et de Biologie Cellulaire** 27:33–36.

Morton, D.B. (1976). Lysosomal enzymes in mammalian spermatozoa. In: Dingle, J.T. & Dean, R.T., eds. **Lysosomes in Biology and Pathology**, vol. 5, Amsterdam, North Holland Publ. Co.

Novikoff, A.B. (1963). Lysosomes in the physiology and pathology of cells. Contribution of staining methods. In: de Reuck, A.V.S. & Cameron, M.P., eds. **Ciba Foundation Symposium on Lysosomes.** Boston, Little, Brown Co. pp.36–77.

Novikoff, A.B. & Goldfischer, S. (1961). Nucleoside diphosphatase activity in the Golgi apparatus and its usefulness for cytological studies. **Proceedings of the National Academy of Sciences USA** 47:802–810.

Rambourg, A. & Clermont, Y. (1986). Tridimensional structure of the Golgi apparatus in type A ganglion cells of the rat. **The American Journal of Anatomy** 167:393–409.

Rambourg A., Clermont, Y. & Hermo, L. (1979). Three-dimensional architecture of the Golgi aparatus in Sertoli cells of the rat. **The American Journal of Anatomy** 154:455–476.

Rambourg, A., Clermont, Y. & Marraud, A. (1974). Three-dimensional structure of the osmium impregnated Golgi apparatus as seen in the high voltage electron microscope. **The American Journal of Anatomy** 140:27–46.

Rambourg, A., Segretain, D. & Clermont, Y. (1984). Tridimensional architecture of the Golgi apparatus in the atrial muscle cell of the rat. **The Journal of Anatomy** 170:163–179.

Rambourg, A., Clermont, Y., Hermo, L. & Segretain, D. (1987). Tridimensional architecture of the Golgi apparatus and its components in mucous cells of Brunner's glands of the mouse. **The American Journal of Anatomy** 179:95–107.

Sandoz, D. (1970). Evolution des ultrastructures au cours de la formation de l'acrosome chez la souris. **Journal de Microscopie** 9:535–558.

Sandoz, D. (1972). Etude radioautographique de l'incorporation *in vitro* du galactose-^{3}H dans les spermatides de souris. **Journal de Microscopie** 15:403–408.

Sandoz, D. & Roland, J.C. (1976). Radioautographic study of glycoproteins and proteoglycans. **Journal de Microscopie et de Biologie Cellulaire** 27:207–214.

Smith, C.E. (1980). Ultrastructural localization of nicotinamide adenine dinucleotide phosphatase (NADPase) activity to the intermediate saccules of the Golgi apparatus in rat incisor ameloblasts. **The Journal of Histochemistry and Cytochemistry** 28:16–26.

Susi, F.R., Leblond, C.P. & Clermont, Y. (1971). Changes in the Golgi apparatus during spermiogenesis in the rat. **The American Journal of Anatomy** 130;251–278.

Tang, X.M., Lalli, M.F. & Clermont, Y. (1982). A cytochemical study of the Golgi apparatus of the spermatid during spermiogenesis in the rat. **The American Journal of Anatomy** 163:283–293.

Golgi complex: integrity, composition, genesis and homeostasis

ALAN M. TARTAKOFF

Institute of Pathology, Case Western Reserve University, 2085 Adelbert Road, Cleveland, OH 44106, USA.

Abstract

The known constituents of the Golgi complex are synthesized in other cytoplasmic compartments. In an attempt to explain their transit to and residence in the Golgi, this chapter will:

(1) Summarize present knowledge on Golgi protein and lipid composition,
(2) Describe those situations in which structurally 'aberrant' proteins overaccumulate in the Golgi, and
(3) Review studies of conditions and agents which disturb intracellular transport at the level of the Golgi complex or alter its organization.

Key words

Golgi complex, protein transport, cisternae, vesicles, GERL, acid hydrolases, glycosylation, temperature manipulation, colchicine, taxol, nocodazole, monensin, anti-clathrin antibodies.

Introduction

Several outstanding features of the Golgi complex have brought it to the forefront of interest in membrane cell biology.

(1) The characteristic stacked cisternae of the Golgi complex are found in all eukaryotic cells, including yeast.
(2) The Golgi complex appears to be a site of convergence of the secretory and endocytic paths. In addition to being an obligatory way-station for the passage of soluble and membrane-associated molecules from their site of synthesis in the rough endoplasmic reticulum (RER) toward more distal destinations, the Golgi has been increasingly implicated as an intermediate along the endocytic pathway. For example, selected soluble tracers and ligands taken up by receptor-mediated endocytosis move through Golgi cisternae and post-cisternal elements (Farquhar, 1985; Snider & Rogers, 1986; Fishman & Fine, 1987). This intersection between the secret-

ory and endocytic pathways poses complex questions with regard to the intrinsic identity and dynamics of these membranes.

(3) In the context of the secretory path (and this may prove also to be true for Golgi participation in the endocytic pathway), the Golgi, defined in the broadest sense, is a sorting center. By 'broadest sense' is meant that the Golgi should be considered to include the characteristic stacked cisternae, associated vesicles, and also those associated cisternal or tubular structures variously referred to as 'GERL' or the 'trans-Golgi network' (Novikoff, 1976; Simons & Griffiths, 1986). Several of these structures bear a coating of adherent material on the endodomain of their membranes. In this sense it may be inappropriate to refer to these membranes as being altogether smooth.

The sorting functions which are attributable to the Golgi defined in this inclusive sense are as follows:

(a) The differential transport of acid hydrolases which delivers them to lysosomes rather than to the plasma membrane. In selected, but certainly not in all cells, these events are ascribable to the addition of the mannose-6-phosphate signal at the level of the Golgi (e.g., von Figura and Hasilik, 1986; Chapter by Hasilik & Von Figura in this volume).

(b) The division between the so-called 'constitutive' and 'regulated' secretory pathways, i.e., the choice of whether a newly-synthesized macromolecule will be continuously conducted *via* small vesicles to the cell surface *vs.* the option of being conducted to storage granules whose exocytosis will occur only upon appropriate hormonal or electrical stimulation (e.g., Kelly, 1985).

(c) The choice, which pertains to many polarized cells which have characteristically distinct apical as opposed to basolateral membrane domains, as to whether a given newly-synthesized secretory or membrane protein will be conducted to one or the other of these compositionally and functionally distinct plasma membrane domains (Simons & Fuller, 1985; Fuller *et al.*, 1985).

The basis of our understanding of Golgi structure and function comes from an evolving set of conceptual and technical approaches. For example, cytochemical studies of the Golgi established some time ago that this organelle is compositionally heterogeneous, but it is only much more recently that distinct functions have been ascribed (at least provisionally) to its subcompartments. At the moment, the best prospect for appreciating the extent of this heterogeneity comes from the use of cytochemistry and immunocytochemistry (Chicheportiche &Tartakoff, 1987); however, considering that Golgi-enriched subcellular fractions have been available for some time, and that a modest degree of isopycnic or electrophoretic subfractionation has been obtained (Deutscher *et al.*, 1983; Dunphy & Rothman, 1983; Morré *et al.*, 1983; Goldberg & Kornfeld, 1983; Lodish *et al.*, 1987), a further appreciation of the biochemical heterogeneity of the Golgi will surely be forthcoming.

Composition

The composition of the Golgi complex is only incompletely known. The best characterized resident components are integral membrane proteins (Table 1). Certain of these have served as cytochemical markers; others have received a great deal of attention with regard to their metabolic roles. This is the case for the diverse family of sugar transferases, glycosidases and sugar nucleotide transporters responsible for the maturation of O- and N-glycans, glycosaminoglycans and glycolipids (Tartakoff, 1983b; Hirschberg & Snider, 1987). Available information indicates, moreover, that a number of the transferases are not uniformly distributed across the Golgi stack, although the data appear to be somewhat variable from one cell type to the next (Roth *et al.*, 1986). The gist of the information as it now stands, however, is that the linear proximal-to-distal axis of transport of newly synthesized macromolecules across the Golgi corresponds to the distribution of oligosaccharide processing enzymes – those which act first are characteristic of the more proximal, whereas those which act later are characteristic of more distal cisternae (e.g., Dunphy & Rothman, 1985).

A further Golgi integral membrane glycoprotein has been shown, by microinjection of the mRNA from a hybridoma which secretes a corresponding monoclonal antibody, to be of functional significance for intracellular transport (Burke & Warren, 1984).

Table 1. *Golgi associated enzyme activities*

Name	Sub-Golgi localization
N-glycan trimming mannosidases	throughout (?)
N-glycan sugar transferases	distal or post cisternal
O-glycan sugar transferases	?
Glycosaminoglycan sugar transferases, sulfotransferases	?
Sugar nucleotide permeases	?
Endoprotease (dibasic amino acid specific)	condensing vacuoles
Nucleoside diphosphatase	distal
NADPase	medial
Acid phosphatase	post cisternal
Proton pump (?)	distal, post cisternal (?)
Mannose-6-phosphate signal synthetic enzymes	early or medial
Acyl transferase	early (?)

See text for references.

A second class of, at least potential, Golgi constituents is the soluble content itself. In the case of the RER there is a limited number of soluble proteins which for unknown reasons are retained within the cisternal space (egasyn, immunoglobulin-binding protein, a 'stress' protein, disulfide interchange enzyme: Edman *et al.*, 1985; Medda & Swank, 1985; Bole *et al.*, 1986; Munro & Pelham, 1986). No such proteins are yet characterized for the Golgi complex, but the osmiophilic (or reducing) substance characteristic of the cisternal space of the most proximal cisternae may be a candidate (Friend & Murray, 1959).

A third class of Golgi components pertains to the endodomain of these membranes, i.e., the surface in contact with the cell sap (Table 2). Here, what is surprising and possibly quite important from the point of view of understanding function, is that proteins have recently been recognized which, although highly Golgi-specific, are nevertheless bound only superficially, and may be released by salt washing of Golgi-enriched subcellular fractions. Among these proteins are certain antigens of undefined function (Chicheportiche & Tartakoff, 1984, 1987), one of the regulatory subunits of adenylcyclase (Nigg *et al.*, 1985a, b; de Camilli *et al.*, 1986), a protein which shares structural features with MAP-2, one of the microtubule-associated proteins (Allan & Kreis, 1986), a 'coating' substance (Griffiths *et al.*, 1985; Orci *et al.*, 1986a), and clathrin (Griffiths *et al.*, 1981; Merisko *et al.*, 1986a, b; Tooze & Tooze, 1986). Further superficial components are materials responsible for the close apposition of Golgi cisternae and for creating a 'zone of exclusion' around the Golgi (Mollenhauer & Morré, 1978). Cell-free reconstitution studies have also implicated several cytosolic proteins in proximal to distal Golgi transport (Paquet *et al.*, 1986; Dunphy *et al.*, 1986; Wattenberg & Rothman, 1986; Wattenberg *et al.*, 1986).

With regard to smaller molecules (Table 3), those which have been characterized are several vitamins (Crane & Morré, 1977; Nyquist *et al.*, 1970, 1971a, b), phosphoadenosine phosphosulfate (PAPS) and several sugar

Table 2. *Proteins associated with the endodomain of Golgi elements*

Name	Molecular weight
GC I	54 000, 86 000
GC II	44 000
MAP2 homologue	110 000?
clathrin	180 000
coat material	?
intercisternal 'glue'	?
tubulin (?)	55 000
fibrillar material (zone exclusion)	?

See text for references.

Table 3. *Non-protein components*

Component
cholesterol-sphingomyelin-rich
vitamin A, K, ubiquinone, Coenzyme Q
osmiophilic material
sugar nucleotides, PAPS

See text for references.

nucleotides and their hydrolysis products (Hirschberg & Snider, 1987). There is also limited information with regard to lipid composition. At least in the hepatocyte and the pancreatic exocrine cell, in moving along the secretory pathway one encounters a progressive increase in both cholesterol and sphingomyelin content (rough microsomes vs. Golgi enriched fractions vs. secretion granule fractions; Meldolesi *et al.*, 1971). Moreover, cytochemical studies employing filipin to label what appear to be cholesterol-rich membrane domains indicate that the cholesterol content increases proximal to distal across the Golgi stack (Orci *et al.*, 1981).

A further compositional characteristic of the (distal) Golgi appears to be the presence of a proton pump, which at first approximation, resembles that which has been more extensively characterized in both secretory granules and endocytic compartments (endosomes, lysosomes). The information here is far from complete. What has been observed at the immunocytochemical level (Anderson & Pathak, 1985; Schwartz *et al.*, 1985; Orci *et al.*, 1986b) is that acidotropic weak bases (in addition to being concentrated within lysosomes, endosomes and immature and mature secretion granules) accumulate in distal Golgi elements of constitutive secretory cells. The extent of accumulation within the Golgi is considerably less than what is observed among endocytic compartments. This cytochemical approach is unfortunately only qualitative.

In addition, biochemical studies (Barr *et al.*, 1984; Glickman *et al.*, 1983) suggest that either ATP or NADH may power a proton pump recovered in Golgi-enriched subcellular fractions. It remains to be proved to what extent these fractions are free from endocytic contaminants.

Constructs – mutants that localize to the Golgi complex

Few Golgi proteins have been obtained in highly purified form. Therefore, it cannot be determined whether common structural features might explain either why the integral membrane proteins of the Golgi come to reside within the Golgi as a whole or within subcompartments, or why endodomain components show a Golgi preference.

The protein which is best characterized is β-1-4-galactosyl-transferase, a glycophosphoprotein which bears both N and O glycans (Strous *et al.*, 1985b). A curious characteristic of this protein, which might be pursued to ascertain what features are responsible for its Golgi localization, is that in selected cell types it is associated with the cell surface (Roth *et al.*, 1985). Presumably in these cases there has been a structural alteration or post-synthetic modification which explains its altered subcellular localization.

A further naturally occurring situation, essentially the reverse of the above, is encountered in human patients suffering from a mislocalization of sucrase-isomaltase (Hauri *et al.*, 1985). In at least one case this protein, which normally is a plasma membrane constituent of intestinal epithelial cells, resides at high concentration within the distal Golgi and is absent at the cell surface. The underlying structural alteration is unknown.

A major research effort has recently allowed a systematic study of covalent changes which cause proteins normally destined for delivery to the plasma membrane to lodge instead in Golgi or Golgi-associated compartments (Fig. 1). Such mis-localization, in contrast to retention in the RER, cannot simply be ascribed to protein precipitation, inability to diffuse in the plane of the membrane, etc. The common protocol for such experiments has involved introducing amino acid substitutions into viral envelope glycoproteins (and their derivatives) and subsequently to cause transient expression of these proteins in appropriate host cells. The host cells most frequently employed are simian COS cells because of their compatibility with SV4O-based expression vectors.

A serious reservation with regard to this line of experimentation is that one is interrupting the transport of a protein which is normally destined for the plasma membrane: it is not clear to what extent this interruption of transport may reflect biologically significant 'sorting signals' comparable to those which normally govern Golgi-specific localization. It would, perhaps, be more meaningful to inquire into which changes in the structure of a normal Golgi resident would abrogate its Golgi localization. Nevertheless, in one striking study in which a recombinant fusion protein (between growth hormone and the carboxy terminal portion of influenza hemagglutinin) has been expressed (Fig. 1, construct 1), the fusion protein localizes to the Golgi area, as judged by immunofluorescence. In fact the compartment in question has been shown by electron microscopy not to be Golgi cisternae themselves, but a pre-Golgi compartment which retains continuity with the RER. In a further high-resolution study (Fig. 1, construct 5), a variant of the G protein has been shown to move just far enough through the Golgi to acquire branched GlcNAc residues, but not galactose.

The principal biochemical tools used here to evaluate the progress of transport are the study of the degree of maturation of N-glycans and fatty acid addition. Unfortunately, most of these investigations have used neither subcellular fractionation nor electron microscopic immunocytochemistry to establish with greater certainty the subcellular localization of the products under study.

Taken as a whole, the constructs illustrate that alterations of either the ectodomain (Fig. 1, constructs 1–2, 4, 8), the transmembrane domain (Fig. 1, constructs 6,7) or the endodomain (Fig. 1, construct 5) may interrupt transport in the Golgi area. For the moment, most of these covalent changes cannot be interpreted in terms of the three-dimensional structure of the proteins in question. The structural alterations which result in interruption in the Golgi therefore do not point with any unanimity at one structural feature which might itself be responsible for Golgi localization.

Considering that the Golgi is a center for such a multiplicity of post-synthetic modifications of proteins, one might nevertheless anticipate that a characteristic post-synthetic modification itself might result in sufficient conformational change to cause arrest within the Golgi area. Several recent studies have pointed to the importance of trimerization of G and HA and structural 'maturation' of secretory proteins as a prerequisite for exit from the RER (Rome *et al.*, 1983; Lemansky *et al.*, 1984; Lodish & Kong, 1984; Copeland *et al.*, 1986; Gething *et al.*, 1986; Kreis & Lodish, 1986).

Perturbation

A number of drugs interrupt the progress of intracellular transport (Table 4). In addition, temperature manipulation has identified at least two intermediates along the secretory path within the Golgi area. Each Golgi perturbant studied to date has distinct anatomical consequences. Apart from the studies described below, other investigations show that many of the same agents and conditions also inhibit the endocytic paths.

Treatment of cells with the tubulin perturbants colchicine, taxol or nocodazole or injection of an antibody specific for tyrosylated tubulin leads to fragmentation and dispersal of small Golgi stacks (islands [Wehland & Sandoval, 1983; Rogalski *et al.*, 1984; Sandoval *et al.*, 1984; Thyberg & Moskalewski, 1985]). Despite this gross structural disorganization, which suggests an intimate structural relationship between microtubules and the Golgi, the Golgi often remains functionally competent. It continues to conduct proteins to the cell surface, matures their N-glycans and causes accurate apical vs. basolateral sorting of at least some viral envelope glycoproteins (Rogalski *et al.*, 1984; Salas *et al.*, 1986; Rindler *et al.*, 1987). In other cases, interruption of transport has been observed (Dehm & Prockop, 1972; Ehrlich *et al.*, 1974; Redman *et al.*, 1978; Oda & Ikehara, 1981; Thyberg & Moskalewski, 1985).

A second class of agents known to interrupt intracellular transport are those which perturb ion balance. The prototype here is the carboxylic ionophore, monensin, which carries sodium and protons (and, to a lesser extent, potassium) across membranes (Tartakoff, 1983a). In many animal

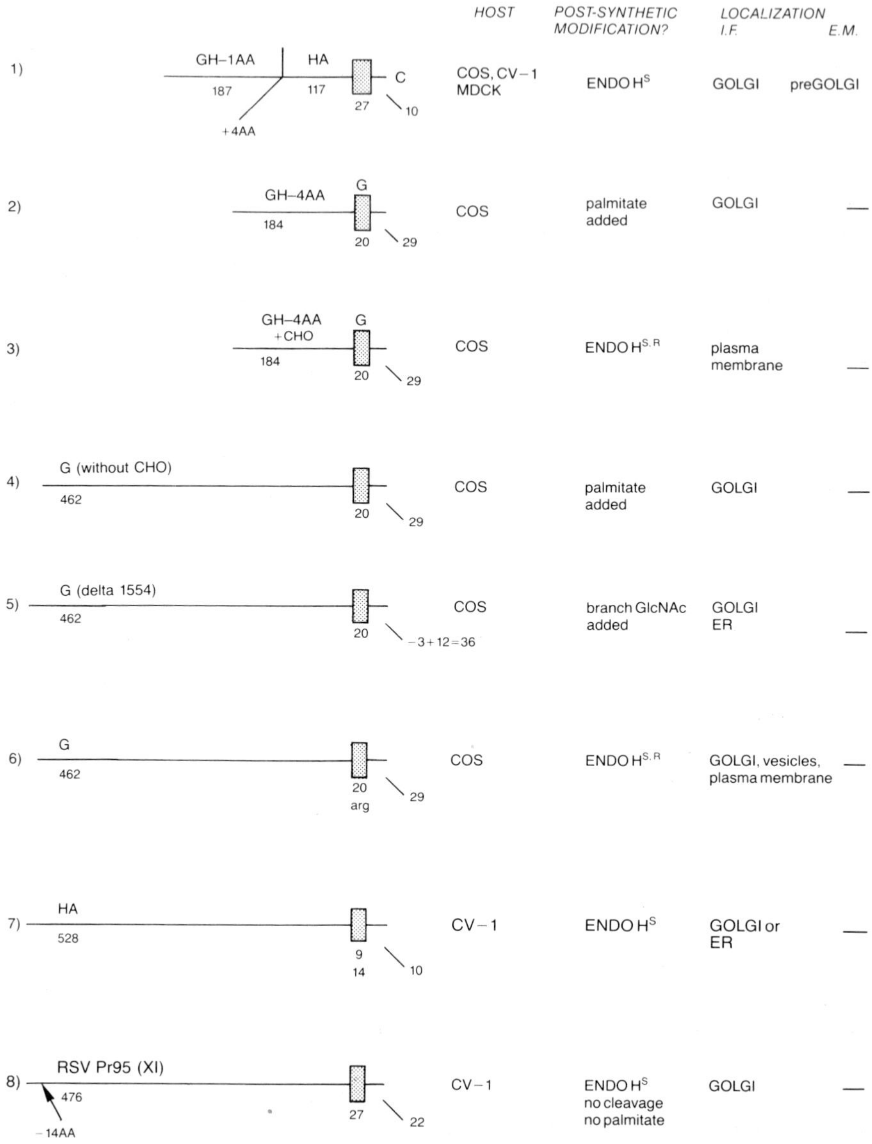

Fig. 1 Golgi-arrest constructs. In each case the N-terminus is toward the left, the transmembrane domain is designated by a filled rectangle, and the C-terminus is toward the right. The numbers beneath the polypeptides indicate the total length of the segments in question. The entry Postsynthetic Modification indicates to what extent N-glycan maturation has occurred (reflected by acquisition of resistance to Endoglycosidase H (ENDO H^R)), palmitic acid has been added, or proteolytic cleavage has occurred. ENDO H^R is usually characteristic of arrival in medial Golgi cisternae, palmitic acid is added either just before or just after exit from

and plant cells the Golgi becomes rapidly dilated and intracellular transport is slowed if not stopped altogether. Moreover, treatment with agents such as monensin (or acidotropic amines) (1) diverts lysosomal enzymes to the secretory pathway (the secreted forms of these hydrolases have acquired the mannose-6-phosphate signal [Gonzalez-Noriega *et al.*, 1980; Pohlman *et al.*, 1984]), (2) interrupts N-glycan maturation of secretory and plasma membrane glycoproteins (Tartakoff, 1983a,b; Strous *et al.*, 1985a; Thorens & Vassalli, 1986), and (3) deviates newly synthesized proteins from the regulated to the constitutive route (Moore *et al.*, 1983; Wagner *et al.*, 1986). The polarity of apical vs. basolateral sorting appears not to be upset by acidotropic amines (Matlin, 1986).

To rationalize the slowing of transport, (1) and (3), one might suppose that what is occurring is a neutralization of the normally somewhat acidic

Caption for Fig. 1 (cont.)

the RER and proteolysis is a distal and/or post-Golgi event. The abbreviations I.F. and E.M. designate observation by immunofluorescence or electron microscopy.

Construct (1) This construct includes a slightly truncated form of growth hormone (lacking the most C-terminal amino acid), 4 extra amino acids and a total of 154 amino acids from the carboxy terminus of hemagglutinin [HA] (Rizzolo *et al.*, 1985).

Construct (2) This construct includes a slightly truncated form of growth hormone [GH] (lacking the 4 most C-terminal amino acids), and a total of 49 amino acids from the carboxy terminus of the G protein [G] of vesicular stomatitis virus (Guan & Rose, 1984).

Construct (3) This construct is identical to (2) but includes N-glycans linked to amino acids 12, 95 or both. Growth hormone itself lacks carbohydrate. These variants were produced by site-directed mutagenesis (Guan & Rose, 1985).

Construct (4) Variants of G were constructed lacking one or both glycosylation sites (residues 179, 336). When both are absent, no transport to the cell surface occurs (Machamer *et al.*, 1985).

Construct (5) The delta 1554 construct lacks the 3 most C-terminal amino acids of the cytoplasmic domain of G. Twelve foreign amino acids are added in their place. Alternative constructs in which only the 3 foreign amino acids are added allow transport to the cell surface (Rose & Bergmann, 1983).

G was replaced by either glutamine or arginine. Transport continues in the former case but is interrupted in at least half of the cells in the latter case (Adams & Rose, 1985).

Construct (7) Truncation of the transmembrane domain (normally 27 amino acids) to 9 or 14 amino acids leads to arrest of transport and rapid degradation (Doyle *et al.*, 1986).

Construct (8) In the mutant illustrated, 14 amino acids (Nos. 4–17) have been deleted. Normally the precursor Pr95 is cleaved to generate gp37 (a transmembrane fragment) and gp85 (an extracellular fragment) after exit from the Golgi (Wills *et al.*, 1983).

Table 4. *Impact of microtubule-active drugs, monensin, acidotropic bases and temperature reduction*

Event	Impact				
	Microtubule-active drugs	Monensin	Acidotropic bases	Temperature reduction	Mitosis
Golgi coherence	fragmentation (islands)	normal	normal	some fragmentation at 15°C	fragmentation (vesicles)
Golgi dilation	modest	rapid, extensive	modest	none	?
Protein transport to cell surface	variable	slowed/ stopped	often slowed	interrupted (10–20°C)	stopped
Oligosaccharide maturation	normal	often interrupted	often interrupted	interrupted at 10–15°C; complete at 20°C	—
Sulfation	?	interrupted	?	?	—
Proteolysis	normal	interrupted	often slowed	?	—
Acid hydrolase secretion	?	dramatic	dramatic	?	—
Delivery of 'regulated' secretory products to granules	variable	interrupted	deviated to constitutive path	interrupted	—
Polar insertion of membrane glycoproteins and protein secretion	disputed	?	slow but accurate	slowed	—

See text for references.

cisternal space of distal Golgi elements, thereby incapacitating pH-dependent receptor-mediated sorting operations. The only experimental work pertaining to such receptors concerns the 215 kD mannose-6-phosphate receptor, which is trapped in endosomes after chloroquine treatment (Brown *et al.*, 1986). More indirect explanations might also be invoked. The impact of monensin may be more complex than that of the amines since this ionophore facilitates exchange of protons, sodium and potassium ions across membranes. Many of its effects on the Golgi may be simulated by ouabain (Whetsell & Bunge, 1969; Tartakoff & Vassalli, 1977).

Golgi function is also radically interrupted during mitosis. Here the interruption is both anatomical and functional: exit from the RER and exocytosis are interrupted, and fragmentation of the RER and Golgi is extreme; stacked cisternae can no longer be located (e.g., Warren, 1985). How reassembly of the Golgi occurs at the end of mitosis is totally unknown; however, as with the nuclear envelope (and, presumably, the endoplasmic reticulum) membrane–cytoskeletal interactions may play an important stabilizing role.

Several studies have explored the impact of temperature reduction on Golgi function and structure. For example, immunocytochemical studies of virally-infected tissue culture cells and the exocrine pancreas show that temperature reduction to 15°C allows exit of newly-synthesized proteins from the RER and results in their overaccumulation within a compartment interposed between the transitional elements of the RER and the most proximal Golgi cisterna (Saraste & Kuismanen, 1984; Saraste *et al.*, 1986). Alternatively when the temperature is held at 20°C, the viral glycoproteins pass further along the secretory pathway and attain the 'trans-Golgi region' where they overaccumulate in a tubular structure which is coated with material that is not recognized by anti-clathrin antibodies (Griffiths *et al.*, 1985). Studies of the maturation of the N-glycans of these envelope glycoproteins indicate that within this distal compartment they already have completed the acquisition of terminal sugars (Matlin & Simons, 1983).

Further studies of the exocrine pancreas indicate that temperature reduction to 10°C can be used to interrupt exit from the RER, whereas temperature reduction to 20°C allows exit from the RER to continue and results in overaccumulation of the exocrine proteins within the Golgi, as determined by subcellular fractionation and electron microscopic autoradiography (Tartakoff, 1986; Saraste *et al.*, 1986). Incubation at 10°C has a striking anatomical consequence; the transitional elements of the RER become much more numerous and conspicuous than in control tissue maintained and fixed at 37°C. Although it is not known why temperature reduction interrupts vesicular transport, alterations in membrane fluidity and/or metabolic (e.g., proton pump) activity may be involved.

Temperature manipulation also regulates Golgi traffic in the temperature sensitive yeast *sec* mutants (Schekman, 1985). In the mutants *sec* 7 and 14, Golgi function in transport, glycosylation and phosphorylation of secretory and plasma membrane proteins and at least one vacuolar hydrolase is reversibly interrupted at 37°C. The corresponding gene products have not yet been identified, but at 37°C a most impressive stack of Golgi cisternae is found. For unknown reasons, in low glucose medium the characteristic stack structure is anatomically modified.

A further means of interrupting Golgi function involves manipulation of the ATP supply. Radical reduction of ATP blocks exit from the RER and eliminates transitional elements (Jamieson & Palade, 1968; Balch & Keller, 1986; Merisko *et al.*, 1986b; Tartakoff, 1986). ATP is also required for transport of proteins out of the Golgi (Orci *et al.*, 1985; Schekman, 1985; Tartakoff, 1986).

Conclusion

The significance of the information now available with regard to (1) structural characteristics of Golgi residents, and (2) agents which perturb transport, cannot be completely evaluated. Ultimately, our understanding of the mechanisms operating in Golgi traversal (reflected by the effects of perturbants) will allow prediction of the fate of individual proteins delivered from the RER to the Golgi. These mechanisms in rudimentary form have been frequently discussed (Tartakoff, 1982; Dunphy & Rothman, 1985), but not adequately resolved. We therefore will continue to see interaction between studies focused on the structure of the substrates of transport and studies of perturbation as well as investigation of transport mutants and attempts at cell-free reconstitution.

References

Adams, G. & Rose, J. (1985). Incorporation of a charged amino acid into the membrane-spanning domain blocks transport but not membrane anchoring of a viral glycoprotein. **Molecular and Cellular Biology** 5:1442–1448.

Allan, V.J. & Kreis, T.E. (1986). A microtubule-binding protein associated with membranes of the Golgi apparatus. **Journal of Cell Biology** 103:2229–2239.

Anderson, R.G. & Pathak, R.K. (1985). Vesicles and cisternae in the trans Golgi apparatus of human fibroblasts are acidic compartments. **Cell** 40:635–643.

Balch, W. & Keller, D. (1986). ATP coupled transport of VSV G protein. **Journal of Biological Chemistry** 261:14690–14696.

Barr, R., Safranski, K., Sun, I.L., Crane, F.L. & Morré, D.J. (1984). A electrogenic proton pump associated with the Golgi apparatus of mouse liver driven by NADH and ATP. **Journal of Biological Chemistry** 259:14064–14067.

Bole, D.G., Hendershot, L.M. & Kearney, J.F. (1986). Post-translational association of immunoglobulin heavy chain binding protein with nascent heavy chains in nonsecreting and secreting hybridomas. **Journal of Cell Biology** 102:1558–1566.

Brown, W., Goodhouse, J. & Farquhar, M. (1986). Mannose-6-phosphate receptors for lysosomal enzymes cycle between the Golgi complex and endosomes. **Journal of Cell Biology** 103:1235-1248.

Burke, B. & Warren, G. (1984). Microinjection of mRNA coding for an anti-Golgi antibody inhibits the intracellular transport of a viral membrane protein. **Cell** 36:847-856.

Chicheportiche, Y. & Tartakoff, A. (1984). Characterization of cytoplasmically oriented Golgi proteins with a monoclonal antibody. **Journal of Cell Biology** 99:2200-2210.

Chicheportiche, Y. & Tartakoff, A. (1987). The use of antibodies for analysis of the secretory and endocytic paths of eukaryotic cells. **Subcellular Biochemistry** 12:234–275.

Copeland, C.S., Doms, R.W., Bolzau, E.M., Webster, R.G. & Helenius, A. (1986). Assembly of influenza hemagglutinin trimers and its role in intracellular transport. **Journal of Cell Biology** 103:1179-1191.

Crane, F.L. & Morré, D.J. (1977). Evidence for coenzyme Q function in Golgi membranes. In: Folkers, K. & Yamamura, Y., eds. **Biomedical and Clinical Aspects of Coenzyme Q**. Amsterdam, Elsevier, pp.3-14.

De Camilli, P., Moretti, M., Donini, S.D., Walter, U. & Lohmann, S.M. (1986). Heterogenous distribution of the cAMP receptor protein RII in the nervous system: evidence for its intracellular accumulation on microtubules, microtubule-organizing centers, and in the area of the Golgi complex. **Journal of Cell Biology** 103:189-203.

Dehm, P. & Prockop, D.J. (1972). Time lag in the secretion of collagen by matrix-free tendon cells and inhibition of the secretory process by colchicine and vinblastine. **Biochimica et Biophysica Acta** 264:375-382.

Deutscher, S.L., Creek, K.E., Merion, M. & Hirschberg, C.B. (1983). Subfractionation of rat liver Golgi apparatus: separation of enzyme activities involved in the biosynthesis of the phosphomannosyl recognition marker in lysosomal enzymes. **Proceedings of the National Academy of Sciences USA** 80:3938-3942.

Doyle, C., Sambrook, J. & Gething, M-J. (1986). Analysis of progressive deletions of the transmembrane and cytoplasmic domains of influenza hemagglutinin. **Journal of Cell Biology** 103:1193-1204.

Dunphy, W. & Rothman, J. (1983). Compartmentation of asparagine-linked oligosaccharide processing in the Golgi apparatus. **Journal of Cell Biology** 97:270–275.

Dunphy, W. & Rothman, J. (1985). Compartmental organization of the Golgi stack. **Cell** 42:13–21.

Dunphy, W.G., Pfeffer, S.R., Clary, D.O., Wattenberg, B.W., Glick, B.S. & Rothman, J.E. (1986). Yeast and mammals utilize similar cytosolic components to drive protein transport through the Golgi complex. **Proceedings of the National Academy of Science USA** 83:1622–1626.

Edman, J.C., Ellis, L., Blacher, R.W., Roth, R.A. & Rutter, W.J. (1985). Sequence of protein disulphide isomerase and implications of its relationship to thioredoxin. **Nature** 317:267–270.

Ehrlich, H.P., Ross, R. & Bornstein, P. (1974). Effects of antimicrotubular agents on the secretion of collagen. **Journal of Cell Biology** 62:390–404.

Farquhar, M. (1985). Progress in unraveling pathways of Golgi traffic. **Annual Review of Cell Biology** 1:447–488.

Fishman, J.B. & Fine, R.E. (1987). A *trans* Golgi-derived exocytic coated vesicle can contain both newly synthesized cholinesterase and internalized transferrin. **Cell** 48:157–164.

Friend, D.S. & Murray, M.J. (1959). Osmium impregnation of the Golgi apparatus. **American Journal of Anatomy** 117:135–150.

Fuller, S.D., Bravo, R. & Simons, K. (1985). An enzymatic assay reveals that proteins destined for the apical or basolateral domains of an epithelial cell line share the same late Golgi compartments. **The EMBO Journal** 4:297–307.

Gabel, C. & Bergmann, J.(1985). Processing of the asparagine-linked oligosaccharide of secreted and intracellular forms of the VSVG protein: *in vivo* evidence of Golgi apparatus compartmentalization. **Journal of Cell Biology** 101:460–469.

Gething, M-J., McCammon, K. & Sambrook, J. (1986). Expression of wild-type and mutant forms of influenza hemagglutinin: The role of folding in intracellular transport. **Cell** 46:939–950.

Glickman, J., Croen, K., Kelly, S. & Al-Awqati, Q. (1983). Golgi membranes contain an electrogenic H^+ pump in parallel to a chloride conductance. **Journal of Cell Biology** 97:1303–1308.

Goldberg, D.E. & Kornfeld, S. (1983). Evidence for extensive subcellular organization of asparagine-linked oligosaccharide processing and lysosomal enzyme phosphorylation. **The Journal of Biological Chemistry** 258:3159–3165.

Gonzalez-Noriega, A., Grubb, J.H., Talkad, V. & Sly, W.S. (1980). Chloroquine inhibits lysosomal enzyme pinocytosis and enhances lysosomal enzyme secretion by impairing receptor recycling. **Journal of Cell Biology** 85:839–852.

Griffiths, G., Warren, G., Stuhlfauth, I. & Jockusch, B.M. (1981). The role of clathrin-coated vesicles in acrosome formation. **European Journal of Cell Biology** 26:52–60.

Griffiths, G., Pfeiffer, S., Simons, K. & Matlin, K. (1985). Exit of newly synthesized membrane proteins from the *trans* cisterna of the Golgi complex to the plasma membrane. **Journal of Cell Biology** 101:949–964.

Guan, J-L. & Rose, J. (1984). Conversion of a secretory protein into a trans-membrane protein results in its transport to the Golgi complex but not to the cell surface. **Cell** 37:779–787.

Guan, J-L., Machamer, C. & Rose, J. (1985). Glycosylation allows cell-surface transport of an anchored secretory protein. **Cell** 42:489–496.

Hardwick, J., Shaw, K., Wills, J. & Hunter, E. (1986). Amino-terminal deletion mutants of the Rous sarcoma virus glycoprotein do not block signal cleavage but can block intracellular transport. **Journal of Cell Biology** 103:829–838.

Hauri, H-P., Roth, J., Sterchi, E.E. & Lentze, M.J. (1985). Transport to cell surface of intestinal sucrase-isomaltase is blocked in the Golgi apparatus in a patient with congenital sucrase-isomaltase deficiency. **Proceedings of the National Academy of Science USA** 82:4423–4427.

Hirschberg, C.B. & Snider, M.D. (1987). Topography of glycosylation in the rough endoplasmic reticulum and Golgi apparatus. **Annual Review of Biochemistry**, 56:63–87.

Jamieson, J. & Palade, G. (1968). Intracellular transport of secretory proteins in the pancreatic exocrine cell. IV. Metabolic requirements. **Journal of Cell Biology** 39:589–603.

Kelly, R.B. (1985). Pathways of protein secretion in eukaryotes. **Science** 230:25–32.

Kreis, T.E. & Lodish, H.F. (1986). Oligomerization is essential for transport of vesicular stomatitis viral glycoprotein to the cell surface. **Cell** 46:929–937.

Lemansky, P., Gieselmann, V., Hasilik, A. & von Figura, K. (1984). Cathepsin D and β-hexosaminidase synthesized in the presence of 1-deoxynojirimycin accumulate in the endoplasmic reticulum. **Journal of Biological Chemistry** 259:10129–10135.

Lodish, H.F. & Kong, N. (1984). Glucose removal from N-linked oligosaccharides is required for efficient maturation of certain secretory glycoproteins from the rough endoplasmic reticulum to the Golgi complex. **Journal of Cell Biology** 98:1720–1729.

Lodish, H.F., Kong, N., Hirami, S. & Rasmussen, J. (1987). A vesicular intermediate in the transport of hepatoma secretory proteins from the RER to the Golgi complex. **Journal of Cell Biology** 104:221–230.

Machamer, C., Florkiewicz, R. & Rose, J. (1985). A single N- linked oligosaccharide at either of the two normal sites is sufficient for transport of VSV G protein to the cell surface. **Molecular and Cellular Biology** 5:3074–3083.

Matlin, K. & Simons, K. (1983). Reduced temperature prevents transfer of a membrane glycoprotein to the cell surface but does not prevent terminal glycosylation. **Cell** 34:233–243.

Matlin, K.S. (1986). Ammonium chloride slows transport of the influenza virus hemagglutinin but does not cause mis-sorting in a polarized epithelial cell line. **Journal of Biological Chemistry** 261:15172–15178.

Medda, S. & Swank, R.T. (1985). Egasyn, a protein which determines the subcellular distribution of β-glucuronidase, has esterase activity. **Journal of Biological Chemistry** 260:15802–15808.

Meldolesi, J., Jamieson, J.D. & Palade, G.E. (1971). Composition of cellular membranes in the pancreas of the guinea pig. **Journal of Cell Biology** 49:130–149.

Merisko, E.M., Farquhar, M.G. & Palade, G.E. (1986a). Redistribution of clathrin heavy and light chains in anoxic pancreatic acinar cells. **Pancreas** 1:110–123.

Merisko, E.M., Farquhar, M.G. & Palade, G.E. (1986b). The reorganization of the Golgi complex in anoxic pancreatic acinar cells. **Pancreas** 1:95–109.

Mollenhauer, H.H. & Morré, D.J. (1978). Structural compartmentation of the cytosol: zones of exclusion, zones of adhesion, cytoskeletal and intercisternal elements. **Subcellular Biochemistry** 5:327–359.

Moore, H-P., Gumbiner, B. & Kelly, R.B. (1983). Chloroquine diverts ACTH from a regulated to a constitutive secretory pathway in AtT-20 cells. **Nature** 302:434–436.

Morré, D.J., Morré, D.M. & Heidrich, H-G. (1983). Subfractionation of Golgi apparatus by free-flow electrophoresis. **European Journal of Cell Biology** 31:263–274.

Munro, S. & Pelham, H.R.B. (1986). An Hsp7O-like protein in the ER: identity with the 78 kd glucose-regulated protein and immunoglobulin heavy chain binding protein. **Cell** 46:291– 300.

Nigg, E.A., Hilz, H., Eppenberger, H.M. & Dutly, F. (1985a). Rapid and reversible translocation of the catalytic subunit of cAMP-dependent protein kinase type II from the Golgi complex to the nucleus. **The EMBO Journal** 4:2801–2806.

Nigg, E.A., Schafer, G., Hilz, H. & Eppenberger, H.M. (1985b). Cyclic-AMP-dependent protein kinase type II is associated with the Golgi complex and with centrosomes. **Cell** 41:1039–1051.

Novikoff, A.B. (1976). The endoplasmic reticulum: a cytochemist's view (A Review). **Proceedings of the National Academy of Sciences USA** 73:2781–2787.

Nyquist, S.E., Barr, R. & Morré, D.J. (1970). Ubiquinone from rat liver Golgi apparatus fractions. **Biochimica et Biophysica Acta** 208:532–534.

Nyquist, S.E., Crane, F.L. & Morré, D.J. (1971a). Vitamin A: concentration in the rat liver Golgi apparatus. **Science** 173:939–941.

Nyquist, S.E., Matschiner, J.T. & Morré, D.J. (1971b). Distribution of vitamin K among rat liver cell fractions. **Biochimica et Biophysica Acta** 244:645–649.

Oda, K. & Ikehara, Y. (1981). Inhibitory effect of colchicine on translocation of alkaline phosphatase to the plasma membrane concomitant to its induction in rat liver. **Biochimica et Biophysica Acta** 640:398–408.

Orci, L., Montesano, R., Meda, P., Malaisse-Lagae, F., Brown, D., Perrelet, A. & Vassalli, P. (1981). Heterogeneous distribution of filipin-cholesterol complexes across the cisternae of the Golgi apparatus. **Proceedings of the National Academy of Science USA** 78:293–297.

Orci, L., Ravazzola, M., Amherdt, M., Madsen, O., Vassalli, J-D. & Perrelet, A. (1985). Direct identification of prohormone conversion site in insulin-secreting cells. **Cell** 42:671–681.

Orci, L., Glick, B. & Rothman, J. (1986a). A new type of coated vesicular carrier that appears not to contain clathrin. **Cell** 46:171–184.

Orci, L., Ravazzola, M., Amherdt, M., Madsen, O., Perrelet, A., Vassalli, J-D. & Anderson, R.G.W. (1986b). Conversion of proinsulin to insulin occurs coordinately with acidification of maturing secretory vesicles. **Journal of Cell Biology** 103:2273–2281.

Paquet, M.R., Pfeffer, S.R., Burczak, J.D., Glick, B. S. & Rothman, J.E. (1986). Components responsible for transport between successive Golgi cisternae are highly conserved in evolution. **Journal of Biological Chemistry** 261:4367–4370.

Pohlman, R., Krauger S., Hasilik, A. & von Figura, K. (1984). Effect of monensin on intracellular transport and receptor-mediated endocytosis of lysosomal enzymes. **Biochemical Journal** 217:649–658.

Redman, C.M., Banerjee, D., Manning, C., Huang, C.Y. & Green, K. (1978). *In vivo* effect of colchicine on hepatic protein synthesis and on the conversion of proalbumin to serum albumin. **Journal of Cell Biology** 77:400–416.
Rindler, M.J., Ivanov, I.E. & Sabatini, D.D. (1987). Microtubule-acting drugs lead to the nonpolarized delivery of the influenza hemagglutinin to the cell surface of polarized Madin-Darby canine kidney cells. **Journal of Cell Biology** 104:231–241.
Rizzolo L., Finidori, J., Gonzalez, A., Arpin, M., Ivanov, I., Adesnik, M. & Sabatini, D. (1985). Biosynthesis and intra-cellular sorting of growth hormone-viral envelope glycoprotein hybrids. **Journal of Cell Biology** 101:1351–1362.
Rogalski, A.A., Bergmann, J.E. & Singer, S.J. (1984). Effect of microtubule assembly status on the intracellular processing and surface expression of an integral protein of the plasma membrane. **Journal of Cell Biology** 99:1101–1109.
Rome, H., Ockling, C., Wiman, K., Rask, L., Obrink, B. & Peterson, P.A. (1983). Ligand-dependent regulation of intracellular protein transport: Effect of vitamin A on the secretion of the retinol-binding protein. **Journal of Cell Biology** 96:907–910.
Rose, J. & Bergmann, J. (1983) Altered cytoplasmic domain affect intracellular transport of the VSV glycoprotein. **Cell** 34:513–524.
Roth, J., Lentze, M.J. & Berger, E.C. (1985). Immunocytochemical demonstration of ecto-galactosyltransferase in absorptive intestinal cells. **Journal of Cell Biology** 100:118–125.
Roth, J., Tantjes, D.J., Weinstein, J., Paulson, J.C. & Greenwell, P. (1986). Differential subcompartmentation of terminal glycosylation in the Golgi apparatus of intestinal absorptive and goblet cells. **The Journal of Biological Chemistry** 261:14307–14312.
Salas, P.J.I., Misek, D.E., Vega-Salas, D.E., Gundersen, D., Cereijido, M. & Rodriguez-Boulan, E. (1986). Microtubules and actin filiments are not critically involved in the biogenesis of epithelial cell surface polarity. **Journal of Cell Biology** 102:1853–1867.
Sandoval, I.V., Bonifacino, J.S., Klausner, R.D., Henkart, M. & Wehland, J. (1984). Role of microtubules in the organization and localization of the Golgi apparatus. **Journal of Cell Biology** 99:113s–118s.
Saraste, J. & Kuismanen, E. (1984). Pre- and post-Golgi vacuoles operate in the transport of Semliki Forest virus membrane glycoproteins to the cell surface. **Cell** 38:535–549.
Saraste, J., Palade, G.E., Farquhar, M.G. (1986). Temperature-sensitive steps in the transport of secretory proteins through the Golgi complex in exocrine pancreatic cells. **Proceedings of the National Academy of Sciences USA** 83:6425–6429.
Schekman, R. (1985). Protein localization and membrane traffic in yeast. **Annual Review of Cell Biology** 1:115–143.
Schwartz, A.L., Strous, G.J.A.M., Slot, J.W. & Geuze, H.J. (1985). Immunoelectron microscopic localization of acidic intracellular compartments in hepatoma cells. **The EMBO Journal** 4:899–904.
Simons, K. & Fuller, S. (1985). Cell surface polarity in epithelia. **Annual Review of Cell Biology** 1:243–288.
Simons, K. & Griffiths, G. (1986). The *trans* Golgi network: sorting at the exit site of the Golgi complex. **Science** 234:438–443.
Snider, M.D. & Rogers, O. (1986). Membrane traffic in animal cells: Cellular glycoproteins return to the site of Golgi mannosidase I. **Journal of Cell Biology** 103:265–275.
Strous, G.J., du Maine, A., Zijderhand-Bleekmolen, J.E., Slot, J.W. & Schwartz, A.L. (1985a). Effect of lysosomotropic amines on the secretory pathway and on

the recycling of the asialoglycoprotein receptor in human hepatoma cells. **Journal of Cell Biology** 101:531–539.

Strous, G.J., van Kerkhof, P., Willemsen, R., Slot, J.W. & Geuze, H.J. (1985b). Effect of monensin on the metabolism, localization, and biosynthesis of N- and O- linked oligosaccharides of galactosyltransferase. **European Journal of Cell Biology** 36:256–262.

Tartakoff, A.M. (1982). Simplifying the complex Golgi. **Trends in Biochemical Sci-** 7:174–176.

Tartakoff, A.M. (1983a). Perturbation of vesicular traffic with the carboxylic ionophore monensin. **Cell** 32:1026–1028.

Tartakoff, A.M. (1983b). The confined function model of the Golgi complex: center for ordered processing of biosynthetic products of the rough endoplasmic reticulum. **International Review of Cytology** 85:221–252.

Tartakoff, A.M. (1986). Temperature and energy dependence of secretory protein transport in the exocrine pancreas. **The EMBO Journal** 5:1477–1482.

Tartakoff, A.M. & Vassalli, P. (1977). Plasma cell immunoglobulin secretion. **Journal of Experimental Medicine** 146:1332–1345.

Thorens, B. & Vassalli, P. (1986). Chloroquine and ammonium chloride prevent terminal glycosylation of immunoglobulins in plasma cells without affecting secretion. **Nature** 321:618–620.

Thyberg, J. & Moskalewski, S. (1985). Microtubules and the organization of the Golgi complex. **Experimental Cell Research** 159:1–16.

Tooze, J. & Tooze, S.A. (1986). Clathrin-coated vesicular transport of secretory proteins during the formation of ACTH-containing secretory granules in AtT2O cells. **Journal of Cell Biology** 103:839–850.

von Figura, K. & Hasilik, A. (1986). Lysosomal enzymes and their receptors. **Annual Review of Biochemistry** 55:167–193.

Wagner, D.D., Mayadas, T. & Marder, V.J. (1986). Initial glycosylation and acidic pH in the Golgi apparatus are required for multimerization of von Willibrand factor. **Journal of Cell Biology** 102:1320–1324.

Warren, G. (1985). Membrane traffic and organelle division. **Trends in Biochemical Science** 10:439–443.

Wattenberg, B.W., Balch, B. & Rothman, J. (1986). A novel prefusion complex formed during protein transport between Golgi cisternae in a cell-free system. **Journal of Biological Chemistry** 251:2202–2207.

Wattenberg, B.W. & Rothman, J.E. (1986). Multiple cytosolic components promote intra-Golgi protein transport. **Journal of Biological Chemistry** 261:2208–2213.

Wehland, J. & Sandoval, I. (1983). Cells injected with guanosine 5′(α, β-methylene) triphosphate, an α, β non-hydrolyzable analog of GTP, show anomalous patterns of tubulin polymerization affecting cell translocation, intracellular movement and the organization of Golgi elements. **Proceedings of the National Academy of Sciences, USA** 80:1938–1941.

Whetsell, W. & Bunge, R. (1969). Reversible alterations in the Golgi complex of cultured neurons treated with an inhibitor of active Na and K transport. **Journal of Cell Biology** 42:490–500.

Wills, J., Hardwick, M., Shaw, K. & Hunter, E. (1983). Alterations in the transport and processing of Rous sarcoma virus envelope glycoproteins mutated in the signal and anchor regions. **Journal of Cell Biochemistry** 23:81–91.

Lysosomal enzymes and the Golgi apparatus

ANDREJ HASILIK AND KURT VON FIGURA

Physiologisch-Chemisches Institut, Universität Münster, Waldeyerstrasse 15, D-4400 Muenster, FRG.

Abstract

The phosphorylation dependent segregation of lysosomal enzyme precursors is one of the best studied paradigms of protein sorting in eukaryotic cells. Early during the passage through the Golgi apparatus, soluble lysosomal enzyme precursors acquire mannose 6-phosphate residues which are the recognition markers for the segregation of these enzyme precursors from the secretory pathway. The recognition marker mediates binding of the precursors to mannose 6-phosphate specific precursors, which deliver the bound ligands to a prelysosomal compartment. A 215 kD receptor is involved in both intracellular sorting and uptake of lysosomal enzymes from the extracellular fluid. The findings on the activity of several lysosomal hydrolases in the reproductive tract and their release during sperm capacitation should stimulate research on the role of lysosomal enzymes in reproduction.

Key words

Lysosomal enzymes, Golgi apparatus, mannose 6-phosphate, mannose 6-phosphate receptor, coated vesicles, cathepsin D.

Introduction

Since the discovery of lysosomes by Christian De Duve (reviewed by Bainton, 1981) biochemistry has been facing the question of how lysosomal enzymes are delivered into lysosomes. There are two aspects to the transport of lysosomal enzymes, one founded in cell biology and the other in biochemistry. The former aspect focuses on the organelles involved, mainly the Golgi apparatus (Farquhar & Palade, 1981) and the latter on the recognition of lysosomal enzymes by specific receptors.

The Golgi apparatus is a polarized organelle, which plays a pivotal role in the modification and segregation of lysosomal enzymes. Various modification reactions occur in different subcompartments of the Golgi-apparatus, which are referred to as cis-, mid- and trans-Golgi. Segregation takes place in the distal Golgi area, which is characterized by a network of tubules and cisternae. This area was recognized as an organelle by the late Alex B. Novikoff, who referred to it by the acronym GERL, Golgi–Endoplasmic

Reticulum–Lysosome, and proposed that it is specialized for the biogenesis of lysosomes and secretory granules (Novikoff, 1976). These functions are accepted in the current views of the Golgi complex (Griffiths & Simons, 1986), although the segregation organelle is referred to as the trans-Golgi network or trans-Golgi reticulum rather than GERL.

Studies on the biochemical aspects of the transport of lysosomal enzymes were initiated by the discovery of a recognition marker in lysosomal enzymes by Hickman & Neufeld (1972). Studies in a number of laboratories contributed to the identification of the marker, a phosphorylated oligosaccharide, and its receptors.

Several recent reviews deal in detail with the phosphorylation, transport and maturation of lysosomal enzymes, with mannose 6-phosphate specific receptors and with alternative pathways of the transport of lysosomal enzymes (Creek & Sly, 1984; Goldberg *et al.*, 1984; Kornfeld & Kornfeld, 1985; Sahagian, 1984; von Figura & Hasilik, 1986; Kornfeld, 1986). The aim of this chapter is to survey the maturation of carbohydrate side chains in lysosomal enzymes in the Golgi apparatus and to assess the importance of their recognition in the segregation of lysosomal enzymes.

Modifications of the carbohydrate in lysosomal enzymes

Status of the oligosaccharides entering the Golgi apparatus

All lysosomal enzymes analyzed to date contain asparagine (Asn)-linked oligosaccharides. In general, a precursor of such oligosaccharides is transferred from a lipid intermediate, Glc_3-Man_9-$GlcNAc_2$-diphosphodolichol, to certain Asn-residues in nascent glycoproteins within the rough endoplasmic reticulum. The Asn-residues that are able to be glycosylated are recognized as part of an amino acid sequence comprising Asn-X-Ser/Thr (Marshall, 1972). Prior to delivery to the Golgi apparatus, the glucose residues and at least one mannose are hydrolyzed from the side chains. Inhibition of the trimming glucosidases by deoxynojirimycin results in an inhibition of the transport and accumulation of the glucosylated glycoproteins in the endoplasmic reticulum (ER) or in the transitional elements (TE). This has been shown for a variety of glycoproteins including the lysosomal marker cathepsin D (Lemansky *et al.*, 1984; Nauerth *et al.*, 1985). Transport from TE to Golgi is mediated by vesicles (Lodish *et al.*, 1987) and it is possible that the packaging in these vesicles depends on deglucosylation due to specific receptors that recognize either the glucosylated (to be retained) or the deglucosylated (to be transported) glycoproteins. Alternatively, the deglucosylation may be required for the proper folding and solubilization of the glycoproteins after their synthesis.

High mannose and complex oligosaccharides

All known modifications of lysosomal enzymes in the Golgi apparatus concern their carbohydrate moiety. These modifications result in a variety of carbohydrate structures that have been identified in lysosomal enzymes and are listed in Table 1. In lysosomal membrane proteins few side chains other than those of the complex type have been identified (Takasaki *et al.*, 1984; Lewis *et al.*, 1985). In contrast, soluble lysosomal enzymes contain a rather small proportion of complex oligosaccharides as determined by either analysis of purified enzymes, e.g. β-hexosaminidase (Freeze *et al.*, 1979; Tucker *et al.*, 1980; Overdijk *et al.*, 1985), α-fucosidase (Chien & Dawson, 1980), β-glucuronidase (Byrd & Touster, 1981; Mizuochi *et al.*, 1981; Howard *et al.*, 1982), several cathepsins (Takahashi *et al.*, 1983 and 1984; Nakao *et al.*, 1984; Taniguchi *et al.*, 1985), α-mannosidase (Mathur *et al.*, 1984) and α-glucosidase (Mutsaers *et al.*, 1987), or of immunoprecipitated metabolically labeled enzymes such as β-glucuronidase, cathepsin D, β-hexosaminidase, arylsulfatase B and α-galactosidase, which contain some tri- or tetra-antennary oligosaccharides (Goldberg & Kornfeld, 1981; Hasilik & von Figura, 1981; Steckel *et al.*, 1983; LeDonne *et al.*, 1983). From the extent of the resistance to endoglucosaminidase H, an enzyme which cleaves most high mannose and hybrid, but not complex oligosaccharides (Kobata, 1979), and from labeling with [^{3}H]galactose, it has been concluded that in fibroblasts newly synthesized cathepsin D has at the most 30% of the oligosaccharide side chains processed to the complex type (Hasilik & von Figura, 1981; Gieselmann *et al.*, 1983).

The synthesis of complex oligosaccharides depends on an ordered sequence of trimming and glycosylating reactions which was discussed in detail by Kornfeld & Kornfeld (1985). Briefly, the two α1-2-linked mannose residues, which became uncovered by the action of the glucosidases are

Table 1. *Oligosaccharide side chains in lysosomal enzymes*

Oligosaccharide	Lysosomal enzymes	
	soluble	membrane
1 high mannose	+	+/−
2 high mannose phosphorylated	+	−
3 hybrid	+/−	−
4 hybrid phosphorylated	+/−	−
5 complex	+/−	+
6 complex sulfated	+/−	?

The symbols represent regular (+), circumstantial (+/−) and unsearched for (?) occurrence or absence (−) of the oligosaccharides.

trimmed by the action of mannosidase I. This results in uncovering of an α3-linked mannose residue, which is then substituted by a GlcNAcβ 1-2 residue due to GlcNAc-transferase I. The product of this reaction is further trimmed by mannosidase II to yield GlcNAcβl-2Manαl-3(Manαl-6)Manβl-4GlcNAcβ l-4GlcNAcβ, which is then elongated by an ordered action of several enzymes that transfer GlcNAc, Gal, and neuraminic acid residues to the oligosaccharide.

Phosphorylated oligosaccharides

As mentioned above, most of the oligosaccharides present in soluble lysosomal enzymes belong to the high mannose type. Some of these oligosaccharides are phosphorylated. The phosphorylated oligosaccharides in lysosomal enzymes represent the recognition marker, which in 1972 was postulated (Hickman & Neufeld, 1972) to mediate endocytosis of lysosomal enzymes. In 1977 Kaplan *et al.*, reported on inhibition of the endocytosis of β-glucuronidase by mannose 6-phosphate and by the pretreatment of the enzymes with alkaline phosphatase, thus indicating that the marker may contain mannose 6-phosphate residues. These observations have been confirmed for other lysosomal enzymes (Sando & Neufeld, 1977; Ullrich *et al.*, 1978) and were soon followed by demonstration of the presence of mannose 6-phosphate in hydrolysates of lysosomal enzymes or in oligosaccharides cleaved from these enzymes by endoglucosaminidase H (Natowicz *et al.*, 1979; Hasilik & Neufeld, 1980b). Phosphate was found in oligosaccharides in nearly all soluble lysosomal enzymes examined (Tabas & Kornfeld 1980; Myerowitz & Neufeld, 1981; Goldberg & Kornfeld, 1981; Hasilik *et al.*, 1982; Sweeley *et al.*, 1983; Steckel *et al.*, 1983; Nakao *et al.*, 1984; Mainferme *et al.*, 1985; Gal *et al.*, 1985; Reuser *et al.*, 1985) including the classical marker for lysosomes, acid phosphatase (Lemansky *et al.*, 1985) and the azurophilic granule enzyme myeloperoxidase (Hasilik *et al.*, 1984; Stromberg *et al.*, 1986).

Phosphorylated oligosaccharides in lysosomal enzymes contain 5–9, predominantly 6–8, mannose residues. The largest derivative is the typical product of the ER α-mannosidase, which trims the central branch of the oligosaccharide such that it terminates in the α1-3 linked residue:

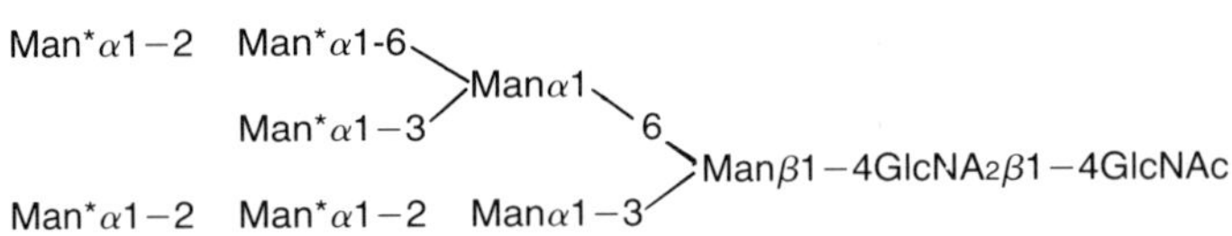

and the smaller species may be formed due to trimming in the Golgi apparatus of some of the terminal residues. The residues marked by the asterisks in the above structure represent those which may carry phosphate groups (Varki & Kornfeld, 1980b), the inner core Manα 1-3(Man 1-6)Manβ is not phosphorylated. The phosphorylated oligosaccharides carry at a maximum two phosphate groups (Hasilik *et al.*, 1980; Varki & Kornfeld, 1980b, 1983; Goldberg & Kornfeld, 1981; Natowicz *et al.*, 1982).

Synthesis of mannose 6-phosphate residues

Mannose 6-phosphate residues are formed in lysosomal enzyme precursors in two reactions:

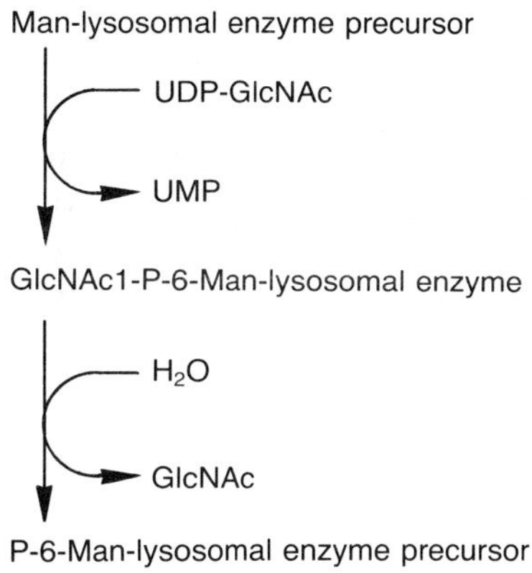

In the first reaction N-acetylglucosaminyl-1-phosphate is transferred to the C-6 hydroxyl of a peripheral mannose residue. The enzyme catalyzing this reaction, UDP-N-acetylglucosamine: lysosomal enzyme N-acetylglucosaminyl-1-phosphotransferase, has been found in a variety of cells and tissues (Reitman & Kornfeld, 1981a; Hasilik *et al.*, 1981a; Waheed *et al.*, 1982b) and partially purified from rat liver (Waheed *et al.*, 1982). This enzyme phosphorylates high mannose oligosaccharides in lysosomal enzymes, but not (or inefficiently) in other glycoproteins (Reitman & Kornfeld, 1981b; Waheed *et al.*, 1982a). The enzyme also phosphorylates simple carbohydrates such as α-methylmannoside, but with much lower efficiency than lysosomal enzymes (Reitman & Kornfeld, 1981 b; Waheed *et al.*, 1982). In mannobiose derivatives the enzyme phosphorylates preferentially the non-reducing terminal residue (Madiyalakan *et al.*, 1987). The phosphorylation of lysosomal enzymes is abolished after their denaturation (Reitman & Kornfeld, 1981b; Waheed *et al.*, 1982; Lang *et al.*, 1984). The reaction appears to depend on recognition by the transferase of a structural signal in lysosomal enzymes. The phosphorylation of lysosomal enzymes is inhibited in the presence of deglycosylated lysosomal enzymes (Lang *et al.*, 1984). The primary structure of several lysosomal enzymes has been deduced from the nucleotide sequence of their cDNA's. However, the signal for phosphorylation has not been elucidated. It is likely that the signal depends on the tertiary structure of the lysosomal enzymes.

The transfer of N-acetylglucosaminyl 1-phosphate residues to lysosomal enzymes is followed by removal of the N-acetyl-glucosaminyl residues. The enzyme catalyzing this reaction N-acetylglucosamine 1-phosphate diester α-N-acetylglucosaminidase has been partially purified from several tissues and has been shown to be distinct from the lysosomal α-N-acetyl-glucosaminidase (Varki & Kornfeld, 1980, 1981; Waheed *et al.*, 1981). It is a glycosidase rather than a phosphodiesterase (Varki *et al.*, 1983). Both the transferase and the N-acetylglucosamine 1-phosphodiester α-N-acetylglucosaminidase fractionate with Golgi membranes (Waheed *et al.*, 1981; Varki & Kornfeld, 1980a, 1981; Pohlmann *et al.*, 1982). Goldberg & Kornfeld (1983) have shown that these enzymes can be separated from α-mannosidase II and galactosyltransferase and localized them to the cis-region of the Golgi complex. The polarized distribution of the processing enzymes in the Golgi apparatus seems to determine the direction of the processing of lysosomal enzymes. It appears that the phosphorylation protects mannose residues from trimming and thus, prevents the conversion of the oligosaccharides to the complex type. Oligosaccharides which do not become phosphorylated on the branch linked α1-3 to the β-mannosyl residue are subject to further processing and may give rise to phosphorylated hybrid oligosaccharides (Varki & Kornfeld, 1983; Yamamoto *et al.*, 1985). In mutant cells unable to phosphorylate lysosomal enzymes an increased proportion of the oligosaccharide side chains are converted to the complex type (Vladutiu & Rattazzi, 1975; Hasilik & von Figura, 1981; Vladutiu, 1983; Kress et al., 1980; Miller et al., 1981).

Recently, we have observed that certain lysosomal enzymes can be labeled with sulfate in cultured fibroblasts (Braulke *et al.*, 1987). [^{35}S]Sulfate was incorporated into cathepsin D, arylsulfatase A and α-chain of β-hexosaminidase. The β-chain of β-hexosaminidase was not labeled. The radioactivity was associated with complex type oligosaccharides. It was sensitive to endoglycosidase F and the labeling was inhibited in the presence of deoxynojirimycin, which is known to inhibit ER glucosidases and thus, to interfere with the formation of complex oligosaccharides (Saunier *et al.*, 1982; Peyrieras *et al.*, 1983). The function of the sulfate label in lysosomal enzymes is not known. Previously, sulfation of complex oligosaccharides in various non-lysosomal glycoproteins has been reported (reviewed by Kornfeld & Kornfeld, 1985). The sulfate label is lost from the lysosomal enzymes within a few hours. In contrast, the phosphate label, which has also been studied in human fibroblasts persists in lysosomal enzymes for more than 1 day (Hasilik & Neufeld, 1980b). In accordance with this observation, Gabel & Foster (1986) have shown that upon endocytosis of β-glucuronidase in mouse L-cells the mannose 6-phosphate marker persists in this enzyme even after the delivery to lysosomes.

Mannose 6-phosphate-dependent segregation of lysosomal enzymes and its efficiency

From the Golgi apparatus the lysosomal enzymes are transported to the trans-Golgi network. Here, the phosphorylated lysosomal enzymes are segregated from other biosynthetic products by virtue of binding to mannose 6-phosphate specific receptors. Two mannose 6-phosphate receptors have been identified: a 215 kD receptor, first isolated by Sahagian *et al.* (1981), and a 46 kD receptor, which was discovered by Hoflack & Kornfeld (1985a,b).

The 215 kD receptor (von Figura *et al.*, 1985; Sahagian & Steer, 1985) and the 46 kD receptor as well (Stein, unpublished observations) are transmembrane glycoproteins that are oriented with the N-terminus to the extracytosolic side. Both the 215 kD receptor (Gabel *et al.*, 1982; Fischer *et al.*, 1982) and the 46 kD receptor (Hoflack *et al.*, 1987) bind preferably oligosaccharides containing two phospho-monoesters and dissociate from them at pH < 6. The 215 kD receptor is involved in the transport of endogenous and exogenous lysosomal enzymes (Gonzalez-Noriega *et al.*, 1980), while the 46 kD receptor participates in endogenous transport only (Stein, unpublished observations). The 215 kD receptor is present in the trans-Golgi area (Sahagian, 1984; Geuze *et al.*, 1984, 1985), in the endocytic compartment (Sahagian, 1984; Geuze *et al.*, 1984; Brown *et al.*, 1986) and in coated vesicles associated with these compartments. The subcellular distribution of the 46 kD receptor is similar to that of the 215 kD receptor (Geuze, personal communication). *In vitro*, the 215 kD receptor is incorporated together with clathrin into coated vesicles (Pearse, 1985).

It appears that the segregation of lysosomal enzymes involves their binding to mannose 6-phosphate specific receptors and packaging of these complexes in coated vesicles (Schulze-Lohoff *et al.*, 1985; Tuemmers *et al.*, 1983; Campbell & Rome, 1983; Kedersha & Rome, 1986). There is ample cytological evidence that coated vesicles are involved in the transport of lysosomal enzymes from the periphery of the Golgi apparatus toward lysosomes (Novikoff, 1976; Holtzman *et al.*, 1967; Bentfeld-Barker & Bainton, 1982; Wehland *et al.*, 1982; Griffiths & Simons, 1986) or the acrosome (Griffiths *et al.*, 1981; Werner, 1986).

With the coated vesicles lysosomal enzymes are delivered to a prelysosomal compartment where a proteolytic processing of the precursors is initiated (Gieselmann *et al.*, 1983). The pathway of this delivery is not clear. It has been demonstrated that the precursor of cathepsin D is associated with both exo- and endocytic coated vesicles (Lemansky *et al.*, 1987). These results are compatible with models of the transport of lysosomal enzymes *via* an intracellular route (Gonzalez-Noriega *et al.*, 1980), or *via* the cell surface (von Figura & Weber, 1978). If the transport occurs *via* the cell surface then the final segregation of receptor-bound lysosomal enzymes from the secretory products entering the non-regulated secretory pathway may take place at the plasma membrane in the period between the fusion of the exocytotic vesicles with the plasma membrane and the reinternalization of the receptor-bound components.

To date it is a matter of discussion whether the transport of the plasma membrane proteins involves coated vesicles (Wehland *et al.*, 1982; Strous *et al.*, 1983; Kinnon & Owen, 1983; Benson *et al.*, 1985; Griffiths & Simons, 1986). In the regulated pathway the nascent secretory granules contain coated membrane areas, but only at an immature stage (Orci *et al.*, 1985; Tooze & Tooze, 1986). While the constitutive secretion may proceed in the absence of clathrin (Payne & Schekman, 1985), there is evidence that coated buds in the trans Golgi area contain lysosomal enzymes and the 215 kD receptor and albumin as well (Geuze *et al.*, 1985). Therefore, it appears that secretory products are not completely excluded from coated vesicles, in which lysosomal enzymes are packaged. By analyzing the contents of exocytic and endocytic coated vesicles for lysosomal enzymes and secretory products it should be possible to differentiate between the alternative pathways.

Mannose 6-phosphate-dependent and independent segregation of lysosomal enzymes

I-cell fibroblasts lack N-acetylglucosaminyl-1-phospho-transferase (Hasilik *et al.*, 1981a, b; Reitman *et al.*, 1981; Hasilik *et al.*, 1981a; Owada & Neufeld, 1982) and do not phosphorylate lysosomal enzymes (Bach *et al.*, 1979; Hasilik & Neufeld, 1980b; Hasilik *et al.*, 1982; Lemansky *et al.*, 1985). As a consequence these cells secrete lysosomal enzymes (Hickman & Neufeld, 1972). In affected patients (I-cell disease) enhanced concentrations of lysosomal enzymes in serum are observed (Wiesmann *et al.*, 1971). Phosphorylation of β-galactosidase is impaired in certain G_{M1}-gangliosidosis patients (Hoogeveen *et al.*, 1986) and the defect results in an enhanced secretion of the precursor of this enzyme in fibroblasts. In certain mutations of β-hexosaminidase (Proia & Neufeld, 1982; Proia *et al.*, 1984; D'Azzo *et al.*, 1984) and α-glucosidase (Reuser *et al.*, 1985) lack of phosphorylation is likely to result from a defect in the transport of the enzyme precursors into the Golgi apparatus. In many cell types other than fibroblasts the segregation of lysosomal enzymes is to a large extent independent of the phosphorylation. Alternative mechanisms for segregation of lysosomal enzymes have not been elucidated to date (see discussion in reviews by Kornfeld, 1986; von Figura & Hasilik, 1986).

Little is known about the transport of lysosomal membrane proteins. Lysosomal membrane proteins do not bind to mannose 6-phosphate receptor (Krentler *et al.*, 1986). β-Glucocerebrosidase (Erickson *et al.*, 1985) and other lysosomal membrane glycoproteins (Lewis *et al.*, 1985; D'Souza

& August, 1986) contain mostly complex oligosaccharides. It has been reported that β-glucosidase is not transported to lysosomes if the conversion of high mannose to the complex oligosaccharides is inhibited (Aerts *et al.*, 1986). In a study on several lysosomal membrane proteins Barriocanal *et al.* (1986) observed that the glycosylation was necessary for stabilization of these proteins rather than for their transport to lysosomes.

Non-segregated lysosomal enzymes

In various cells a characteristic proportion of lysosomal enzymes is subjected to secretion. In fibroblasts secreted lysosomal enzymes are represented by precursors that had not been segregated into the lysosomes (reviewed by Hasilik & von Figura, 1984). Precursors of lysosomal enzymes are secreted also by endothelial cells (Hasilik *et al.*, 1981b). However, macrophages secrete mature forms of lysosomal enzymes in addition to the precursors (Skudlarek & Swank, 1981; Imort *et al.*, 1983; Skudlarek *et al.*, 1984). In human urine both precursor and mature forms of the lysosomal enzymes are found and in human plasma the lysosomal enzymes exist predominantly in the precursor form (Zuehlsdorf *et al.*, 1983). Little is known about the molecular forms of lysosomal enzymes in the seminal plasma which contains rather high concentrations of certain lysosomal enzymes. Caput and cauda epididymal cells in culture have been shown to secrete predominantly the mature form of α-galactosidase (Skudlarek & Orgebin-Crist, 1986). The mature forms of lysosomal enzymes are likely to originate from lysosomes, although extracellular processing cannot be excluded. For a cathepsin-related protein a much higher rate of secretion has been observed in transformed 3T3-cells than in non-transformed cells (Gal *et al.*, 1985). The secreted protein is phosphorylated and the mechanism of the enhancement of secretion is not known. Secretion of precursor cathepsin D in human 3T3 monocytes is enhanced upon preincubation of these cells with 1,25-dihydroxy vitamin D_3 (Stein *et al.*, 1987). In cells, in which segregation of lysosomal enzymes depends on mannose 6-phosphate receptors, such as fibroblasts, secretion of lysosomal enzyme precursors can be stimulated by weak bases and monensin (Gonzalez-Noriega *et al.*, 1980; Hasilik & Neufeld, 1980a). These drugs interfere with the reutilization of the mannose 6-phosphate receptor (Gonzalez-Noriega *et al.*, 1980; Braulke, unpublished observations). It should be of interest to study the mechanisms, which might be utilized in tissues to regulate the extent of the secretion of lysosomal enzymes.

Acknowledgement

The authors wish to dedicate this chapter to the memory of Dr. Alex B. Novikoff.

Note added in proof

From molecular cloning experiments it has been concluded that the 215 kD mannose 6-phosphate receptor is identical to insulin-like growth factor II receptor (Morgan, D.O. *et al.*, **Nature** 329 [1987] 301–307). Therefore, the current literature refers to this receptor as the Man 6-P/IGF-II receptor.

References

Aerts, J.M.F.G., Brul, S., Donker-Koopman, W.E., van Weely, S., Murray, G.J., Barranger, J.A., Tager, J.M. & Schram, A.W. (1986). Efficient routing of glucocerebrosidase to lysosomes requires complex oligosaccharide chain formation. **Biochemical and Biophysical Research Communications** 141:452–458.

Bach, G., Bargal, R. & Cantz, M. (1979). I-cell disease: Deficiency of extracellular hydrolase phosphorylation. **Biochemical and Biophysical Research Communications** 91: 976–981.

Bainton, D.F. (1981). The discovery of lysosomes. **The Journal of Cell Biology** 91:66s–76s.

Barriocanal, J.G., Bonifacino, J.S., Yuan, L. & Sandoval, I.V. (1986). Biosynthesis, glycosylation, movement through the Golgi system, and transport to lysosomes by an N-linked carbohydrate-independent mechanism of three lysosomal integral membrane proteins. **The Journal of Biological Chemistry** 261:16755–16763.

Benson, R.J.J., Porter-Jordan, K., Buoniconti, P. & Fine, R.E. (1985). Biochemical and cytochemical evidence indicates that coated vesicles in chick embryo myotubes contain newly synthesized acetylcholinesterase. **The Journal of Cell Biology** 101:1930–1940.

Bentfeld-Barker, M.E. & Bainton, D.F. (1982). Identification of primary lysosomes in human megakaryocytes and platelets. **Blood** 59:472–481.

Braulke, T., Hille, A., Huttner, W.B., Hasilik, A. & von Figura, K. (1987). Sulfated oligosaccharides in human lysosomal enzymes. **Biochemical and Biophysical Research Communications**, in the press.

Brown, W.J., Goodhouse, J. & Farquhar, M.G. (1986). Mannose-6- phosphate receptors for lysosomal enzymes cycle between the Golgi complex and endosomes. **The Journal of Cell Biology** 103:1235– 1247.

Byrd, J.C. & Touster, O. (1981). Isolation and partial characterization of the oligosaccharide moieties of rat preputial α-glucuronidase. **Biochimica et Biophysica Acta** 677:69–78.

Campbell, C.H. & Rome, L.H. (1983). Coated vesicles from rat liver and calf brain contain lysosomal enzymes bound to mannose 6-phosphate receptors. **The Journal of Biological Chemistry** 258:13347–13352.

Chien, S.F. & Dawson, G. (1980). Purification and properties of two forms of human α-L-fucosidase. **Biochimica et Biophysica Acta** 614:476–488.

Creek, K.E. & Sly, W.S. (1984). The role of the phosphomannosyl receptor in the transport of acid hydrolases to lysosomes. In: Dingle, J.T., Dean, R.T. & Sly, W., eds. **Lysosomes in Biology and Pathology**. Elsevier, pp. 63–82.

D'Azzo, A., Proia, R.L., Kolodny, E.H., Kaback, M.M. & Neufeld, E.F. (1984). Faulty association of α- and β-subunits in some forms of β-hexosaminidase A deficiency. **The Journal of Biological Chemistry** 259:11070–11074.

D'Souza, M.P. & August, J.T. (1986). A kinetic analysis of biosynthesis and localization of a lysosome-associated membrane glycoprotein. **Archives of Biochemistry and Biophysics** 249:522–532.

Erickson, A.H., Ginns, E.I. & Barranger, J.A. (1985). Biosynthesis of the lysosomal enzyme glucocerebrosidase. **The Journal of Biological Chemistry** 260:14319–14324.

Farquhar, M.G. & Palade, G.E. (1981). The Golgi apparatus (complex)-(1954–1981)- from artifact to center stage. **The Journal of Cell Biology** 91:77s–103s.

Fischer, H.D., Creek, K.E. & Sly, W.S. (1982). Binding of phosphorylated oligosaccharides to immobilized phosphomannosyl receptors. **The Journal of Biological Chemistry** 257:9938–943.

Freeze, H., Geiger, B. & Miller, A.L. (1979). Carbohydrate composition of human placental N-acetylhexosaminidase A and B. **Biochemical Journal** 177:749–752.

Gabel, C.A., Goldberg, D.E. & Kornfeld, S. (1982). Lysosomal enzyme oligosaccharide phosphorylation in mouse lymphoma cells: specificity and kinetics of binding to the mannose 6-phosphate receptor *in vivo*. **The Journal of Cell Biology** 95: 536–542.

Gabel, C.A. & Foster, S.A. (1986). Mannose 6-phosphate receptor-mediated endocytosis of acid hydrolases: internalization of α-glucuronidase is accompanied by a limited dephosphorylation. **The Journal of Cell Biology** 103:1817–1827.

Gal, S., Willingham, M.C. & Gottesman, M.M. (1985). Processing and lysosomal localization of a glycoprotein whose secretion is transformation stimulated. **The Journal of Cell Biology** 100:535–544.

Geuze, H.J., Slot, J.W., Strous, G.J.A.M., Peppard, J., von Figura, K., Hasilik, A. & Schwartz, A.L. (1984). Intracellular receptor sorting during endocytosis: Comparative immunoelectron microscopy of multiple receptors in rat liver. **Cell** 37:195–204.

Geuze, H.J., Slot, J.W., Strous, G.J.A.M., Hasilik, A. & von Figura, K. (1985). Possible pathways for lysosomal enzyme delivery. **The Journal of Cell Biology** 101:2253–2262.

Gieselmann, V., Pohlmann, R., Hasilik, A. & von Figura, K. (1983). Biosynthesis and transport of cathepsin D in cultured human fibroblasts. **The Journal of Cell Biology** 97:1–5.

Goldberg, D.E. & Kornfeld, S. (1981). The phosphorylation of α-glucuronidase oligosaccharides in mouse P388D_1 cells. **The Journal of Biological Chemistry** 256:13060–13067.

Goldberg, D.E. & Kornfeld, S. (1983). Evidence for extensive subcellular organization of asparagine-linked oligosaccharide processing and lysosomal enzyme phosphorylation. **The Journal of Biological Chemistry** 258:3159–3165.

Goldberg, D., Gabel, C. & Kornfeld, S. (1984). Processing of lysosomal enzyme oligosaccharide units. In: Dingle, J.T., Dean, R.T., Sly, W., eds. **Lysosomes in Biology and Pathology**. Elsevier, pp. 45–62.

Gonzalez-Noreiga, A., Grubb, J.H., Talkad, V. & Sly, W.S. (1980). Chloroquine inhibits lysosomal enzyme pinocytosis and enhances lysosomal enzyme secretion by impairing receptor recycling. **Journal of Cell Biology** *85:839–852.*

Griffiths, G., Warren, G., Stuhlfauth, I. & Jockusch, B.M. (1981). The role of clathrin-coated vesicles in acrosome formation. **European Journal of Cell Biology** 26:52–60.

Griffiths, G. & Simons, K. (1986). The trans Golgi network: sorting at the exit site of the Golgi complex. **Science** 234:438–443.

Hasilik, A. & Neufeld, E.F. (1980a). Biosynthesis of lysosomal enzymes in fibroblasts. **The Journal of Biological Chemistry** 255:4937–4945.

Hasilik, A. & Neufeld, E.F. (1980b). Biosynthesis of lysosomal enzymes in fibroblasts. **The Journal of Biological Chemistry** 255:4946–4950.

Hasilik, A., Klein, U., Waheed, A., Strecker, G. & von Figura, K. (1980). Phosphorylated oligosaccharides in lysosomal enzymes: Identification of α-N-acetylglucosamine(l)phospho(6)mannose diester groups. **Proceeding of the National Academy of Sciences USA** 77: 7074–7078.

Hasilik, A. & von Figura, K. (1981). Oligosaccharides in lysosomal enzymes. **European Journal of Biochemistry** 121:125–129.

Hasilik, A. & von Figura, K. (1984). Processing of lysosomal enzymes in fibroblasts. In: Dingle, J.T., Dean, R.T., Sly, W., eds. **Lysosomes in Biology and Pathology**. Elsevier, pp. 3–16.

Hasilik, A., Waheed, A. & von Figura, K. (1981a). Enzymatic phosphorylation of lysosomal enzymes in the presence of UDP-N-acetylglucosamine. Absence of the activity in I-cell fibroblasts. **Biochemical and Biophysical Research Communications** 98:761–767.

Hasilik, A., Voss, B. & von Figura, K. (1981b). Transport and processing of lysosomal enzymes by smooth muscle cells and endothelial cells. **Experimental Cell Research** 133:23–30.

Hasilik, A., Waheed, A., Cantz, M. & von Figura, K. (1982). Impaired phosphorylation of lysosomal enzymes in fibroblasts of patients with mucolipidosis III. **European Journal of Biochemistry** 122:119–123.

Hasilik, A., Pohlmann, R., Olsen, R.L. & von Figura, K. (1984). Myeloperoxidase is synthesized as larger phosphorylated precursor. **The EMBO Journal** 3:2671–2676.

Hickman, S. & Neufeld, E.F. (1972). A hypothesis for I-cell disease: defective hydrolases that do not enter lysosomes. **Biochemical and Biophysical Research Communications** 49:992–999.

Hoflack, B. & Kornfeld, S. (1985a). Lysosomal enzyme binding to mouse P388D_1 macrophage membranes lacking the 215– kDa mannose 6-phosphate receptor: evidence for the existence of a second mannose-6-phosphate receptor. **Proceedings of the National Academy of Science, USA** 82:4428–4432.

Hoflack, B. & Kornfeld, S. (1985b). Purification and characterization of a cation-dependent mannose 6-phosphate receptor from murine P388D_1 macrophages and bovine liver. **The Journal of Biological Chemistry** 22:12008–12014.

Hoflack, B., Fujimoto, K. & Kornfeld, S. (1987). The interaction of phosphorylated oligosaccharides and lysosomal enzymes with bovine liver cation-dependent mannose 6-phosphate receptor. **The Journal of Biological Chemistry** 262:123– 129.

Holtzman, E., Novikoff, A.B. & Villaverde, H. (1967). Lysosomes and GERL in normal and chromatolytic neurons of the rat ganglion nodosum. **The Journal of Cell Biology** 33:419–435.

Hoogeveen, A.T., Reuser, A.J.J., Kroos, M. & Galjaard, H. (1986). G_{M1}-Gangliosidosis. Defective recognition site on α-galactosidase precursor. **The Journal of Biological Chemistry** 261:5702–5704.

Howard, D.R., Natowicz, M. & Baenziger, J.U. (1982). Structural studies of the endoglycosidase H-resistant oligosaccharides present on human α-glucuronidase. **The Journal of Biological Chemistry** 257:10861–10868.

Imort, M., Zuehlsdorf, M., Feige, U., Hasilik, A. & von Figura, K. (1983). Biosynthesis and transport of lysosomal enzymes in human monocytes and macrophages. **Biochemical Journal** 214:671–678.

Kaplan, A., Achord, D.T. & Sly, W.S. (1977). Phosphohexosyl components of a lysosomal enzyme are recognized by pinocytosis receptors on human fibroblasts. **Proceedings of the National Academy of Sciences USA** 74:2026–2030.

Kedersha, N.L. & Rome, L.H. (1986). Intracellular transport and secretion. **Microbiology** 00 :307–311.

Kinnon, C. & Owen, M.J. (1983). Clathrin-coated vesicles purified from a human B lymphocyte cell line contain newly synthesized glycoproteins. **The Journal of Biological Chemistry** 258:8470–8476.

Kobata, A. (1979). Use of endo- and exoglycosidases for structural studies of glycoconjugates. **Analytical Biochemistry** 100:1–14.

Kornfeld, R. & Kornfeld, S. (1985). Assembly of asparagine-linked oligosaccharides. **Annual Review of Biochemistry** 54:631–640.

Kornfeld, S. (1986). Trafficking of lysosomal enzymes in normal and disease states. **Journal of Clinical Investigation** 77:1–6.

Krentler, C., Pohlmann, R., Hasilik, A. & von Figura, K. (1986). Lysosomal membrane proteins do not bind to mannose-6- phosphate-specific receptors. **Biological Chemistry Hoppe-Seyler** 367:141–145.

Kress, B.C., Freeze, H.H., Herd, J.K., Alhadeff, J.A. & Miller, A.L. (1980). Purification and characterization of I-cell disease α-L-fucosidase. **The Journal of Biological Chemistry** 255:955–961.

Lang, L., Reitman, M., Tang, J., Roberts, R.M. & Kornfeld, S. (1984). Lysosomal enzyme phosphorylation. **The Journal of Biological Chemistry** 23:14663–14671.

Ledonne, N.C. Jr, Fairley, J.L. & Sweeley, C.C. (1983). Biosynthesis of α-galactosidase A in cultured chang liver cells. **Archives of Biochemistry and Biophysics** 224:186–195.

Lemansky, P., Gieselmann, V., Hasilik, A. & von Figura, K. (1984). Cathepsin D and α-hexosaminidase synthesized in the presence of 1-deoxynojirimycin accumulate in the endoplasmic reticulum. **The Journal of Biological Chemistry** 259:10129–10135.

Lemansky, P., Gieselmann, V., Hasilik, A. & von Figura, K. (1985). Synthesis and transport of lysosomal acid phosphatase in normal and I-cell fibroblasts. **Journal of Biological Chemistry** 260:9023–9030.

Lemansky, R., Hasilik, A., von Figura, K., Helmy, S., Fishman, J., Fine, R.E., Kedersha, N.L. & Rome, L.H. (1987). Lysosomal enzyme precursors in coated vesicles derived from the exocytic and endocytic pathways. **The Journal of Cell Biology**, in the press.

Lewis, V., Green, S.A., Marsh, M., Vihko, P., Helenius, A. & Mellman, I. (1985). Glycoproteins of the lysosomal membrane. **The Journal of Cell Biology** 100:1839–1847.

Lodish, H.F., Kong, N., Hirani, S. & Rasmussen, J. (1987). A vesicular intermediate in the transport of hepatoma secretory proteins from the rough endoplasmic reticulum to the Golgi complex. **The Journal of Cell Biology** 104:221–230.

Madiyalakan, R., Jain, R.K. & Matta, K.L. (1987). Phosphorylation of the 1,2-linked mannosylsaccharide by N-acetylglucosamine-1-phosphotransferase from fibroblasts occurs at the terminal mannose. **Biochemical and Biophysical Research Communications** 142:354–358.

Mainferme, F., Wattiaux, R. & von Figura, K. (1985). Synthesis, transport and processing of cathepsin C in Morris Lepatoma 7777 cells and rat hepatocytes. **European Journal of Biochemistry** 153:211–216.

Marshall, R.D. (1972). Glycoproteins. **Annual Review of Biochemistry** 41:673–702.

Mathur, R., Alvares, K., Balasubramanian, A.S. (1984). Two forms of acid α-D-mannosidase in monkey brain: Evidence for the co-existence of high mannose and complex oligosaccharides in one form. **Biochemical and Biophysical Research Communications** 123:1185–1193.

Miller, A.L., Kress, B.C., Stein, R., Kinnon, C., Kern, H., Schneider, J.A. & Harms, E. (1981). Properties of N-acetyl- α-D-hexosaminidase from isolated normal and I-cell lysosomes. **The Journal of Biological Chemistry** 256:9352–9362.

Mizuochi, T., Nishimura, Y., Kato, K. & Kobata, A. (1981). Comparative studies of asparagine-linked oligosaccharide structures of rat liver microsomal and lysosomal α-glucuronidases. **Archives of Biochemistry and Biophysics** 209:298–303.

Mutsaers, J.H.G.M., van Halbeek, H., Vliegenthart, J.F.G., Tager, J.M., Reuser,

A.J.J., Kroos, M. & Galjaard, H. (1987). Determination of the structure of the carbohydrate chains of acid α-glucosidase from human placenta. **Biochimica et Biophysica Acta** 911:244–251.

Myerowitz, R.& Neufeld, E.F. (1981). Maturation of a-L-Iduronidase in cultured human fibroblasts. **The Journal of Biological Chemistry** 256:3044–3048.

Nakao et al., (1983).

Nakao, Y., Kozutsumi, Y., Kawasaki, T., Yamashina, I., van Halbeek, H. & Vliegenthart, J.F.G. (1984). Oligosaccharides on cathepsin D from porcine spleen. **Archives of Biochemistry and Biophysics** 229:43–54.

Natowicz, M.R., Chi, MM-Y., Lowry, O.H. & Sly, W.S. (1979). Enzymatic identification of mannose 6-phosphate on the recognition marker for receptor-mediated pinocytosis of β-glucuronidase by human fibroblasts. **Proceedings of the National Academy of Sciences USA** 79:4322–4326.

Natowicz, M., Baenziger, J.U. & Sly, W.S. (1982). Structural studies of the phosphorylated high mannose-type oligosaccharides on human β-glucuronidase. **The Journal of Biological Chemistry** 257:4412–4420.

Nauerth, A., Lemansky, P., Hasilik, A., von Figura, K., Bause, E. & Legler, G. (1985). Cell type dependent inhibition of transport of cathepsin D in HepG2 cells and fibroblasts exposed to deoxy-manno-nojirimycin and deoxynojirimycin. **Biological Chemistry Hoppe-Seyler** 366:1009–1016.

Novikoff, A.B. (1976). The endoplasmic reticulum: A cytochemist's view (A review). **Proceedings of the National Academy of Sciences USA** 73:2781–2787.

Orci, L., Ravazzola, M., Amherdt, M., Madsen, O., Vassalli, J-D. & Perrelet, A. (1985). Direct identification of prohormone conversion site in insulin-secreting cells. **Cell** 42:671–681.

Overdijk, B., Beem, E.P., van Steijn, G.J., Trippelvitz, L.A.W., Lisman, J.J.W., Pazparente, J., Cardon, P., Leroy, Y., Fournet, B., van Halbeek, H., Mutsaers, J.H.G.M. & Vliegenthart, J.F.G. (1985). Structural analysis of the carbohydrate chains of β-N-acetylhexosaminidases from bovine brain. **Biochemical Journal** 232:637–641.

Owada, M. & Neufeld, E.F. (1982). Is there a mechanism for introducing acid hydrolases into liver lysosomes that is independent of mannose 6-phosphate recognition? Evidence from I-cell disease. **Biochemical and Biophysical Research Communications** 105:814–820.

Payne, G.S. & Schekman, R. (1985). A test of clathrin function in protein secretion and cell growth. **Science** 230:1009–1013.

Pearse, B.M.F. (1985). Assembly of the mannose-6-phosphate receptor into reconstituted clathrin coats. **The EMBO Journal** 4:2457–2460.

Peyrieras, N., Bause, E., Legler, G., Vasilov, R., Claesson, L., Peterson, P. & Ploegh, H. (1983). Effects of the glucosidase inhibitors nojirimycin and deoxynojirimycin on the biosynthesis of membrane and secretory glycoproteins. **The EMBO Journal** 2:823–832.

Pohlmann, R., Waheed, A., Hasilik, A. & von Figura, K. (1982). Synthesis of phosphorylated recognition marker in lysosomal enzymes is located in the cis part of Golgi apparatus. **The Journal of Biological Chemistry** 257:5323–5325.

Proia, R.L. & Neufeld, E.F. (1982). Synthesis of ß-hexosaminidase in cell-free translation and in intact fibroblasts: An insoluble precursor a chain in a rare form ofTay-Sachs disease. **Proceedings of the National Academy of Science USA** 79:6360–6364.

Proia, R.L., D'Azzo, A. & Neufeld, E.F. (1984). Association of α- and β- subunits during the biosynthesis of β-hexosaminidase in cultured human fibroblasts. **The Journal of Biological Chemistry** 259:3350–3354.

Reitman, M.L. & Kornfeld, S. (1981a). UDP-N-Acetylglucos-amine:Glycoprotein N-Acetylglucosamine-1-phosphotransferase. **The Journal of Biological Chemistry** 256:4275–4281.
Reitman, M.L. & Kornfeld, S. (1981b). Lysosomal enzyme targeting. **The Journal of Biological Chemistry** 256:11977–11980.
Reitman, M.L., Varki, A. & Kornfeld, S. (1981). Fibroblasts from patients with I-cell disease and pseudo-Hurler poly-dystrophy are deficient in uridine 5'-diphosphate-N-acetylglucosamine:glycoprotein N-acetylglucosaminyl-phosphotransferase activity. **Journal of Clinical Investigation** 67:1574–1579.
Reuser, A.J.J., Kroos, M., Oude Elferink, R.P.J. & Tager, J.M. (1985). Defects in synthesis, phosphorylation, and maturation of acid α-glucosidase in glycogenosis type II. **The Journal of Biological Chemistry** 260:8336–8341.
Sahagian, G.G. (1984). The mannose 6-phosphate receptor: function, biosynthesis and translocation. **Biology of the Cell** 51:207–214.
Sahagian, G.G., Distler, J. & Jourdian, G.W. (1981). Characterization of a membrane-associated receptor from bovine liver that binds phosphomannosyl residues of bovine testicular α-galactosidase. **Proceedings of the National Academy of Sciences USA** 78:4289–4293.
Sahagian, G.G. & Steer, C.J. (1985). Transmembrane orientation of the mannose-6-phosphate receptor in isolated clathrin-coated vesicles. **The Journal of Biological Chemistry** 260:9838–9842.
Sando, G.N. & Neufeld, E.F. (1977). Recognition and receptor-mediated uptake of a lysosomal enzyme, α-iduronidase, by cultured human fibroblasts. **Cell** 12:619–627.
Saunier, B., Kilker, R.D., Tkacz, J.S., Quaroni, A. & Herscovics, A. (1982). Inhibition of N-linked complex oligosaccharide formation by 1-deoxynojirimycin, an inhibitor of processing glucosidases. **The Journal of Biological Chemistry** 257:14155–14161.
Schulze-Lohoff, E., Hasilik, A. & von Figura, K. (1985). Cathepsin D precursors in clathrin-coated organelles from human fibroblasts. **The Journal of Cell Biology** 101:824–829.
Skudlarek, M.D. & Swank, R.T. (1981). Turnover of two lysosomal enzymes in macrophages. **The Journal of Biological Chemistry** 256:10137–10144.
Skudlarek, M.D., Novak, E.K. & Swank, R.T. (1984). Processing of lysosomal enzymes in macrophages and kidney. In: Dingle, J.T., Dean, R.T. & Sly, W., eds. **Lysosomes in Biology and Pathology**. Elsevier, pp. 17–43.
Skudlarek, M.D. & Orgebin-Crist, M.C. (1986). Glycosidases in cultured rat epididymal cells: enzyme activity, synthesis and secretion. **Biology of Reproduction** 35:167–178.
Steckel, F., Hasilik, A. & von Figura, K. (1983). Biosynthesis and maturation of arylsulfatase B in normal and mutant cultured human fibroblasts. **The Journal of Biological Chemistry** 258:14322–14326.
Stein, M., Braulke, T., von Figura, K. & Hasilik, A. (1987). Effects of differentiation inducing agents on synthesis, maturation and secretion of cathepsin D in U937 and HL-60 cells. **Biological Chemistry Hoppe-Seyler**, in the press.
Stromberg, K., Persson, A.M. & Olsson, I. (1986). The processing and intracellular transport of myeloperoxidase – modulation by lysosomotropic agents and monensin. **European Journal of Cell Biology** 39:424–31.
Strous, G.J.A.M., Willemsen, R., van Kerkhof, P., Slot, J.W., Geuze, H.J. & Lodish, H.F. (1983). Vesicular stomatitis virus glycoprotein, albumin, and transferrin are transported to the cell surface *via* the same Golgi vesicles. **The Journal of Cell Biology** 97:1815–1822.

Sweeley, C.C., Ledonne, N.C. Jr. & Robbins, P.W. (1983). Post-translational processing reactions involved in the biosynthesis of lysosomal α-N-acetylgalactosaminidase in cultured human fibroblasts. **Archives of Biochemistry and Biophysics** 223:158–165.

Tabas, I. & Kornfeld, S. (1980). Biosynthetic intermediates of α-glucuronidase contain high mannose oligosaccharides with blocked phosphate residues. **The Journal of Biological Chemistry** 255:6633–6639.

Takahashi, T., Schmidt, P.G. & Tang, J. (1983). Oligosaccharide units of lysosomal cathepsin D from porcine spleen. **The Journal of Biological Chemistry** 258:2819–2830.

Takahashi, T., Schmidt, P.G. & Tang, J. (1984). Novel carbohydrate structures of cathepsin B from porcine spleen. **The Journal of Biological Chemistry** 259:6059–6062.

Takasaki, S., Murray, G.J., Furbish, F.S., Brady, R.O., Barranger, J.A. & Kobata, A. (1984). Structure of the N-asparagine-linked oligosaccharide units of human placental β-glucocerebrosidase. **The Journal of Biological Chemistry** 259:10112–10117.

Taniguchi, T., Mizuochi, T., Towatari, T., Katunuma, N. & Kobata, A. (1985). Structural studies on the carbohydrate moieties of rat liver cathepsins B and H. **Journal of Biochemistry** 97:973–976.

Tooze, J. & Tooze, S.A. (1986). Clathrin-coated vesicular transport of secretory proteins during the formation of ACTH-containing secretory granules in AtT2O cells. **The Journal of Cell Biology** 103:839–850.

Tucker, S.M., Pierce, R. & Price, R.G. (1980). Characterisation of human N-acetyl-α-D-glucosaminidase isoenzymes as an indicator of tissue damage in disease. **Clinica Chimica Acta** 102:29–40.

Tümmers, S., Zuehlsdorf, M., Robenek, H., Hasilik, A. & von Figura, K. (1983). Molecular forms of cathepsin D in coated vesicle preparations. **Hoppe-Seyler's Zeitschrift für Physiologische Chemie** 364:1287–1295.

Ullrich, K., Mersmann, G., Weber, E. & von Figura, K. (1978). Evidence for lysosomal enzyme recognition by human fibroblasts *via* a phosphorylated carbohydrate moiety. **The Biochemical Journal** 170:643–650.

Varki, A. & Kornfeld, S. (1980a). Identification of a rat liver α-N-acetylglucosaminyl phosphodiesterase capable of removing 'blocking' α-N-acetylglucosamine residues from phosphorylated high mannose oligosaccharides of lysosomal enzymes. **The Journal of Biological Chemistry** 255:8398–8401.

Varki, A. & Kornfeld, S. (1980b). Structural studies of phosphorylated high mannose-type oligosaccharides. **Journal of Biological Chemistry** 255:10847–10858.

Varki, A. & Kornfeld, S. (1981). Purification and characterization of rat liver α-N-acetylglucosaminyl phosphodiesterase. **The Journal of Biological Chemistry** 256:9937–9943.

Varki, A. & Kornfeld, S. (1983). The spectrum of anionic oligosaccharides released by endo-β-N-acetylglucosaminidase H from glycoproteins. **The Journal of Biological Chemistry** 258:2808–2818.

Varki, A., Sherman, W. & Kornfeld, S. (1983). Demonstration of the enzymatic mechanisms of α-N-acetyl-D-glucosamine-1-phosphodiester N-acetylglucosaminidase (formerly called α-N-acetylglucosaminylphosphodiesterase) and lysosomal α-N-acetylglucosaminidase. **Archives of Biochemistry and Biophysics** 222:145–149.

Vladutiu, G.D. (1983). Effect of the co-existence of galactosyl and phosphomannosyl residues on α-hexosaminidase on the processing and transport of the enzyme in mucolipidosis I fibroblasts. **Biochimica et Biophysica Acta** 760:363–370.

Vladutiu, G.D. & Rattazzi, M.C. (1975). Abnormal lysosomal hydrolases excreted by cultured fibroblasts in I-cell disease (mucolipidosis II). **Biochemical and Biophysical Research Communications** 67:956–964.

von Figura, K. & Weber, E. (1978). An alternative hypothesis of cellular transport of lysosomal enzymes in fibroblasts. **Biochemical Journal** 176:943–950.

von Figura, K., Gieselmann, V. & Hasilik, A. (1985). Mannose 6-phosphate-specific receptor is a transmembrane protein with a C-terminal extension oriented towards the cytosol. **The Biochemical Journal** 225:543–547.

von Figura, K. & Hasilik, A. (1986). Lysosomal enzymes and their receptors. **Annual Review of Biochemistry** 55:167– 193.

Waheed, A., Hasilik, A. & von Figura, K. (1981). Processing of the phosphorylated recognition marker in lysosomal enzymes. **The Journal of Biological Chemistry** 256:5717–5721.

Waheed, A., Hasilik, A. & von Figura, K. (1982a). UDP-N-acetylglucosamine: lysosomal enzyme precursor N-acetylglucosamine-1-phosphotransferase. **The Journal of Biological Chemistry** 257:12322–12331.

Waheed, A., Pohlmann, R., Hasilik, A., von Figura, K., van Elsen, A. & Leroy, J.G. (1982b) Deficiency of UDP-N-acetylglucosamine; lysosomal enzyme N-acetylglucosamine-1-phosphotransferase in organs of I-cell patients. **Biochemical and Biophysical Research Communication** 105:1052–1058.

Wehland, J., Willingham, M.C., Gallo, M.G. & Pastan, I. (1982). The morphologic pathway of exocytosis of the vesicular stomatitis virus G protein in cultured fibroblasts. **Cell** 28:831–841.

Werner, G. (1986). Changes in the Golgi apparatus during spermiogenesis in the backswimmer, Notonecta glauca. **Biology of the Cell** 57:169–180.

Wiesmann, U., Vassella, F. & Herschkowitz, N. (1971). 'I-cell' disease: leakage of lysosomal enzymes into extracellular fluids. **New England Journal of Medicine** 285:1091– 1095.

Yamamoto, K., Tsuji, T., Tarutani, O. & Osawa, T. (1985). Phosphorylated high mannose-type and hybrid-type oligosaccharide chains of human thyroglobulin isolated from malignant thyroid tissue. **Biochimica et Biophysica Acta** 838:84–91.

Zuehlsdorf, M., Imort, M., Hasilik, A. & von Figura, K. (1983). Molecular forms of α-hexosaminidase and cathepsin D in serum and urine of healthy subjects and patients with elevated activity of lysosomal enzymes. **Biochemical Journal** 213:733–740.

Pharmacological intervention in Golgi function in MDCK cells

J. MICHAEL EDWARDSON, PHILLIPA U. DANIELS
AND CAROLE Y. NADIN

Department of Pharmacology, University of Cambridge, Hills Road, Cambridge CB2 2QD, UK.

Abstract

The effects of the carboxylic ionophores monensin and nigericin and the weak bases chloroquine and ammonium chloride have been examined on the intracellular processing and transport, in MDCK (Madin–Darby canine kidney) cell monolayers, of the two influenza virus envelope proteins, haemagglutinin and neuraminidase, and of a group of secretory proteins. In the presence of the carboxylic ionophores, the glycosylation of haemagglutinin is interrupted, apparently at the level of the *medial* Golgi, while in the presence of the weak bases, glycosylation proceeds almost to completion. Both groups of compounds delay the delivery of haemagglutinin and neuraminidase to the cell surface, although the effects of the two groups on the transport of neuraminidase are different. Similar effects are seen with the secretory proteins: processing is affected only by monensin, although both monensin and chloroquine delay secretion.
It is concluded that :

(1) All proteins so far studied that are destined for the apical surface of MDCK cell monolayers show delayed, but not blocked, intracellular transport in the presence of the carboxylic ionophores.
(2) Monensin and nigericin produce similar effects on the processing and transport of the influenza virus envelope proteins.
(3) The carboxylic ionophores and the weak bases produce different effects on both processing and transport. The effects of the bases are consistent with an action in a late Golgi compartment, whereas the ionophores appear to act earlier in the transport pathway.

The possible mechanisms of action of these agents are discussed.

Key words

carboxylic ionophores, weak bases, intracellular transport, Golgi complex, epithelia, influenza, monensin, nigericin, chloroquine, ammonium chloride, glycosylation, haemagglutinin, neuraminidase.

Introduction

Two groups of compounds have relatively selective effects on the function of the Golgi complex, with little effect on protein synthesis. These are the carboxylic ionophores, such as monensin and nigericin, and the 'acidotropic' weak bases, such as chloroquine and primaquine (Fig. 1).

Tartakoff & Vassalli (1977) originally showed that, in plasma cells, monensin caused a dramatic swelling of the normally compressed cisternae of the Golgi complex and a concomitant block in the intracellular transport of immunoglobulin within the Golgi complex. In addition, the immunoglobu-

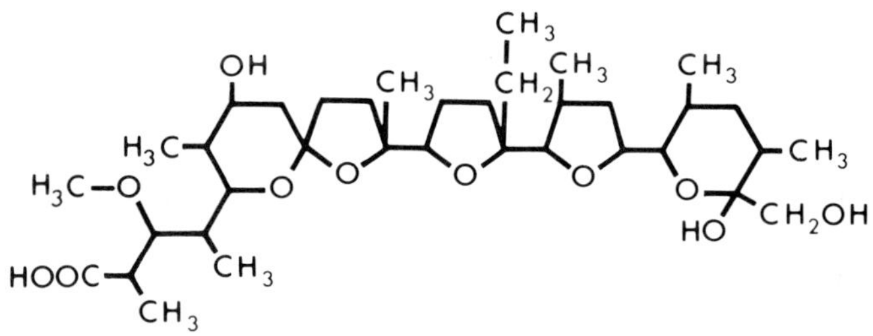

monensin

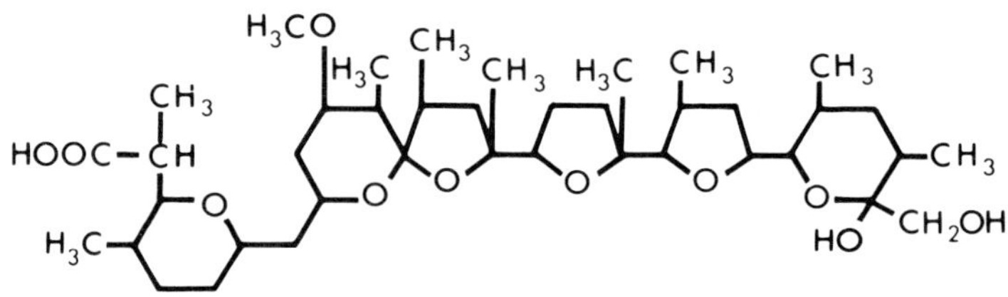

nigericin

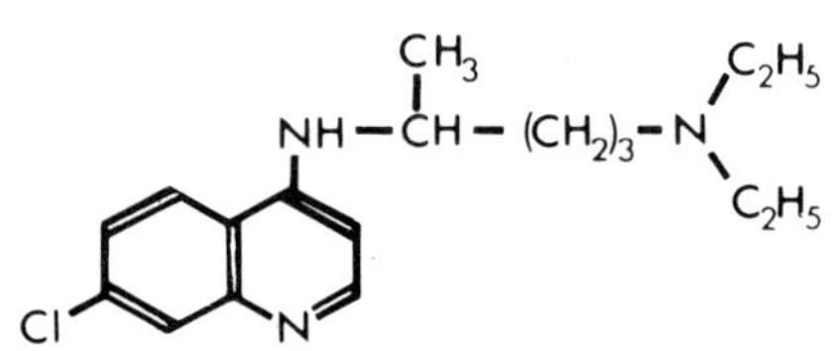

chloroquine

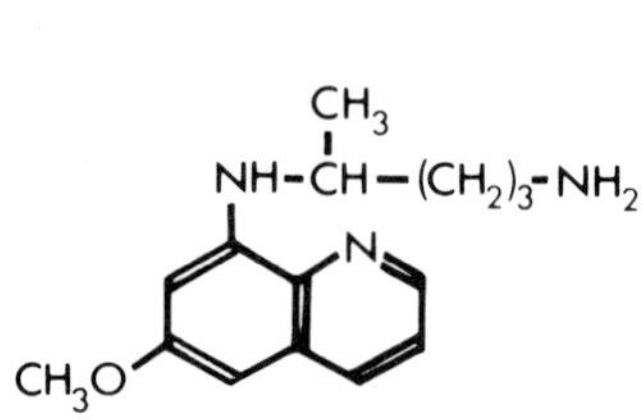

primaquine

Fig. 1 Chemical structures of typical carboxylic ionophores and acidotropic weak bases.

lin failed to receive terminal galactose and sialic acid residues, which indicated that the monensin-induced block was at the level of the *medial* cisternae of the Golgi complex. Results similar to these have since been obtained for a variety of membrane and secretory proteins in many cell types, and monensin and nigericin have been used widely as tools to study the intracellular transport of newly synthesised proteins (Tartakoff, 1983). However, the effects of monensin are not always predictable. For example, the mouse macrophage F_c receptor reaches the plasma membrane in monensin-treated cells, but is not galactosylated (Green *et al.*, 1985). Hence, monensin appears not to exert a single well-defined block in the intracellular transport pathway.

The acidotropic weak bases, also, cause dilation of Golgi cisternae and a dramatic slowing in the intracellular transport of secretory and membrane proteins (Strous *et al.*, 1985). However, the effects of the weak bases may differ from those of the carboxylic ionophores. For example, terminal glycosylation of the envelope protein G of vesicular stomatitis virus (VSV) in hepatoma cells is blocked by monensin but not by primaquine (Strous *et al.*, 1985), despite the fact that both agents block the transport of G to the plasma membrane. The form of G produced in the presence of primaquine is consistent with a block in transport on the *trans* side of the Golgi complex. Another interesting effect of the weak bases is their interference with the intracellular sorting of newly synthesised secretory proteins. For example, in the pituitary cell line AtT-20, chloroquine causes the diversion of ACTH from the regulated to the constitutive secretory pathway (Moore *et al.*, 1983).

The carboxylic ionophores are open-chain compounds containing oxygenated heterocyclic rings and a single terminal carboxyl group (Pressman, 1976). In their deprotonated anionic form they form rings *via* head-to-tail hydrogen bonding and ligand alkali metal cations through their internally-disposed oxygen atoms. The resulting neutral complex has a high lipid solubility because of the shielding of its polar interior by the hydrocarbon exterior. In cellular membranes, the ionophore facilitates the electroneutral exchange of Na^+, K^+ and H^+ and causes the dissipation of transmembrane gradients of these ions. Unlike the carboxylic ionophores, the weak bases would be expected only to dissipate pH gradients. The neutral forms of these compounds rapidly cross membranes, becoming protonated and trapped within acidic compartments. As the intra-compartmental concentration of the base increases, water will also enter, leading to osmotic swelling. This 'base capture' principle has been exploited recently by Anderson & Pathak (1985), who used 3-(2,4-dinitroanilino)-3′-amino-N-methyl-dipropylamine (DAMP), to demonstrate that a compartment on the *trans* side of the Golgi complex, that has been called the *trans* Golgi network, or TGN (Griffiths & Simons, 1986), is mildly acidic. Since the carboxylic ionophores and the weak bases share the ability to dissipate transmembrane pH gradients, it is tempting to speculate that they exert their principal effect on this compartment. However, it must be remembered that monensin and primaquine have different effects on the processing of the same protein in a single cell type. Hence, other effects of monensin must also be involved.

In ion transporting epithelia, functional asymmetry depends on the differential localisation of proteins in the two domains of the plasma membrane, apical and basolateral. For example, in the epithelial MDCK cell line the enzyme aminopeptidase is found only in the apical surface, while Na^+, K^+ -ATPase is found only in the basolateral surface (Louvard, 1980). Investigations of the mechanisms underlying this asymmetric targetting of plasma membrane proteins have used epithelial cell lines infected with enveloped viruses such as influenza and VSV. These viruses bud asymmetrically from epithelial monolayers (influenza apically and VSV basolaterally [Rodriguez-Boulan & Sabatini, 1978]) and the envelope proteins (influenza haemagglutinin and neuraminidase, and VSV G) are inserted exclusively into the surface from which the particular virus will bud (Rodriguez-Boulan & Pendergast, 1980). Interestingly, monensin has a differential effect on the intracellular transport of VSV G and influenza haemagglutinin in MDCK cells, blocking the transport of the former in the Golgi complex (Alonso-Caplen & Compans, 1983) but merely delaying the appearance of the latter at the plasma membrane (Edwardson, 1984). This ability of monensin to discriminate between haemagglutinin and G has been taken to indicate that the pathways taken by the two proteins diverge at a stage prior to the point at which the transport of G is blocked (Alonso-Caplen & Compans, 1983), although other interpretations, such as a differential sensitivity of the two proteins to the cellular effects of the ionophore, are possible. In addition, it has been reported (Alonso-Caplen and Compans, 1983) that nigericin (a potassium-selective ionophore), at a concentration of 1 μM, reduces the yield of influenza virus from infected MDCK cells by about 10-fold, while monensin (a sodium-selective ionophore) at the same concentration has no effect on virus yield. This observation raises the possibility that ionophores of different ion selectivity may have qualitatively different effects on the intracellular transport of newly synthesised proteins.

In this chapter results are reported of experiments on the effects of two carboxylic ionophores (monensin and nigericin) and two weak bases (chloroquine and ammonium chloride) on the intracellular processing and transport, in MDCK cell monolayers, of the two influenza virus envelope proteins, haemagglutinin and neuraminidase, and of a group of endogenous secretory proteins. The aims of this study were to compare the cellular effects of these two groups of compounds and to investigate the role of low pH compartments in the transport and sorting of proteins in this epithelial cell line.

Polarisation of MDCK cell monolayers

Figure 2 shows an electron micrograph of a section through an MDCK cell growing in a monolayer on a nitrocellulose filter. The monolayer was infected with influenza virus and processed 7 h post-infection. Influenza virions are seen budding exclusively from the apical surface of the cell, a result which demonstrates the polarisation of the monolayer.

Processing of haemagglutinin

Haemagglutinin acquires high-mannose oligosaccharides as it is synthesised in the rough endoplasmic reticulum (rER) and these are then processed in the Golgi complex by the removal of glucose and mannose residues and the addition of N-acetylglucosamine, galactose and fucose residues (Nakamura & Compans, 1979). These processing events were seen as an increase in molecular weight and an acquisition of resistance to the action of the enzyme endoglycosidase H (endo H, Fig. 3a). The half-time for acquisition of endo H resistance was about 15 min (not shown). Monensin interrupted

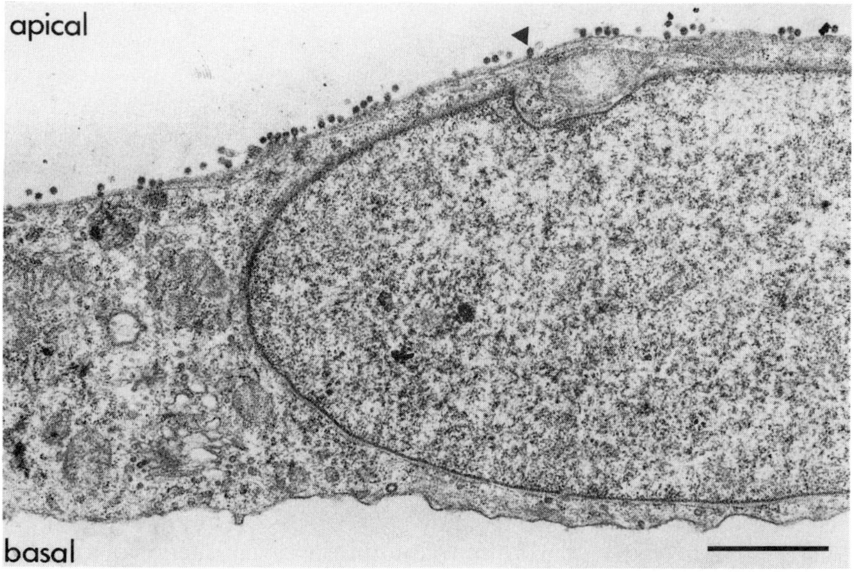

Fig. 2 Asymmetric budding of influenza virions from an infected MDCK cell monolayer.
An MDCK cell monolayer growing on a nitrocellulose filter was infected with influenza virus. At 7 h post-infection the monolayer was fixed and processed for electron microscopy as described in Materials and Methods. The arrowhead shows a budding virion. The bar represents 1 μm.

(a)

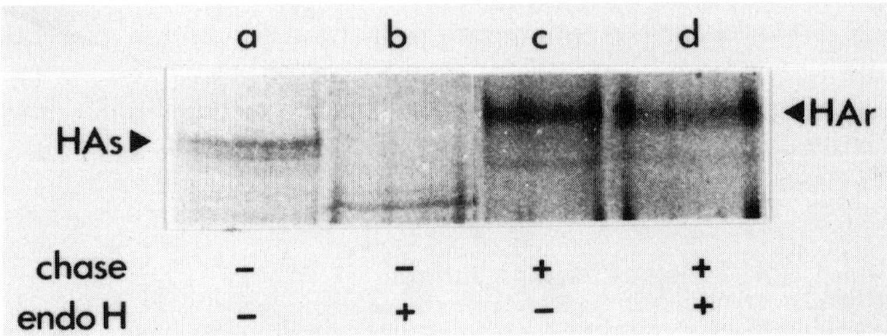

(b)

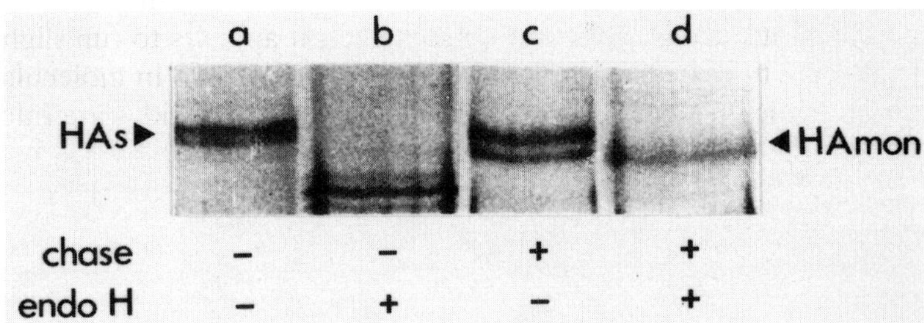

Fig. 3 Effect of carboxylic ionophores on the processing of haemagglutinin. (a) Influenza virus-infected MDCK cell monolayers were labelled 5 h post-infection with [^{35}S]methionine for 10 min and harvested either immediately (a,b) or after a 30 min chase period in the presence of excess unlabelled methionine (c,d). Samples either received no further treatment (a,c) or were exposed to endo H (0.4 mU) for 15 h at 37°C (b,d). Only the part of the gel containing haemagglutinin is shown. Bands are HAs, haemagglutinin sensitive to endo H; HAr, haemagglutinin resistant to endo H.
(b) As for (a) except that post-infection maintenance medium contained 1μM monensin.

Golgi processing of haemagglutinin, resulting in the production of a doublet that migrated slightly more quickly than the normal rER form (Fig. 3b). Endo H converted the doublet into a single band that ran in the same position as the lower band of the doublet. The simplest interpretation of this result is that monensin is revealing forms of haemagglutinin that are normally short-lived processing intermediates. In MDCK cells infected with influenza virus, haemagglutinin possesses six oligosaccharide moieties (Nakamura *et al.*, 1980). Since both forms of haemagglutinin seen here in the presence of monensin are smaller than the rER form, their oligosaccharides must have undergone extensive trimming of glucose and mannose residues without complete addition of terminal sugars. The larger endo H-sensitive form must possess at least one oligosaccharide that contains five or more mannose residues, whereas the smaller endo H-resistant form has only oligosaccharides that contain less than five mannose residues. These characteristics would place the block to processing between the *medial* and the *trans* Golgi. In additional experiments it was found that both monensin and nigericin produce the same effects on the processing of haemagglutinin and that both ionophores cause an extensive vesiculation of the Golgi complex (not shown). Figure 4 shows that in the presence of either ammonium chloride or chloroquine a larger form of haemagglutinin is produced than was seen in the presence of monensin. The mobility of this form is difficult to distinguish from that of the fully processed form seen in control cells, although on some gels it appears to run slightly more quickly. In the presence of endo H, a slight reduction in molecular weight is seen, which indicates that not all of the oligosaccharides are fully processed.

Since neuraminidase co-migrates with nucleoprotein, it was not possible to examine the effects of these agents on its processing using the methods described above.

Delivery of haemagglutinin to the plasma membrane

The delivery of haemagglutinin to the apical surface of MDCK cell monolayers was assayed by measuring its sensitivity to exogenously applied

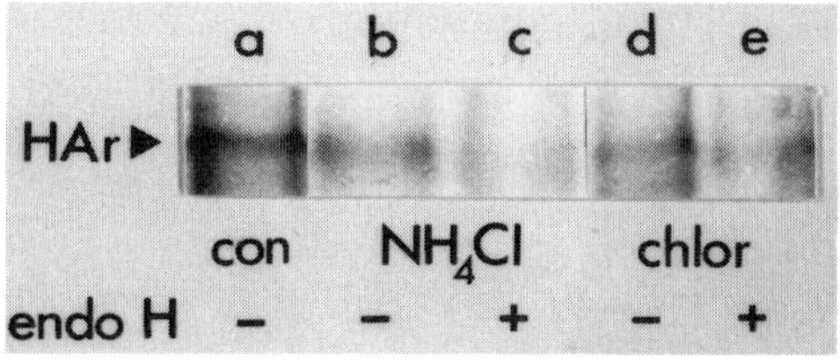

Fig. 4 Effect of weak bases on the processing of haemagglutinin. Influenza virus-infected MDCK cell monolayers were labelled 5 h post-infection with [^{35}S]methionine and chased for 1 h in the presence of excess unlabelled methionine and either no other additives (a), 20 mM ammonium chloride (b,c) or 150 μM chloroquine (d,e). Samples either received no further treatment (a,b,d) or were exposed to endo H (0.4 mU) for 15 h at 37°C (c,e). Only the part of the gel containing haemagglutinin is shown. HAr is haemagglutinin resistant to endo H.

(a)

(b)

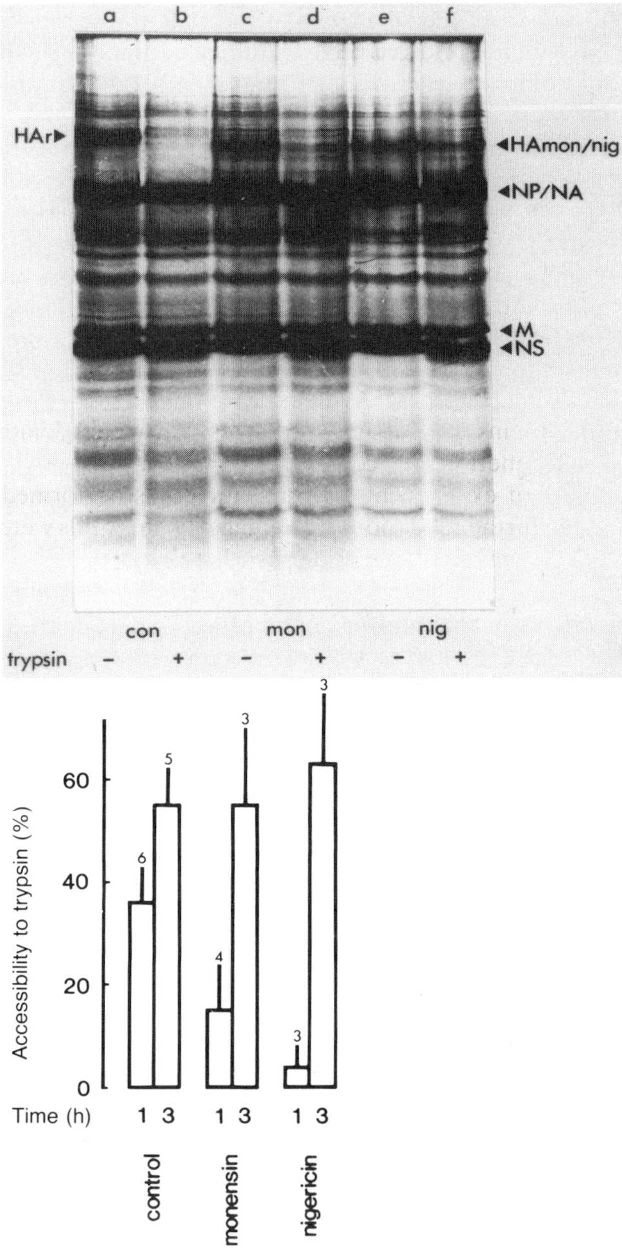

Fig. 5 Effect of carboxylic ionophores on the delivery of haemagglutinin to the plasma membrane.

(a) Post-infection maintenance medium contained either no additives (a,b), 1 μM monensin (c,d) or 1 μM nigericin (e,f). Monolayers were labelled 5 h post-infection with [^{35}S]methionine for 10 min and chased for 1 h in the presence of excess methionine (plus ionophore where trypsin (Matlin & Simons, 1983). In an initial experiment, cells were pulse-labelled with [^{35}S]methionine for 10 min and 'chased' for 1 h in the presence of excess unlabelled methionine. Duplicate monolayers were used as controls or treated with monensin or nigericin. One monolayer of each pair was treated with trypsin. The fluorograph obtained is shown in Fig. 5a. It is clear that the haemagglutinin in the control cells is almost completely sensitive to trypsin after a 1 h chase, while the haemagglutinin in the ionophore-treated cells is only partially sensitive to trypsin. In the absence of EDTA, tight junctions between cells remain intact throughout the trypsinisation (not shown), so that only the apical surface of the monolayer is exposed to trypsin. Hence, the complete digestion of haemagglutinin seen in the control monolayer indicates that this protein is delivered exclusively to the apical surface, in parallel with the polarisation of viral budding. The intracellular viral proteins were resistant to digestion, which demonstrates that trypsin did not penetrate the cells.

In a second series of experiments trypsinisation was performed on ice, in order to block any further transport of haemagglutinin. Cells were chased for either 1 h or 3 h and the haemagglutinin bands on the fluorographs were subjected to densitometric scanning. Values for the percentage accessibility to trypsin (calculated from the percentage loss of haemagglutinin) are shown in Fig. 5b. In control monolayers accessibilities were 36% after a 1 h chase and 55% after 3 h. In the presence of monensin and nigericin, accessibilities were significantly reduced at 1h ($P < 0.05$ and 0.01, respectively), but not significantly different from control at 3 h. At 1 h, accessibility in the monensin-treated monolayers was higher than that in the nigericin-treated monolayers, but not significantly so. Hence, both ionophores delay the appearance of haemagglutinin at the cell surface, although at long chase times externalisation is not reduced. Even after a 3 h chase, however, terminal glycosylation of haemagglutinin remained interrupted in the presence of the ionophores.

Since at 37°C trypsin digests almost all of the haemagglutinin after a 1 h chase, the inability of trypsin to digest all of the haemagglutinin at 4°C represents an inherent inefficiency of the enzyme under these conditions, and not a retention of a proportion of the haemagglutinin within the cell.

Caption for Fig. 5 (cont.)

appropriate). Cells were then harvested either immediately (a,c,e) or after exposure to trypsin (0.5 mg/ml in PBS) for 15 min at 37°C (b,d,f). Bands are HA, haemagglutinin; NP/NA, nucleoprotein/neuraminidase; M, matrix protein; NS, non-structural protein.

(b) Monolayers were treated as above, except that two chase times were used (1 h and 3 h) and trypsinisation was at 4°C. Accessibility to trypsin was calculated from densitometer scans of the haemagglutinin bands. Values are means ± s.e.m. Figures above the bars are the numbers of determinations made.

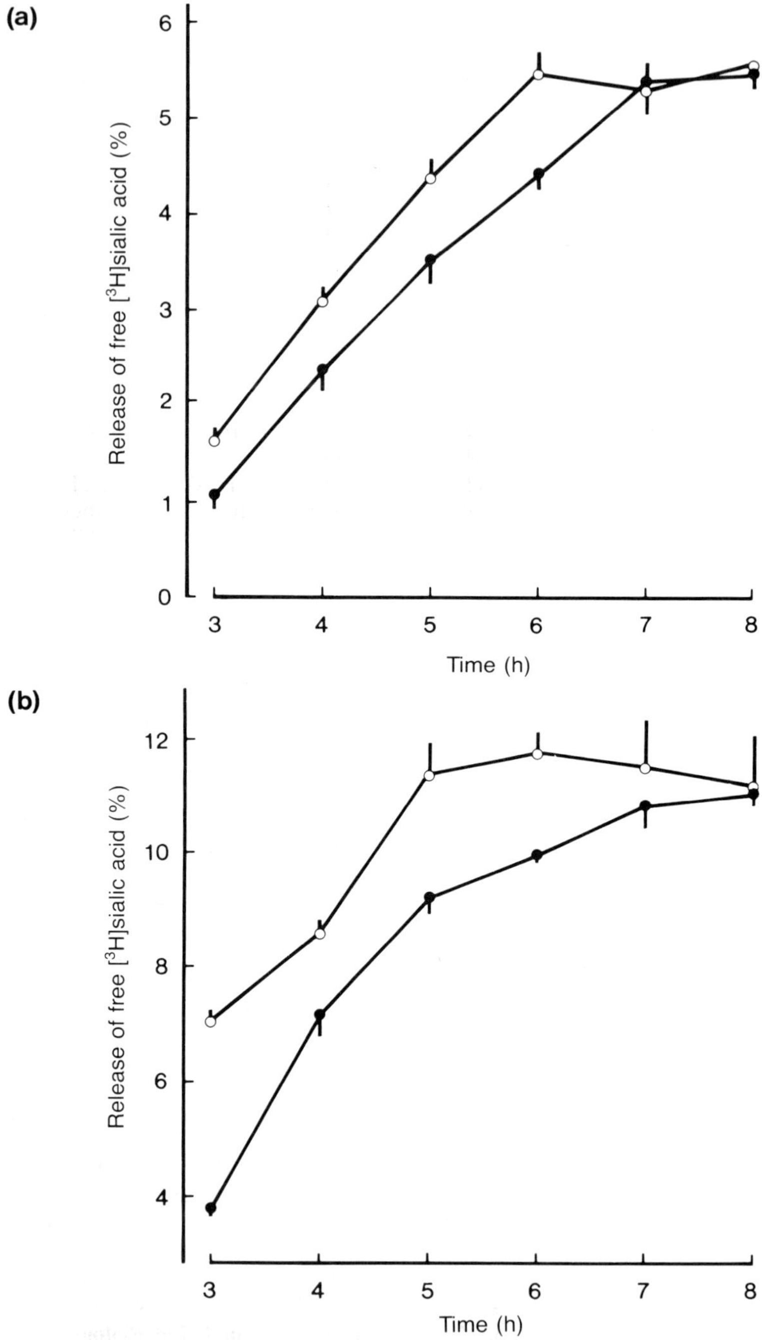

Fig. 6 Effect of carboxylic ionophores on the delivery of neuraminidase to the plasma membrane.

It should also be pointed out that the observation that tryptic digestion of haemagglutinin at long chase times was not affected by the ionophores suggests that no mis-sorting of this protein is induced by these agents.

It has been reported recently (Matlin, 1986) that the delivery of haemagglutinin to the plasma membrane of MDCK cells is delayed by about 20–30 min in the presence of ammonium chloride, but that no mis-sorting is seen. We have found, using the procedure described above, that after a 1 h chase in the presence of ammonium chloride, the accessibility of haemagglutinin to trypsin (at 4°C) is reduced from 68% to 24%. The effect of chloroquine has not yet been examined.

Delivery of neuraminidase to the plasma membrane

To measure the delivery of neuraminidase to the cell surface an assay was developed that exploits the enzyme activity of this protein. At various times after infection MDCK cell monolayers were cooled on ice and the amount of neuraminidase activity present at the cell surface was measured *via* the cleavage of free [^{3}H]sialic acid from the envelope proteins of [^{3}H]sialic acid-labelled SFV added as a suspension above the monolayers. It was established in preliminary experiments that the release of free [^{3}H]sialic acid produced was linear with both time and neuraminidase activity over the relevant ranges. Release occurring over the 30 min period chosen will therefore be proportional to the amount of enzyme present at the cell surface. Release was measured at hourly intervals between 3 h and 8 h post-infection, in the absence and presence of both carboxylic ionophores (Fig. 6). It was found that in control monolayers delivery of neuraminidase to the plasma membrane rose fairly steeply between 3 h and 5 -- 6h and thereafter reached a plateau. The plateau occurs shortly after the time at which viral budding begins (Edwardson, 1984). Hence, it is likely that a steady state is established in which the rate of delivery of neuraminidase to the surface is balanced by the loss of neuraminidase in shed virions. In the presence of the ionophores, the curve relating the surface delivery of neuraminidase to time post-infection is shifted in an approximately parallel

Caption for Fig. 6 (cont.)

(a) Influenza virus-infected monolayers (either control (o) or treated with 1 μM monensin from 1 h post-infection (•)) were cooled on ice at various times, and then exposed to SFV with envelope glycoproteins bearing [^{3}H]sialic acid (approximately 20 000 dpm) in 0.5 ml ice-cold PBS containing 2 mg/ml BSA for 30 min. Protein was precipitated by the addition of 10% (w/v)TCA. Release of free [^{3}H]sialic acid into theTCA supernatant was determined and expressed as a percentage of the total radioactivity added. Values are means ± s.e.m. (n = 3).
(b) As for (a) except that monolayers were either untreated (o) or treated with 1 μM nigericin from 1 h post-infection (•).

manner to a later time. The extent of this shift is about 45 min in the presence of monensin and about 1 h in the presence of nigericin. In both cases, delivery is not significantly different from control after 7 h, presumably when the steady state between delivery and budding is achieved in the presence of the ionophores.

Chloroquine and ammonium chloride had effects on the delivery of neuraminidase to the plasma membrane that differed from those of monensin and nigericin (Fig. 7). In the presence of both bases the delivery of neuraminidase to the cell surface became significantly depressed only at 5 h, whereas with the carboxylic ionophores surface delivery was already depressed at the earliest time point investigated (3 h). In addition, the delivery of enzyme to the cell surface reaches a plateau at a lower level in

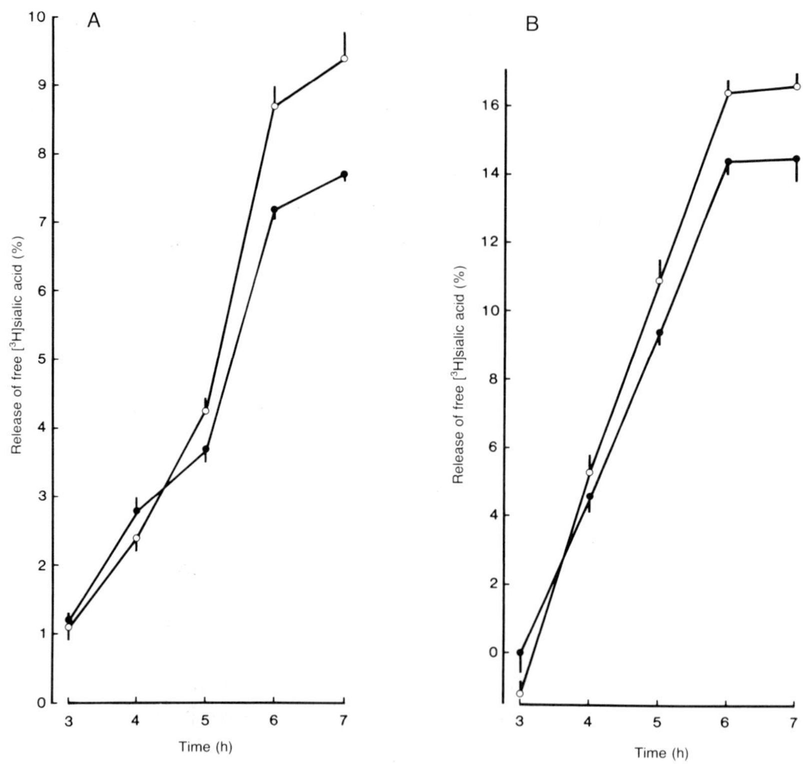

Fig. 7 Effect of weak bases on the delivery of neuraminidase to the plasma membrane.

(a) As for Fig. 6 except that monolayers were either untreated (o) or treated with 150 μM chloroquine from 1 h post-infection (•).

(b) As for (a) except that monolayers were either untreated (o) or treated with 20 mM ammonium chloride from 1 h post-infection (•).

base-treated cells than in control cells. At present we can offer no explanation of these differences. However, they do suggest that the two groups of agents are affecting transport in different ways.

Secretion of endogenous proteins

MDCK cells growing on tissue culture dishes secrete as yet unidentified proteins into the bathing medium. In the presence of monensin, a change in the processing of most of the proteins was observed, along with a reduction in the rate of secretion (Fig. 8). In the presence of chloroquine, on the other hand, the forms of the proteins were all apparently unchanged (Fig. 9a), but the rates of secretion were in all cases reduced (Fig. 9b). Once again, then, the two agents apparently affect intracellular transport differently.

Discussion

It has been shown that both monensin and nigericin interrupt glycosylation of influenza haemagglutinin at a stage that is consistent with a block in transport at the level of the *medial* cisternae of the Golgi complex. However, haemagglutinin still appears at the cell surface in this incompletely glycosylated form. There are two possible explanations of this paradoxical result. Either haemagglutinin is able to by-pass the *trans* Golgi cisternae and so escape an ionophore-induced block of transport, or else it passes through this compartment but is not terminally glycosylated (perhaps because the relevant enzymes no longer recognise haemagglutinin as a substrate). It is impossible, with the available information, to distinguish between these possibilities. However, if a by-pass is present, it is clear that this does not represent the major route normally followed by haemagglutinin, since in untreated cells both haemagglutinin and VSV G pass completely through the Golgi complex into the TGN. Indeed, the TGN is likely to be the site at which these two proteins are segregated (Griffiths & Simons, 1986). Consequently, it is unlikely that the differential effect of monensin on the intracellular transport of haemagglutinin and VSV G reveals the site at which the two transport pathways normally diverge.

After an initial delay, the rate of delivery of neuraminidase to the cell surface in carboxylic ionophore-treated cells is approximately equal to that in control cells. The simplest interpretation of this result is that the transport of the protein out of an as yet unidentified compartment is initially reduced by the action of the ionophore, but is eventually restored by the accumulation of neuraminidase in this compartment. This behaviour is not seen with chloroquine and ammonium chloride, where delivery of neuraminidase to the surface is depressed only late in the infection and remains so even at long times post-infection.

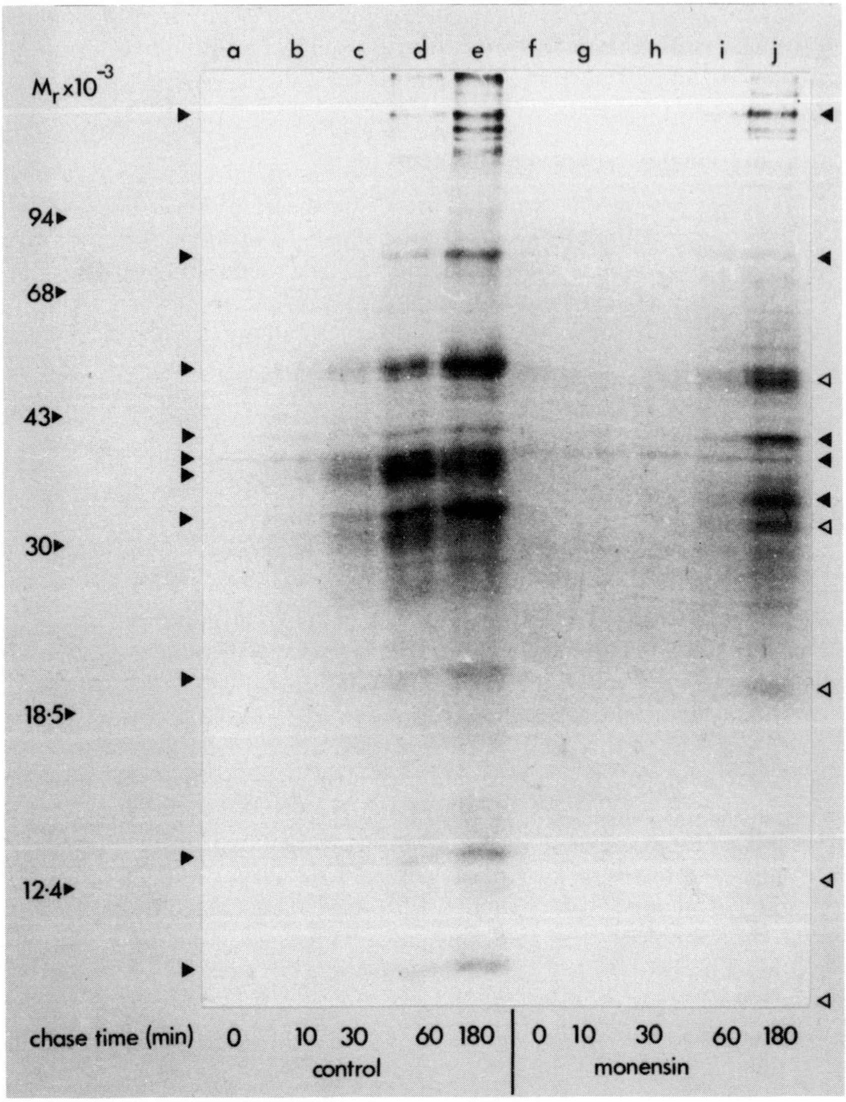

Fig. 8 Effect of monensin on the secretion of endogenous proteins. Monolayers were labelled with [^{35}S]methionine for 10 min and chased for various times in the presence of excess unlabelled methionine. One set of monolayers was treated with 1 μM monensin from 1 h before labelling. After the chase the apical bathing medium was removed and centrifuged to remove cell debris. Soybean trypsin inhibitor (50 μg) was added to the supernatants and protein was precipitated with 10% (w/v) TCA. For the monensin-treated samples open arrowheads depict those proteins which run in different positions relative to the proteins in the control samples. The positions of molecular weight markers are also shown.

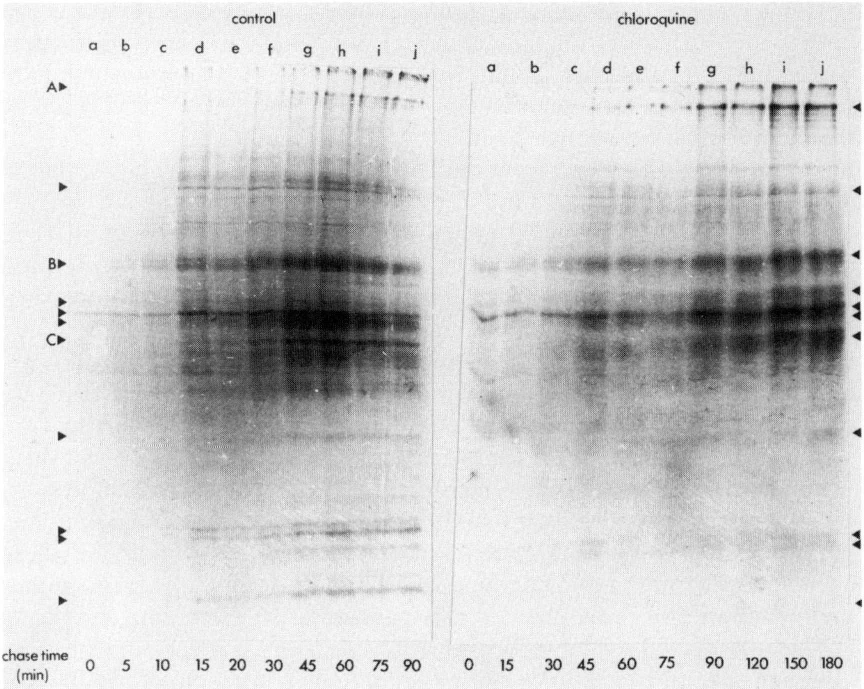

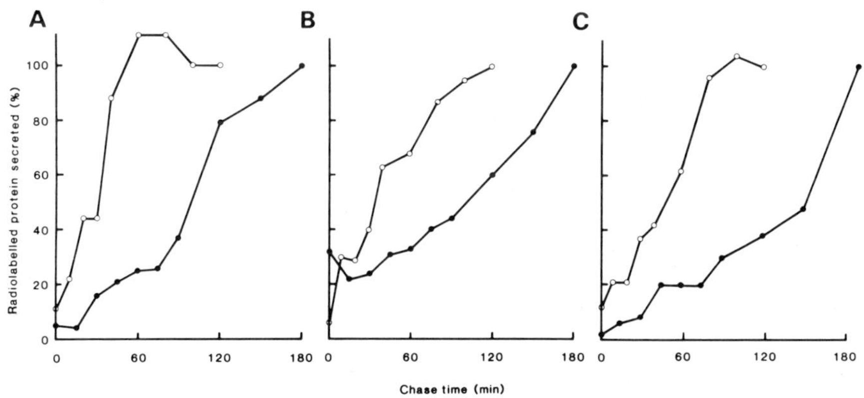

Fig. 9 Effect of chloroquine on the secretion of endogenous proteins. (a) As for Fig. 8 except that one set of monolayers was treated with 150 μM chloroquine during the chase. Arrowheads show the secreted proteins. (b) Time courses of secretion of selected proteins A, B and C from the fluorographs in (a). (o), control; (•) in the presence of chloroquine. Values are corrected for the relative amount of protein secreted at the last time-point in the presence of the weak base.

The secretion of endogenous proteins from MDCK cells is delayed by both monensin and chloroquine, although only the former appears to affect processing of the proteins, leading in several cases to the secretion of species of abnormally low molecular weight. Although the glycosylation states of these proteins have not yet been examined, a reasonable explanation of this result would be that, as in the case of haemagglutinin, glycosylation does not proceed to completion in the presence of monensin.

In MDCK cells, therefore, monensin blocks processing events that are believed to occur normally in the *trans* cisternae of the Golgi complex and the TGN, while chloroquine has a much less dramatic effect on processing. In addition, the observation that monensin delays, but does not block, the delivery of apically-destined proteins to the cell surface in these cells appears to be a general one. Since the major effect of chloroquine is to dissipate transmembrane pH gradients, its effect on intracellular transport may be a result of a selective effect on the exit of proteins from the TGN, which is believed to be a low pH compartment. The carboxylic ionophores, on the other hand, clearly have effects further back in the intra-Golgi transport pathway, which may be a consequence of other known effects of these agents, for example the dissipation of Na^+ and K^+ gradients across either the membranes of the Golgi complex or the plasma membrane.

The role played by the low pH within the TGN is at present obscure, although several possibilities can be identified. For example, the low pH may permit the binding of proteins to be transported out of the TGN to a 'sorting receptor' that cycles between the TGN and the plasma membrane. If this is the case, then one might expect that lowering the pH outside the cell would block the transport of proteins out of the TGN by inhibiting dissociation of proteins from the receptors at the plasma membrane, and that this effect would be 'asymmetric' in MDCK cell monolayers, since a receptor would cycle only through one domain of the plasma membrane. In recent experiments (unpublished results), we have found no such evidence for such a targetting system. Other possibilities are that transport of proteins out of the TGN requires a clustering step and that this in turn requires a low pH, or that the carboxylic ionophores and the weak bases produce their effects by perturbing membrane flow through, and out of, the Golgi complex. Clearly, further study is needed to distinguish between these possibilities, and to elucidate the precise role of intra-compartmental pH in the transport of proteins to the plasma membrane.

Materials and methods

Cells and viruses

MDCK cells were grown as monolayers in Eagle's minimum essential medium (EMEM) supplemented with 10% newborn calf serum, non-essential amino acids and 160 μg/ml gentamicin. Baby hamster kidney (BHK-21) cells were grown in Glasgow's minimum essential medium (GMEM) supplemented with 10% newborn calf serum, 10% tryptose phosphate broth and 160 μg/ml gentamicin. Cells were grown at 37°C in an atmosphere of 5% CO_2 in air. All cells, media and supplements were supplied by Flow Laboratories, Rickmansworth, Herts.

The WSN (H1N1) strain of influenza virus, initially obtained from Dr S.C. Inglis (Department of Pathology, University of Cambridge, U.K.) was grown in fertile hens' eggs, purified by centrifugation on a continuous sucrose density gradient (15–60% [w/v] in 50 mM Tris-HCl, pH 7.4, 100 mM NaCl) for 2 h at 100 000 × g and assayed by plaquing on MDCK cell monolayers. Stacks of virus were stored in liquid nitrogen.

Semliki Forest virus (SFV), initially obtained from Dr T.R. Hesketh (Department of Biochemistry, University of Cambridge, U.K.) was grown on BHK-21 cells. Cells (10^8 at 50% confluence) were washed twice with EMEM containing 0.2% (w/v) bovine serum albumin BSA (Sigma, Poole, Dorset) and 20 mM Hepes, pH 6.8, and infected with 0.01 plaque-forming units (pfu) of SFV/cell in the same medium for 1 h at 37°C. The medium was then replaced with 50 ml complete EMEM. After 15 h at 37°C the medium was harvested and centrifuged at 5 000 × g for 30 min to sediment cell debris. SFV was sedimented by centrifugation at 60 000 × g for 2.5 h and purified by centrifugation for 2 h at 100 000 × g on a gradient consisting of 10–20% (w/v) sucrose in 50 mM Tris-HCl, pH 7.4, 100 mM NaCl above 25–50% (w/v) sucrose in the same buffer. Stocks of virus were stored in liquid nitrogen.

Drugs

Monensin and nigericin (Calbiochem-Behring, Bishops Stortford, Herts) were stored at −20°C as 1 mM solutions in absolute ethanol and used at 1 μM. Chloroquine (Sigma) was stored at −20°C as a 15 mM solution in water and used at 150 μM. Ammonium chloride (Sigma) was stored at −20°C as a 2 M solution in water and used at 20 mM.

Infection

MDCK cells were grown on 3 cm diameter tissue culture dishes until just confluent. Monolayers were washed twice with 2 ml EMEM containing 0.2% (w/v) BSA and infected with 10 pfu of influenza virus in the same medium for 1 h at 37°C. The medium was then replaced with EMEM containing 2% newborn calf serum.

Labelling with [^{35}S]methionine

MDCK cell monolayers, either untreated or 5 h post-infection, were washed twice with 2 ml EMEM lacking methionine and then labelled for 10 min by exposure to 0.5 ml of the same medium containing 60 μCi/ml [^{35}S]methionine (Amersham International, Amersham, Bucks). Further labelling was prevented and incubation continued by addition of 2 ml EMEM containing 10 times the normal level of unlabelled methionine (1.5 mg/ml). To terminate the incubation, samples were washed three times with 2 ml ice-cold phosphate-buffered saline (PBS), containing 137 mM NaCl, 2.7 mM KCl, 8.1 mM Na_2HPO_4, 1.5 mM KH_2PO_4, pH 7.4. Cells were disrupted and cellular DNA was digested by addition of 10 μl 2% (w/v) SDS in 50 mM Tris-HCl, pH 8.8, followed by 2 μl 10 mg/ml DNase I (Sigma Type IV). Cells were scraped from the dishes using a rubber scraper.

Treatment with endoglycosidase H

A sample (10 μl) of cell lysate was taken, 10 μl 2% SDS was added and the mixture boiled for 2 min. After cooling, 200 μl of 100 mM sodium citrate (pH 5.5) containing 0.4 mU endoglycosidase H (endo H; Miles, Slough) and 40 μg/ml phenylmethylsulphonyl fluoride (Sigma) were added and the sample was incubated at 37°C for 15 h. Proteins were precipitated by addition of 200 μl ice-cold acetone and resuspended in 50 μl electrophoresis buffer (see below).

Assay for delivery of haemagglutinin to the plasma membrane

Biosynthetically labelled monolayers were washed twice with 2 ml ice-cold PBS and exposed to 1 ml 0.5 mg/ml trypsin (Sigma Type III) in PBS for 15 min at either 4°C or 37°C. Monolayers were then exposed to 1 ml 0.5 mg/ml soybean trypsin inhibitor (Sigma) in ice-cold PBS for 15 min and then washed twice with 2 ml ice-cold PBS. Cells were harvested as described above.

SDS/polyacrylamide gel electrophoresis

SDS polyacrylamide gel electrophoresis (SDS/PAGE) was carried out as described previously (Edwardson, 1984). After running, gels were fixed for 15 min in 10% (w/v) trichloroacetic acid (TCA) and then immersed for 30 min in AMPLIFY (Amersham). Gels were dried at 80°C under reduced pressure and exposed to X-ray film (Kodak) at −70°C, usually for 1 week. Fluorographs were scanned using a Joyce-Loebl Mark III densitometer scanner.

Preparation of tritiated SFV

Preparation of [^{3}H]SFV was as described for SFV, except that the post-infection maintenance medium contained 5 mCi [^{3}H]N-acetyl-D-mannosamine (30 Ci/mmol; New England Nuclear, Stevenage, Herts). The envelope glycoproteins of shed virions consequently contained [^{3}H]sialic acid (Monaco & Robbins, 1973).

Assay for delivery of neuraminidase to the plasma membrane

At various times after infection, monolayers were washed twice with 2 ml ice-cold PBS containing 0.9 mM $CaCl_2$ and 0.5 mM $MgCl_2$ and then exposed to [^{3}H]SFV (approximately 20 000 dpm) in 0.5 ml ice-cold PBS containing 2 mg/ml BSA for 30 min, with gentle rocking. Protein was precipitated by addition of 10% (w/v) TCA, followed by incubation on ice for 30 min. Protein was pelleted by centrifugation for 2 min in an Eppendorf microfuge. Supernatants were removed and radioactivity counted by liquid scintillation spectrometry, using 0.6% (w/v) butyl-PBD/70% (v/v) toluene/30% (v/v) Triton X-100 as scintillant.

Secretion of proteins from MDCK cells

MDCK cell monolayers were pulse-labelled and chased as described above. At various times of chase the apical bathing medium was removed and centrifuged for 2 min in an Eppendorf microfuge to remove cell debris. Soybean trypsin inhibitor (50 μl of a 1 mg/ml solution in water) was then added to the supernatants and protein precipitated by addition of 10% TCA, as described above. Pellets were then prepared for SDS/PAGE.

Electron microscopy

Monolayers growing on nitrocellulose filters (0.22 μm pore size; Millipore, Harrow, Middx) were fixed for 6 h at 4°C with 2% (v/v) glutaraldehyde, 2% (w/v) paraformaldehyde in 0.1 M Pipes buffer, pH 7.2, containing 0.1 M sucrose. Samples were post-fixed with 1% (w/v) osmic acid, dehydrated, stained with uranyl acetate and lead citrate, and embedded in Spurr resin. Thin sections (60–80 nm) were cut on a Huxley Mark II ultramicrotome, mounted on 400-mesh copper grids and examined in a Philips 300 electron microscope.

Acknowledgements

We are grateful to the Wellcome Trust and the National Kidney Research Fund for financial support. P.U.D. is a Medical Research Council scholar.

References

Alonso-Caplen, F.V. & Compans, R.W. (1983). Modulation of glycosylation and transport of viral membrane glycoproteins by a sodium ionophore. **Journal of Cell Biology** 97: 659–668.

Anderson, R.G.W. & Pathak, R.K. (1985). Vesicles and cisternae in the *trans* Golgi apparatus of human fibroblasts are acidic compartments. **Cell** 40:635–643.

Edwardson, J.M. (1984). Effects of monensin on the processing and intracellular transport of influenza virus haemagglutinin in infected MDCK cells. **Journal of Cell Science** 65: 209–221.

Green, S.A., Plutner, H. & Mellman, I. (1985). Biosynthesis and intracellular transport of the mouse macrophage Fc receptor. **Journal of Biological Chemistry** 260: 9867–9874.

Griffiths, G. & Simons, K. (1986). The *trans* Golgi network: sorting at the exit site of the Golgi complex. **Science** 234: 438–443.

Louvard, D. (1980). Apical membrane aminopeptidase appears at site of cell-cell contact in cultured kidney epithelial cells. **Proceedings of the National Academy of Sciences, U.S.A.** 77:4132–4136.

Matlin, K.S. (1986). Ammonium chloride slows transport of the influenza virus hemagglutinin but does not cause mis-sorting in a polarized epithelial cell line. **Journal of Biological Chemistry** 261:15172–15178.

Matlin, K.S. & Simons, K. (1983). Reduced temperature prevents transfer of a membrane glycoprotein to the cell surface but does not prevent terminal glycosylation. **Cell** 34:233–243.

Monaco, F. & Robbins, J. (1973). Incorporation of N-acetylmannosamine and N-acetylglucosamine into thyroglobulin in rat thyroid *in vitro*. **Journal of Biological Chemistry** 248: 2072–2077.

Moore, H-P., Gumbiner, B. & Kelly, R. (1983). Chloroquine diverts ACTH from α-regulated to a constitutive secretory pathway in AtT-20 cells. **Nature** 302:434–436.

Nakamura, K., Brown, A.S. & Compans, R.W. (1980). Glycosylation sites of influenza virus glycoproteins. Tryptic glycopeptides from the A/WSN (H0Nl) haemagglutinin glycoprotein. **Virology** 107:208–221.

Nakamura, K. & Compans, R.W. (1979). Biosynthesis of the oligosaccharides of influenza viral glycoproteins. **Virology** 93:31–47.

Pressman, B.C. (1976). Biological applications of ionophores. **Annual Reviews of Biochemistry** 45:501–530.

Rodriguez-Boulan, E. & Pendergast, M. (1980). Polarized distribution of viral envelope proteins in the plasma membrane of infected epithelial cells. **Cell** 20:45–54.

Rodriguez-Boulan, E. & Sabatini, D.D. (1978). Asymmetric budding of viruses in epithelial monolayers: a model system for study of epithelial polarity. **Proceedings of the National Academy of Sciences U.S.A.** 75:5071–5075.

Strous, G.J., du Maine, A., Zijderhand-Bleekemolen, J.E., Slot, J.W. & Schwartz, A.L. (1985). Effect of lysosomotropic amines on the secretory pathway and on the recycling of asialoglycoprotein receptor in human hepatoma cells. **Journal of Cell Biology** 101: 531–539.

Tartakoff, A.M. (1983). Perturbation of vesicular traffic with the carboxylic ionophore monensin. **Cell** 32: 1026– 1028.

Tartakoff, A.M. & Vassalli, P. (1977). Plasma cell immunoglobulin secretion. Arrest is accompanied by alterations in the Golgi complex. **Journal of Experimental Medicine** 146: 1332–1345.

The targeting of proteins to specific cell membranes

ANTHONY P. BOULTON

Department of Biochemistry, University College and Middlesex School of Medicine, The Windeyer Building, Cleveland Street, London W1P 6DB, UK.

Abstract

This review deals with the targeting of proteins to specific cell membranes and discusses the nature of targeting signals; the role of N-terminal amino acid sequences in the interaction of proteins with the endoplasmic reticulum and the mitochondrial outer membrane; the roles of localization sequences and proteolytic cleavage sites in the delivery of proteins to other mitrochondrial compartments; and the nature of the signals that direct proteins to the nucleus. The function and intracellular location of casein kinase from the lactating guinea-pig mammary gland are discussed in terms of its suitability as a model for the targeting of proteins to the Golgi apparatus.

Key words

Protein targeting; signal sequences; endoplasmic reticulum; mitochondrial import; nuclear import; mammary gland casein kinase; post-translational modification of milk proteins; oocytes.

Introduction

How do proteins reach their correct destination within the cell? This major question in cell biology is still to be resolved, although much information is accumulating largely due to the use of recombinant DNA technology and immunochemical methods. Consequently, our knowledge of how proteins are directed to the endoplasmic reticulum, how they are imported into mitochondria and chloroplasts, and how lysosomal enzymes are directed to the lysosome has increased dramatically over the past few years. Despite this general increase in understanding, little is known of the precise mechanisms that direct Golgi-specific proteins to the organelle. Our interest lies in this area of investigation and forms the basis of this review in which some of the general principles that have emerged from the study of other systems and then some of our work on casein kinase from the lactating guinea-pig mammary gland will be discussed. Casein kinase is a candidate for study in terms of the biogenesis of the Golgi complex. Reviews on the biogenesis of specific compartments include: the peroxisome (Borst, 1986;

Lazarow & Fujiki, 1985); the mitochondrion (Douglas *et al.*, 1986); and the nucleus (Dingwall & Laskey, 1986).

The scale of the problem of targeting proteins to specific sites within the cell can be envisaged if one considers the eukaryotic cell, bounded by a plasma membrane and containing many membranous structures or subcellular compartments. These include mitochondria, lysosomes, peroxisomes and a nucleus each of which have specific functions within the cell, the metabolic processes of which are therefore compartmentalized such that for example only the mitochondria contain the enzymes that catalyse the oxidative phosphorylation of ADP. The differences in function of the compartments are reflected in variations in both the lipid and protein compositions and as far as the proteins are concerned these can be either structural or catalytic. How is this structural and functional specificity maintained?

Intracellular protein traffic must be controlled to ensure the maintenance of structure and function, and since proteins are synthesized on cytoplasmic ribosomes, either free in the cytosol or attached to the membrane of the endoplasmic reticulum, there must be specific mechanisms for the targeting, transportation and sorting of proteins to their correct compartments or destination. It seems reasonable to suggest that proteins that are to be delivered to a site distal from the site of synthesis carry some signal that will be recognized by a mechanism, possibly a receptor, on the target compartment. Much of the following discussion will be concerned with the nature of such signals.

Targeting signals

Blobel (1980) first suggested that four types of amino acid signal could determine the final destination of a particular protein: signal, stop transfer, insertion and sorting sequences. Signal sequences initiate the transfer of a polypeptide across a specific membrane; stop-transfer sequences interrupt the transfer started by signal sequences; insertion sequences anchor a protein in a membrane, and sorting sequences act as the address on a protein that operates once the protein has been synthesised, and directs it from a donor site to an acceptor. A summary of the main signals used to route proteins is given in Table 1, which classifies the proteins according to their site of synthesis, either on membrane-bound ribosomes or on those free in the cytosol. In general, signals are either internal amino acid sequences or N-terminal sequences. In the case of the N-terminal signals they are either cleaved at some stage or remain as an integral part of the completed protein.

Protein sequencing, either direct or *via* cloned DNA, and the use of site directed mutagenesis has allowed details of the consensus sequences for proteins directed to the same compartment to be built up. Such consensus

Table 1. *Summary of signals used to target proteins*

Proteins synthesized on membrane-bound ribosomes:	
Endoplasmic reticulum –	usually N-terminal cleavable signal sequence
Lysosome Cell membrane	as above but with additional internal sorting signals that specify destination after entry into the endoplasmic reticulum
Proteins synthesized on free ribosomes:	
Nucleus –	internal signal
Chloroplast Mitochondrion	usually N-terminal cleavable signal
Peroxisome –	usually internal signal

sequences are known for proteins of the endoplasmic reticulum (Walter *et al.*, 1984), nuclei (Kalderon *et al.*, 1984) and mitochondria (Schatz & Butow, 1983; Hay *et al.*, 1984) and are summarised in Table 2. To date most experiments have dealt with the question of what is the minimal signal required to direct a protein to a particular compartment, and have been primarily concerned with routing of membrane proteins. The use of site directed and deletion mutagenesis of the signal has led to the principle that alteration of the signal leads to the accumulation of the protein in the cytosol, suggesting the absence of secondary unused signals. Addition mutagenesis where topogenic signals, usually N-terminal signal sequences from one protein are grafted onto a protein not normally found in a particular compartment have confirmed that such signals have a direct role in protein targeting and have demonstrated how powerful such signals can be.

Table 2. *Consensus amino acid signal sequences*

Nucleus –	-Pro-Lys-Lys-Lys-Arg-Lys-(Val)-
Endoplasmic reticulum –	variable (13–16 residues): basic residues in and beyond N-terminal part: central hydrophobic region: C-terminal hydrophilic region
Mitochondrial matrix –	variable: several basic residues: absence of acidic residues: no long hydrophobic stretch

Studies on the accumulation of the large T antigen of SV4O in the nucleus (reviewed in Dingwall & Laskey, 1986) have shown that the process depends on an uncleaved internal sequence of basic amino acids that contains several lysine residues. Substitution of the second lysine in the sequence abolishes nuclear uptake. Attachment, by chemical cross-linking, of the peptide pro-lys-lys-lys-arg-lys-val to bovine serum albumin results in the accumulation of the chimeric protein in the nuclei of *Xenopus* oocytes following microinjection into the cytoplasm (Goldfarb *et al.*, 1986), whilst addition of nuclear localization sequences by recombinant DNA techniques to *Escherichia coli* β-galactosidase leads to the accumulation of the fusion protein in the nuclei of many species (Dingwall & Laskey, 1986). Despite these results and the discovery of homologous sequences in many other viral proteins, particularly those of adenovirus (Smith *et al.*, 1985), no clear consensus sequence has emerged.

In the case of the endoplasmic reticulum, and the targeting of proteins to its membrane, the signal is nearly always at the N-terminal and cleaved co-translationally. There are notable exceptions to this such as ovalbumin where the signal is internal and is not cleaved (Palmiter *et al.*, 1978). Although the primary structures of the N-terminal signals are highly variable in both size and sequence (see Table 3), three general features can be distinguished. Namely, a hydrophobic N-terminal region, a central region highly enriched in hydrophobic residues and a C-terminal part that is less hydrophobic and contains a consensus sequence that is possibly involved in the cleavage of the signal as the protein passes into the endoplasmic reticulum (von Heijne, 1984). Analysis of some 150 signal sequences (von Heijne, 1984) shows that the length varies between 13 and 36 residues which suggests that 13 amino acids should be sufficient to direct a protein to the endoplasmic reticulum. This is borne out by experiments where many proteins that are normally found in the cytosol can be directed to the endoplasmic reticulum as fusion proteins containing N-terminal signals. As might be expected these signals are highly conserved in evolution, even bacterial N-terminal leader sequences are capable of directing β-globin into the endoplasmic reticulum (Lingappa *et al.*, 1984).

The import of proteins into mitochondria is complex and four routes corresponding to the four mitochondrial compartments (the matrix, outer membrane, inner membrane and the intermembrane space) have been distinguished, and although mitochondria contain their own DNA the vast

Table 3. *Examples of amino acid signal sequences for secretory and membrane proteins*

Preproalbumin	met-lys-trp-val-thr-phe-leu-leu-leu-leu-phe-ile-ser-gly-ser-ala-phe-ser*arg
Pre-IgG light chain	met-asp-met-arg-ala-pro-ala-gln-ile-phe-gly-phe-leu-leu-leu-leu-phe-pro-gly-thr-arg- cys*asp
Prelysozyme	met-arg-ser-leu-leu-ile-leu-val-leu-cys-phe-leu-pro-leu-ala-ala-leu-gly*lys
Pre VSV G protein	met-lys-cys-leu-leu-tyr-leu-ala-phe-leu-phe-ile-his-val-asn-cys*lys
Prepenicillinase	met-ser-ile-gln-his-phe-arg-val-ala-leu-pro-phe-phe-ala-phe-cys-leu-pro-val-phe-ala*his

majority of their proteins are synthesized on free ribosomes in the cytosol and are coded for by nuclear genes. The signal that directs proteins to the mitochondrial membrane is N-terminal and is usually cleaved post-translationally by a mitochondrial protease located in the matrix or inter-membrane space. As with the signals that direct proteins to the endoplasmic reticulum the mitochondrial signals are highly variable in amino acid composition and sequence. A common feature however is a high content of basic amino acids which are essential for uptake and processing. In contrast to the endoplasmic reticulum signals those for mitochondrial proteins show a complete absence of acidic residues and hydrophobic stretches (Joh *et al.*, 1985). Presumably these distinctions are of vital importance.

Sorting signals

Sorting signals are those that direct proteins to their destination from either the endoplasmic recticulum or the Golgi complex, which is becoming recognized as the major site of intracellular protein sorting. Most information is available for the lysosomal enzymes which are synthesized on membrane bound ribosomes and are core glycosylated as they enter the endoplasmic reticulum. The sugars are modified in both the endoplasmic reticulum and the Golgi where the phosphorylation of mannose residues at the 6 position is of particular significance (see the chapter by Hasilik & von Figura, in this volume). This acts as a sorting signal that directs the proteins to the lysosome (von Figura & Hasilik, 1986).

Similar, but as yet unknown signals, are presumably present in proteins that are specifically directed from the Golgi complex to either the basolateral or apical membranes of polarised cells (Rindler *et al.*, 1985).

Mechanisms

Despite our knowledge of the composition and structure of signals, little in general is known of how they interact with membranes, with the exception of the interactions that direct proteins to the endoplasmic reticulum and mitochondrial membranes. Although similar details are available for the chloroplast they will not be dealt with here.

At least two components are involved in the import of proteins into the endoplasmic reticulum: a soluble complex, the signal recognition particle (Walter & Blobel, 1982), and an integral membrane protein of the endoplasmic reticulum itself, the signal recognition particle (SRP) receptor or docking protein (Walter & Blobel, 1981). In many *in vitro* systems translating mRNA species coding for high molecular weight proteins, the signal recognition particle binds to the signal peptide as it emerges from the large ribosomal subunit and arrests translation. Interaction with the docking protein relieves the arrest and allows protein synthesis to proceed. This ensures that completion of the polypeptide only occurs on the membrane. Although this is an attractive hypothesis there are exceptions, especially in the translation of low molecular weight proteins, for example, promellitin, where the *in vitro* translation of prepromellitin mRNA is not arrested by signal recognition particle (Zimmermann & Mollay, 1986), and it is suggested that the relative lack of protein folding in such short polypeptide

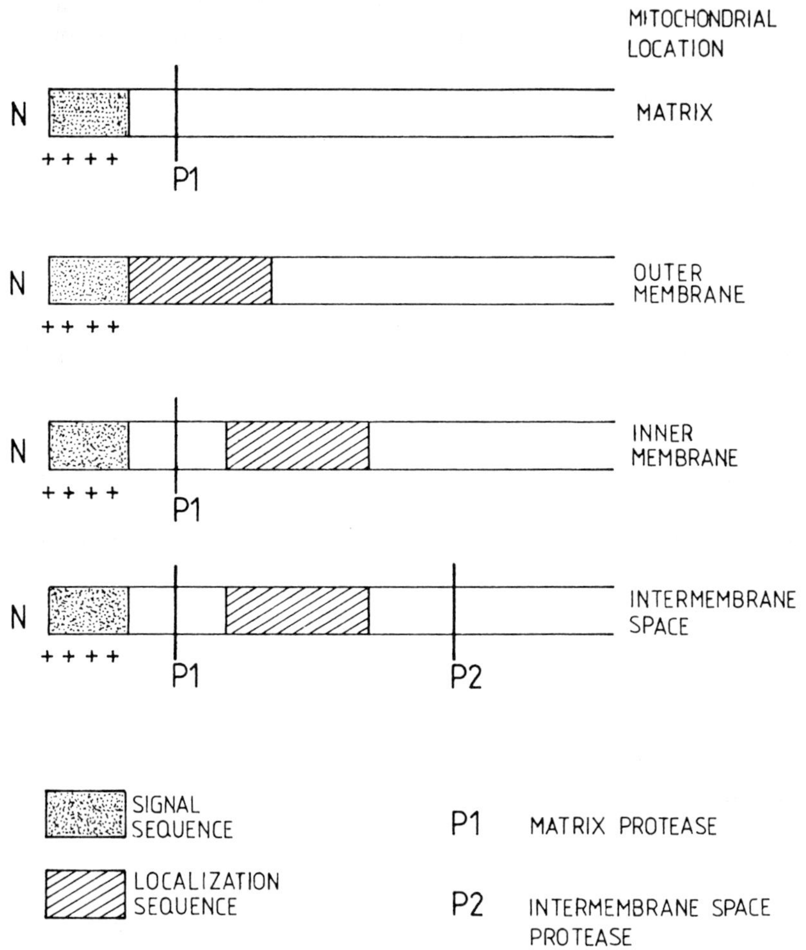

Fig. 1 The relative positions of targeting and processing domains on proteins imported into mitochondria. Targeting of proteins into mitochondria and to the mitochondrial matrix utilises a positively charged hydrophilic sequence (signal sequence). An additional sequence, usually hydrophobic and large enough to span a membrane (localization sequence) anchors the protein in a membrane, the positions of the proteolytic cleavage sites then determine the final destination of the protein (see Fig. 2).

chains obviates the need for an arrest of translation before the signal interacts with the membrane.

As already mentioned there are four routes by which proteins enter the mitochondrion and all rely on an initial recognition of the N-terminal mitochondrial signal sequence. The character and relative position of other signals on the protein then determine its final destination in the mitochondrial subcompartments (see Fig. 1).

A common structural feature exhibited by precursors located outside the mitochondrial matrix is an uninterrupted stretch of hydrophobic amino acids sufficient to span a membrane, the localisation sequence, which anchors the protein in a membrane. On precursors of intermembrane space proteins, for example cytochrome c peroxidase, these hydrophobic stretches are flanked by clusters of basic residues which act as sites of proteolytic cleavage (Kaput *et al.*, 1982). A protein targeted to the outer membrane, such as the 70 kD protein, contains an insertion sequence immediately after the N-terminal signal (Hase *et al.*, 1984), whilst the absence of a localization sequence allows the protein to traverse the two mitochondrial membranes so that it finally resides in the mitochondrial matrix.

The role of these determinants and that of the mitochondrial membrane potential that is required for the import of all but the outer membrane proteins is shown in Fig. 2. Here the role of the membrane potential is shown as being in the formation of a transient fusion of the inner and outer membranes, which allows proteins to cross the intermembrane space (Schleyer & Neupert, 1985). The interaction of the N-terminal signal with the outer membrane is thought of as being energy independent (Frietag *et al.*, 1982; Gasser & Schatz, 1983), and localization of proteins such as the 70 kD protein, to this subcompartment requires only an anchor sequence adjacent to the N-terminal signal (Hase *et al.*, 1984). Precursors to matrix proteins, for example ornithine transcarbamylase, which lack anchor sequences, require the initial interaction of the N-terminal signal with the outer membrane and a membrane potential. The proteins then pass directly into the matrix where the signal is cleaved (Horwich *et al.*, 1985).

The localization of intermembrane space proteins is more complicated. Again, the initial interaction is energy independent but a membrane potential is required to allow insertion of the anchor sequence into the inner membrane. Proteolytic cleavage on both sides of this sequence results in the release of the N-terminal signal into the matrix and of the intermembrane space protein into the correct subcompartment.

Protein targeting to the Golgi

In contrast to the organelles discussed above, there is very little corresponding information on the Golgi complex. The questions are, therefore, what are the signals involved in the targeting of proteins to this organelle and to its constituent compartments and how can they be deciphered?

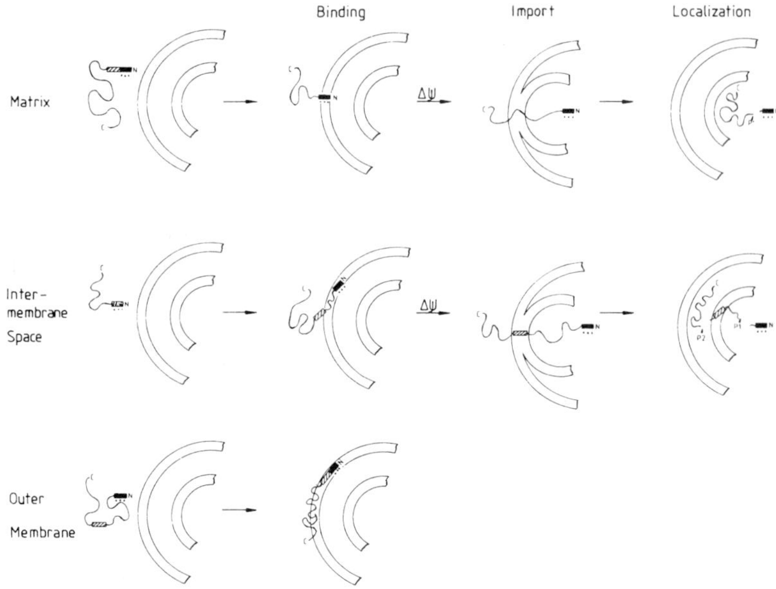

Fig. 2 The role of targeting and processing domains in the import of mitochondrial proteins. In the absence of a localization sequence proteins are directed to the matrix. The initial interaction with the outer mitochondrial membrane is shown as being energy independent, whilst the import of proteins beyond the outer membrane requires an energized membrane. The presence of a transmembrane localization sequence at the inner but not outer membrane locates the protein to the inner membrane. Proteolytic cleavage on either side of the localization sequence releases the protein into the inter-membrane space.

Firstly, a protein is needed that is Golgi specific, preferably tissue specific and, ideally, an integral membrane protein, simple to purify and of known function. Casein kinase from the lactating guinea-pig mammary gland fulfills many of these criteria. Results of our work with this enzyme are summarised below.

The lactating guinea-pig mammary gland synthesises four milk proteins, one acid soluble whey protein, alpha-lactalbumin, and three acid insoluble caseins. The caseins have molecular weights in the range 20–28 kD and are all phosphorylated to varying degrees and to some extent glycosylated (Craig *et al.*, 1978). Experiments with explants from the lactating guinea-pig mammary gland incubated in the presence of [^{35}S]methionine followed by SDS-polyacrylamide gel electrophoresis and fluorography have shown that at short incubation times (e.g., 3 min) incompletely processed caseins are produced that do not migrate with the fully processed proteins secreted by the gland, and that mature caseins that co-migrate with those secreted by the gland are only produced after long incubation times (e.g., 30 min;

Boulton *et al.*, 1984). In contrast, radiolabelled caseins synthesized in the presence of [^{32}P]phosphate consistently migrate with caseins purified from guinea-pig milk irrespective of the length of incubation (Boulton *et al.*, 1984).

Treatment of post-nuclear supernatants (prepared from explants incubated in media containing either [^{35}S]methionine or [^{32}P]phosphate) with proteinase K shows that the caseins are present within a membrane bound environment, since they are protected from proteolytic attack in the absence of detergent, but are completely hydrolysed in the presence of 0.1% (v/v) Triton X-100. Caseins synthesized in explants during a short pulse have identical electrophoretic mobility, when compared with those synthesized and sequestered in mRNA directed cell free protein synthesizing systems in the presence of dog pancreas microsomes, which suggests that the caseins synthesized during a short pulse are sequestered within the endoplasmic reticulum. Confirmation of the timing of the events leading to the synthesis of mature caseins comes from classical pulse-chase experiments using explants in the presence of both radiolabelled methionine and phosphate. Here mature ^{32}P labelled proteins are present immediately after the pulse and remain throughout the chase period of up to 50 min. In contrast, [^{35}S]methionine-labelled mature proteins only start appearing approximately 15 min after the pulse and become progressively more prominent throughout the chase period. These results are consistent with at least two levels of processing leading to the production of the mature proteins, initially sequestration at the level of the rough endoplasmic reticulum with the removal of the signal peptide (Craig *et al.*, 1979), followed some 15 to 20 min later by a second event that leads to the production of caseins that have electrophoretic properties similar to those of caseins isolated from guinea-pig milk. The timing of the second event is consistent with it occurring in the Golgi complex.

Evidence that the second event is a phosphorylation comes from experiments with acid phosphatase. Treatment of caseins isolated from guinea-pig milk results in proteins having identical electrophoretic mobilities to those synthesized in explants during a short pulse with [^{35}S]-methionine. These results suggest that the enzyme that catalyses the phosphorylation of caseins within the mammary gland secretory pathway is located in the Golgi complex, agreeing with the earlier data of Bingham & Farrell (1977), Chew & Mackinlay (1974), and Mackinlay *et al.*, (1977) working with the partially purified enzyme from a variety of species.

Additional evidence for the intracellular localization of casein kinase comes from cell fractionation studies, where in sucrose step gradients it co-separates with galactosyltransferase, a common marker for the Golgi complex (Favard, 1977).

Cell fractionation methods have been developed for the isolation of a Golgi-enriched fraction from the lactating guinea-pig mammary gland and have shown, using trypsin in the presence and absence of both high salt (0.5 M KCl) and detergent (0.1% (v/v) Triton X-100), that casein kinase is probably an integral membrane protein of the complex as the enzymic activity is protected from proteolysis in the absence of detergent, even in the presence of high salt (Pascall *et al.*, 1981). In addition, extraction of the Golgi-enriched fraction with Triton X-114 (Pryde & Phillips, 1986) leads to the enzyme partitioning into the detergent-rich phase, consistent with it being an integral membrane protein (unpublished observations).

Casein kinase thus fulfills two of the criteria expected of a protein that might be of use as a model for an investigation into protein targeting to the Golgi complex. In order to proceed further it was necessary to purify and characterize the protein. Purification was achieved by a combination of sucrose density gradient centrifugation following solubilization of the Golgi-enriched fraction in 0.5 M KCl/0.1% (v/v) Triton X-100, and chromatography on ATP-agarose (Moore *et al.*, 1985). The purified enzyme gave a single band on SDS-polyacrylamide gels stained with silver and had an apparent molecular weight of 74 000.

An antiserum against the purified enzyme raised in rabbits recognized only the membrane-associated enzyme from Golgi-enriched fractions from the lactating guinea-pig mammary gland and not the soluble so called casein kinases present in most tissues, especially liver and reticulocytes (Moore *et al.*, 1985). In these experiments partially purified IgG from the antiserum was preincubated with the protein kinase and followed by incubation with Protein A-sepharose. After centrifugation the residual enzyme activity in the supernatant was assayed. Under these conditions only the mammary gland enzyme was removed from solution. The same antiserum used in immunoblotting procedures recognized the purified enzyme, a protein of identical apparent molecular weight in Golgi-enriched fractions from the lactating guinea-pig mammary gland, but not in similar fractions from guinea-pig liver, spleen, heart or brain. It also failed to recognize the enzyme present in Golgi-enriched fractions from the mammary glands of lactating cow, rat or mouse (Moore *et al.*, 1985). The antiserum was therefore specific to the membrane-associated protein kinase from the lactating guinea-pig mammary gland.

The antiserum has been used to screen a cDNA library constructed in λGt11 using mRNA from the lactating guinea-pig mammary gland (Young & Davis, 1983a, b), and two putative clones (unpublished) which are being characterized, have been isolated.

The question remains as to how a study of Golgi biogenesis might be approached. One possibility depends on the use of the *Xenopus* oocyte, the microinjection of which with mRNA from the lactating guinea-pig mammary gland followed by incubation in media containing [^{35}S]methionine leads to the synthesis and secretion of antibody-precipitable radiolabelled milk proteins. They co-migrate with those synthesized in explants during a short pulse and in microsome supplemented wheat-germ cell-free protein synthesizing systems. These represent those proteins that are synthesized and sequestered in the endoplasmic reticulum and which are unphosphorylated. It appears that although the oocyte contains a functional secretory pathway, including a Golgi complex, it lacks the specific enzyme, casein kinase, that completes the processing of milk proteins. Similar results have been obtained using carp insulin mRNA. In this last case proinsulin rather than insulin is secreted (Rapaport, 1981). Others have used the microinjection of *Xenopus* oocytes to transcribe and translate cDNA (Davey *et al.*, 1985), or have used vectors containing SP6 and/or T7 promoters to synthesize mRNA *in vitro* followed by microinjection into the oocyte (Kreig & Melton, 1984). An approach here would be to use the *Xenopus* oocyte to test for the correct addressing of the protein by microinjection of either the cDNA directly, or of *in vitro* synthesized mRNA and to use cell fractionation to isolate a Golgi-enriched fraction followed by enzyme assays. A functional test would be to follow microinjection of casein kinase sequences with microinjection with milk protein mRNA and look for the secretion of fully mature caseins.

Once having established that casein kinase could be directed to the Golgi complex, *in vitro* mutagenesis could be used to delete potential addressing sequences and determine subsequent localization of the enzyme. Similar studies could be carried out on eukaryotic cells in culture. Polarized cells such as MDCK cells might be of particular use (see the chapter by Edwardson *et al.*, in this volume.)

Acknowledgements

I wish to thank the Medical Research Council for financial support.

References

Bingham, E.W. & Farrell, H.M. (1977). Phosphorylation of casein by the lactating mammary gland: A review. **Journal of Dairy Science** 60:1199–1207.

Blobel, G. (1980). Intracellular protein topogenesis. **Proceedings of the National Academy of Sciences U.S.A.** 77:1496–1500.

Borst, P. (1986). How proteins get into microbodies (peroxisomes, glyoxysomes, glycosomes). **Biochimica et Biophysica Acta** 866:179–203.

Boulton, A.P., Pascall, J.C. & Craig, R.K. (1984). Identification and subsequent phosphorylation of sequestered partially processed caseins in the lactating guinea-pig mammary gland. **Biochemical Journal** 222:501–510.

Boulton, A.P., Lane, C.D., Pascall, J.C. & Craig, R.K. (1985). The absence from the oocyte secretory apparatus of a protein kinase capable of phosphorylating sequestered caseins. **Journal of Applied Chemistry** 7:79–85.

Chew, L.F. & Mackinlay, A.G. (1974). Histone and casein kinases of lactating bovine mammary gland. **Biochimica et Biophysica Acta** 359:73–82.

Craig, R.K., McIlreavy, D. & Hall, R.L. (1978). Separation and partial characterization of guinea-pig caseins. **Biochemical Journal** 173:633–641.

Craig, R.K., Boulton, A.P., Harrison, O.S., Parker, D. & Campbell, P.N. (1979). Studies on the intracellular segregation of polyribosome-associated messenger ribonucleic acid species in lactating guinea pig mammary gland. **Biochemical Journal** 181: 737–756.

Dingwall, C. & Laskey, R.A. (1986). Protein import into the cell nucleus. **Annual Review of Cell Biology** 2:367–390.

Davey, J., Dimmock, N.J. & Colman, A. (1985). Identification of the sequence responsible for the the nuclear accumulation of the influenza virus nucleoprotein in *Xenopus* oocytes. **Cell** 40:667–675.

Douglas, M.O., McCammon, M.T. & Vassarotti, A. (1986). Targeting proteins into mitochondria. **Microbioligical Reviews** 50:166–178.

Favard, P. (1977) The diversity of membranes. In: **Mammalian Cell Membranes**, vol. 2 (G.A. Jamieson and D.M. Robinson, eds.), Butterworths, London, pp. 108–140.

Frietag, H., Janes, M. & Neupert, W. (1982). Biosynthesis of mitochondrial porin and insertion into the outer mitochondrial membrane of *Neurospora crassa*. **European Journal of Biochemistry** 126:197–202.

Gasser, S. & Schatz, G. (1983). Import of proteins into mitochondria. *In vitro* studies on the biogenesis of the outer membrane. **Journal of Biological Chemistry** 258:3427–3430.

Goldfarb, D.S., Gariepy, J., Schoolnik, G. & Kornberg, R.G. (1986). Synthetic peptides as nuclear localization signals. **Nature** 322:641–644.

Hase, T., Muller, U., Riezman, H. & Schatz, G. (1984). A 70 kD protein of the yeast mitochondrial outer membrane is targeted and anchored via its extreme amino terminus. **EMBO Journal** 3:3157–3164.

Hay, R., Bohni, P. & Gasser, S. (1984). How mitochondria import proteins. **Biochimica et Biophysica Acta** 779:65– 87.

Horwich, A., Kalousek, F., Mellman, I. & Rosenberg, I. (1985). A leader peptide is sufficient to direct mitochondrial import of a chimeric protein. **EMBO Journal** 4:1129–1135.

Joh, T., Nomiyama, H., Maeda, S., Shimada, K. & Morino, Y. (1985). Cloning and sequence analysis of a cDNA encoding porcine mitochondrial aspartate aminotransferase. **Proceedings of the National Academy of Sciences USA** 82:6065–6069.

Kalderon, D., Roberts, B.C., Richardson, W.D. & Smith, A.E. (1984). A short amino acid sequence able to specify nuclear location. **Cell** 39:499–509.

Kaput, J., Goltz, S. & Blobel, G. (1982). Nucleotide sequence of the yeast nuclear gene for cytochrome c peroxidase precursor. Functional implications of the pre-sequence for protein transport into mitochondria. **Journal of Biological Chemistry** 257:15054–15058.

Kreig, P.A. & Melton, D.A. (1984). Functional messenger RNAs are produced by SP6 *in vitro* transcription of cloned cDNAs. **Nucleic Acid Research** 12:7057–7070.

Lazarow, P.B. & Fujiki, Y. (1986). Biogenesis of peroxisomes. **Annual Review of Cell Biology** 1:489–530.

Lingappa, V.R., Chaides, J., Spencer-Yost, C. & Hedgpath, J. (1984). Determinants for protein localization. Beta lactamase signal sequence directs globin across microsomal membranes. **Proceedings of the National Academy of Sciences U.S.A.** 81:456–460.

Mackinlay, A.G., West, D.W. & Manson, W. (1977). Specific casein phosphorylation by a casein kinase from lactating bovine mammary gland. **European Journal of Biochemistry** 76:233–243.

Moore, A., Boulton, A.P., Heid, H.W., Jarasch, E-D. & Craig, R.K. (1985). Purification and tissue specific expression of casein kinase from the lactating guinea-pig mammary gland. **European Journal of Biochemistry** 152:729–737.

Palmiter, R.D., Gagnon, J. & Walsh, K.A. (1978). Ovalbumin: A secreted protein without a transient hydrophobic leader sequence. **Proceedings of the National Academy of Sciences U.S.A.** 75:94–98.

Pascall, J.C., Boulton, A.P. & Craig, R.K. (1981). Characterization of a membrane bound serine-specific casein kinase isolated from lactating guinea-pig mammary gland. **European Journal of Biochemistry** 119:91–99.

Pryde, J.G. & Phillips, J.H. (1986). Fractionation of membrane proteins by temperature induced phase separation in Triton X-114. Application to subcellular fractions of the adrenal medulla. **Biochemical Journal** 233:525–533.

Rapaport, T.A. (1981). Intracellular compartmentation and secretion of carp proinsulin synthesized in *Xenopus* oocytes. **European Journal of Biochemistry** 115:665–669.

Rindler, M.J., Ivanov, I.E., Plesken, H. & Sabatini, D.D. (1985). Polarized delivery of viral glycoproteins to the apical and basolateral plasma membranes of MDCK cells infected with temperature sensitive viruses. **Journal of Cell Biology** 100:136–151.

Schatz, G. & Butow, R.A. (1983). How are proteins imported into mitochondria. **Cell** 32:316–318.

Schleyer, M. & Neupert, W. (1985). Transport of proteins into mitochondria. Translocation intermediates spanning contact sites between outer and inner membranes. **Cell** 43:339–350.

Smith, A.E., Kalderon, D., Roberts, B.L., Colledge, W.H., Edge, M., Gillet, P., Paucha, E. & Richardson, W.D. (1985). The nuclear location signal. **Proceedings of the Royal Society of London** Series B 226:43–58.

von Figura, K., & Hasilik, A. (1986). Lysosomal enzymes and their receptors. **Annual Reviews of Biochemistry** 55:167–193.

von Heijne, G. (1984). How signal sequences maintain cleavage specificity. **Journal of Molecular Biology** 173:243–251.

von Heijne, G. (1985). Signal sequences. The limits of variation. **Journal of Molecular Biology** 184:99–105.

Walter, P. & Blobel, G. (1981). Translocation of proteins across the endoplasmic reticulum. III Signal recognition protein (SRP) causes signal sequence dependent and site-specific arrest of chain elongation that is released by micosomal membranes. **Journal of Cell Biology** 91:557–561.

Walter, P. & Blobel, G. (1982). Signal recognition particle contains a 7S RNA essential for protein translocation across the endoplasmic reticulum. **Nature** 299:691–698.

Walter, P., Gilmore, R. & Blobel, G. (1984). Protein translocation across the endoplasmic reticulum. **Cell** 38:5–8.

Young, R.A. & Davis, R.W. (1983a). Efficient isolation of genes using antibody probes. **Proceedings of the National Academy of Sciences U.S.A.** 80:1194–1198.

Young, R.A. & Davis, R.W. (1983b). Yeast RNA polymerase II genes: Isolation by antibody probes. **Science** 222:778– 782.

Zehavi-Willner, T.L. & Lane, C.D. (1977). Subcellular compartmentation of albumin and globin made in oocytes under the direction of injected messenger RNA. **Cell** 11:683–693.

Zimmermann, R. & Mollay, C. (1986). Import of honeybee prepromelittin into the endoplasmic reticulum. Requirements for membrane insertion, processing and sequestration. **Journal of Biological Chemistry** 261:12889–12895.

III Centriole and flagella formation

The mammalian sperm flagellum: structure and development

DAVID M. PHILLIPS

The Population Council, 1230 York Avenue, New York, NY 10021, USA.

Abstract

Mammalian spermatozoa are propelled by a complex motile apparatus composed of a 9+2 flagellum and prominent accessory structures. The largest of these structures are the 9 outer dense fibers which extend for nearly the entire length of the flagellum, and the fibrous sheath which surrounds the dense fibers in the principal piece. Evidence from several laboratories employing different experimental approaches has shown that both dense fibers and the fibrous sheath develop gradually during spermiogenesis, are composed primarily of a few proteins, and function as elastic elements which are involved in altering the shape of the flagellar wave form. In addition to these prominent structures mammalian sperm contain structures which connect the dense fibers to the sperm head, satellite fibers associated with the dense fibers, and an annulus situated between the midpiece and the principal piece. Structurally the flagellar accessory structures appear to be unique to mammalian spermatozoa and some studies suggest that they contain unique proteins.

Is the mammalian flagellum a possible target for a contraceptive agent? The observation that the structures associated with the 9+2 flagellum are unique is encouraging. Perhaps an agent will be found which could interfere with the formation of the dense fibers or fibrous sheath without affecting other cells of the body. However, since these are intracellular structures, an immunoglobulin would presumably not reach them. There is evidence, however, that some internal structures contain proteins which pass through the cell membrane. These look to be more promising targets for immunologically based contraceptives.

Key words

Flagellum, axonemal microtubules, axoneme, annulus, midpiece, principal piece, outer dense fibers, satellite fibers, neck region, connecting piece, proximal centriole.

Introduction

The mammalian sperm flagellum is a large, complex structure that occupies most of the volume of the cell. In addition to the usual 9+2 axonemal

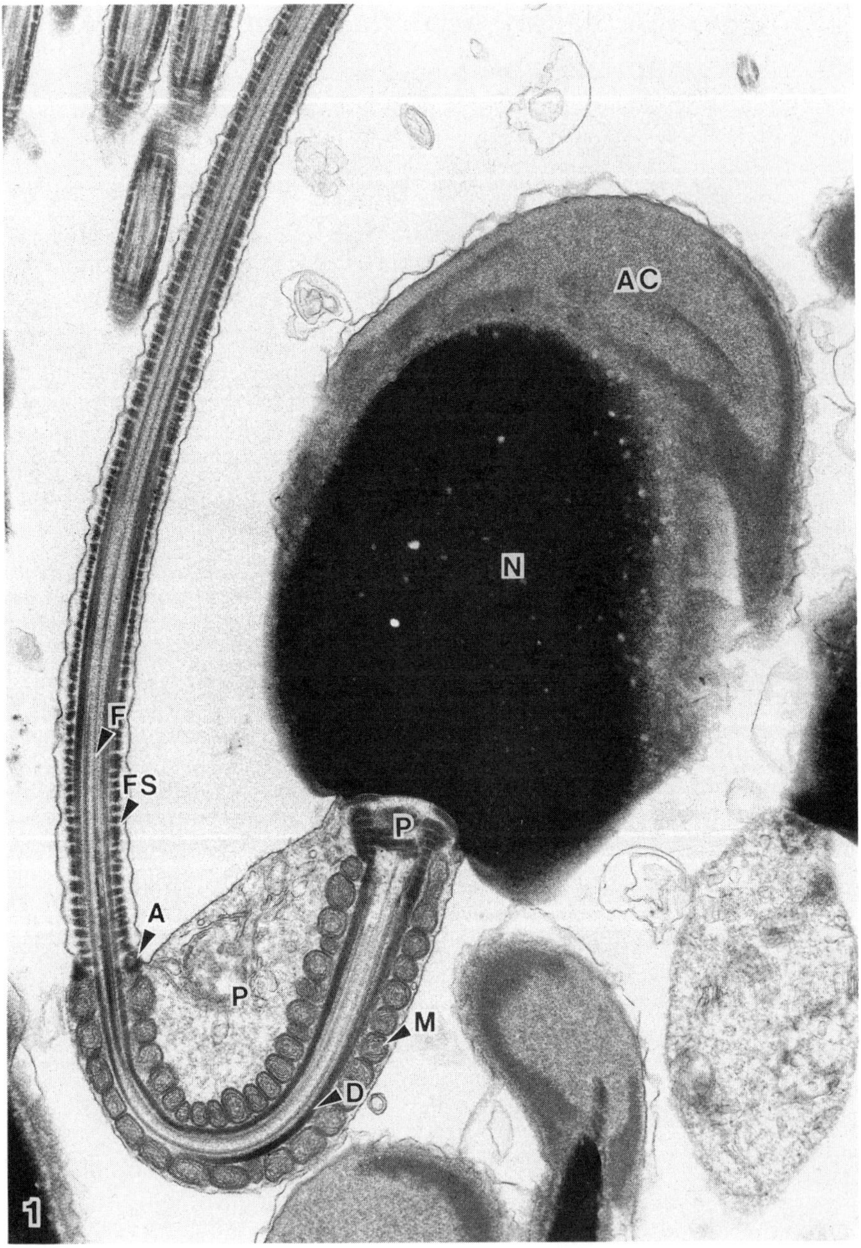

Fig. 1 This spermatozoon of the spiny rat, *Proechimys*, shows the typical features of the mammalian sperm flagellum. The midpiece extends from the proximal centriole (P) to the annulus (A). Dense fibers (D) surround a typical 9+2 flagellum (F). The principal piece, which takes up most of

microtubules, there are prominent organelles which, unlike the flagellar tubules are not observed in any other cell type. The focus of this discussion will be on the structure, function and formation of these organelles. As targets for potential new antifertility agents, prominent unique structures are of particular interest since agents directed against them or their formation might have relatively less detrimental effects on other cells.

Mammalian sperm flagella are broadest at the neck region just behind the sperm head. With the exception of the 9+2 tubules, organelles gradually diminish in size or in amount. Thus the flagellum tapers. Most organelles are situated at precise locations along the length of the flagellum – the columns of the connecting piece are found on the neck region, the annulus lies between the midpiece and the principal piece, mitochondria are restricted to the midpiece (Fig. 1). Outer dense fibers are exceptional in that they are found along almost the entire length of the sperm cell.

Molecular analysis of the organelles which comprise the mammalian sperm flagellum has been difficult, especially in the case of development of sperm structures. Although relatively pure preparations of both fibrous sheath and dense fiber proteins have been analyzed, less prominent structures such as the satellite fibers, annulus, and neck regions have not been isolated from spermatozoa. Isolation of flagellar organelles from spermatids is difficult as the seminiferous epithelium contains spermatids in all stages of development as well as spermatogonia, spermatocytes, myoid and Sertoli cells. Fractionation of stages of spermatids has been accomplished, but it has not led to isolation of flagellar organelles. Thus most of our knowledge of the formation of the mammalian sperm flagellum has been derived from ultrastructural studies.

The neck region

The area at the base of the flagellum is termed the neck region or connecting piece. For a detailed description of the structure and morphogenesis of the neck region, see Fawcett & Phillips (1970). It is assumed that the complex structures in this region serve to anchor the flagellum to the nucleus. Here the nucleus is very precisely shaped. Electron-dense material is observed along the edge of the nucleus and in association with a proximal centriole (Fig. 2). The most prominent structures in the neck are the 9 striated columns of the connecting piece which are continuous anteriorly with the dense material adjacent to the nucleus and posteriorly with the dense fibers. Columns of the connecting piece have alternating electron dense bands of

Caption for Fig. 1 (cont.)

the length of the flagellum, is characterized by the fibrous sheath (FS). Protoplasmic droplet (P), mitochondria (M), nucleus (N), acrosome (AC). × 7 000.

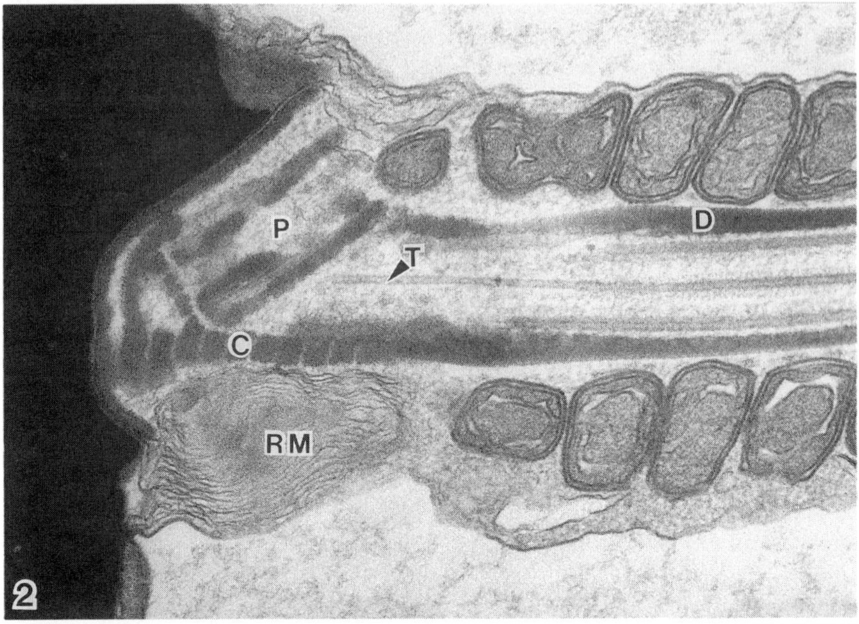

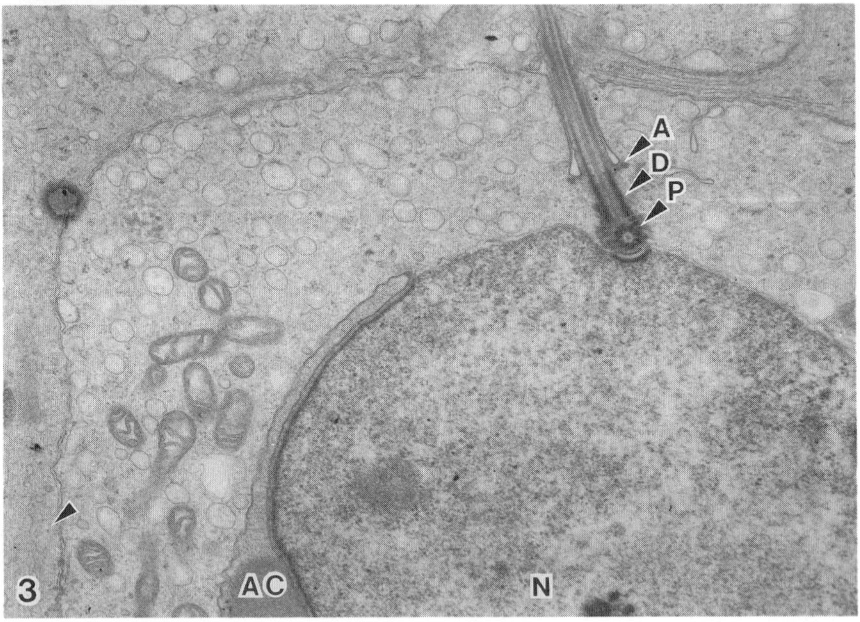

Fig. 2 Neck region of a spermatozoon of a deer mouse, *Peromyscus californicus*. Mature spermatozoa of mammals contain only a proximal centriole (P) as the distal centriole disappears during spermiogenesis. The central pair of flagellar tubules (T) extends into the position where the

65 nm. Each band can be resolved into 9 or 10 bands spaced 5 nm apart. The columns of the connecting piece have been likened to ciliary rootlets, which are striated structures associated with basal bodies and are believed to anchor cilia to cells. Mature mammalian spermatozoa are unique in that the flagellum has no basal body, as this centriole disappears during spermiogenesis (Fig. 2). Thus mammalian sperm generate a beat without a basal body. Bull sperm flagella which are bent and fractured under certain conditions of cryopreservation generate a wave from the distal end of a flagellum which has been fractured in the midpiece (Phillips & Kalay, 1984). Thus, the mammalian sperm flagellum appears to be able to generate a beat from a broken point in the midpiece.

The complex neck region forms gradually throughout spermiogenesis. Very early in spermiogenesis the spermatid flagellum develops from one of a pair of centrioles which are situated adjacent to the plasmalemma, much as in many other cell types (Fig. 3). Even before it associates with the nucleus, dense material begins to form adjacent to the proximal centriole.

Outer dense fibers

Each of the 9 prominent dense fibers is associated with a doublet microtubule. The two dense fibers associated with dense fibers numbers 3 and 8 terminate at the annulus (Fig. 4). Although dense fibers vary greatly in size and shape between species, those associated with doublets 1, 5 and 6 tend to be larger than the others (Fig. 4). Within any species each dense fiber has a characteristic morphology. Viewed in cross-section, morphologically distinct cortical and medullary regions can be seen (Fig. 4). In longitudinal sections or surface replicas, the dense fibers are composed of 40 nm striations (Woolley, 1971; Fawcett, 1975; Phillips, 1975; Olson & Sammons, 1980). Rat and Chinese hamster spermatozoa, which have large dense fibers, have a relatively lower amplitude beat than sperm of species such as rabbit and mouse, which have small dense fibers. This suggests that dense fibers are passive elastic structures (Phillips, 1972). Molecular analysis of dense fibers generally supports this.

There have been several analyses of the polypeptide composition of the isolated outer dense fibers from spermatozoa of rat (Olson & Sammons, 1980; Vera *et al.*, 1984), bull (Brito *et al.*, 1986; Baccetti *et al.*, 1973, 1976), mouse (Bradley *et al.*, 1981), and human (Baccetti *et al.*, 1976). There is

Caption for Fig. 2 (cont.)

distal centriole was. Columns of the connecting piece (C), redundant nuclear membrane (RM), dense fibers (D). × 80 000.

Fig. 3. An early stage in spermiogenesis in the rabbit. Proximal centriole (P), distal centriole (D), annulus (A), acrosome (AC) and nucleus (N). × 14 000.

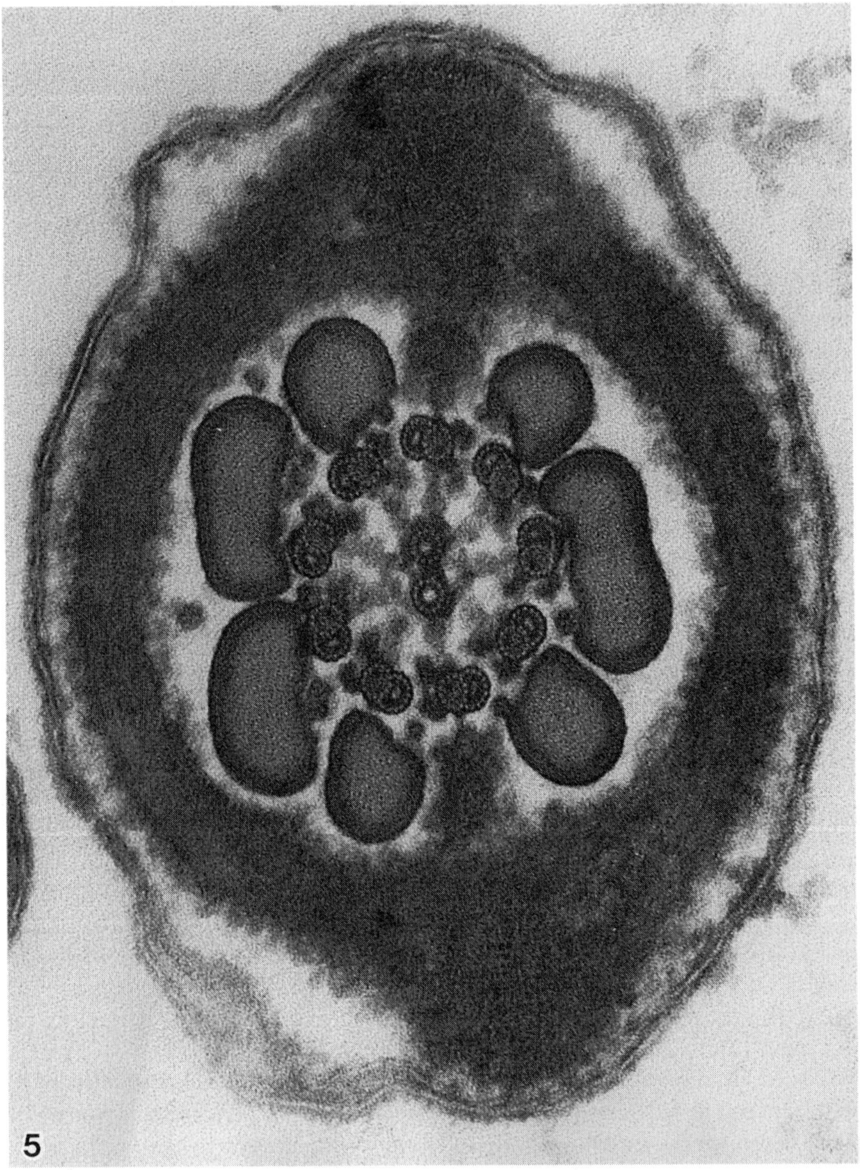

Fig. 4 Cross-section through the principal piece of a rat sperm. × 150 000.

no precise agreement about the number of major polypeptides, their molecular weights or amino acid composition, even within one species. However, all studies have concluded that dense fibers are composed of only 3 to 6 major components. From their amino acid composition and molecular weights it is clear that none of the proteins is tubulin, actin, or myosin, and that at least some of the polypeptides have a high content of cystine. These properties suggest that dense fibers are structural rather than contractile elements. Recently, however, Brito *et al.* (1986) reported that 2 of the 3 major polypeptides of bull sperm outer dense fibers were phosphoproteins. This suggested to these workers that dense fibers might have an active role in motility.

Dense fibers take shape gradually during spermiogenesis. In very early spermatids small dense structures appear between subfiber a and b of each doublet tubule. These continue to enlarge until the spermatids are mature. Using EM radioautography, Irons & Clermont (1982a) found that amino acids were incorporated into dense fibers throughout spermiogenesis. Thus polypeptides which compose dense fibers may be synthesized even in late spermatids as opposed to being assembled from polypeptides which were synthesized earlier.

The fibrous sheath

The fibrous sheath surrounds the dense fibers in the principal piece. Viewed in cross section the fibrous sheath has two thickened longitudinal columns which are in contact with doublets 3 and 8 (Fig. 4). Viewed in longitudinal section the fibrous sheath displays a segmented rib-like appearance (Fig. 5). Biochemical analysis of isolated fibrous sheaths indicates that the fibrous sheath is composed of a single protein of MW 80 kD in rat (Olson *et al.*, 1976) and 74 kD in mouse (O'Brien & Bellvé, 1980). Recently two groups of workers have developed monoclonal antibodies to a major fibrous sheath component (Sakai *et al.*, 1986; Fenderson *et al.*, 1987). Immunocytochemical procedures detected the polypeptide only in late spermatids suggesting that this fibrous sheath component might form after a fibrous sheath framework has already been laid down. The observation of very high disulfide cross-linking in the fibrous sheath suggests that, like the dense fibers, the fibrous sheath may serve a passive elastic function in motility (Bedford & Calvin, 1974; Calvin, 1975).

The fibrous sheath forms gradually during spermiogenesis. In the rat the longitudinal columns appear to be assembled during early spermiogenesis, whereas the ribs arise during late spermatogenesis (Irons & Clermont, 1982b). Thus the rib could be composed of the protein to which the monoclonal antibody has been made since it was detected only in late spermatids.

The annulus

A small electron dense ring, the annulus, lies between the midpiece and the principal piece (Figs. 1 and 5). This structure forms very early in spermiogenesis around the proximal centriole where it is attached to the plasma membrane (Fig. 3). The annulus remains in the neck region until late in

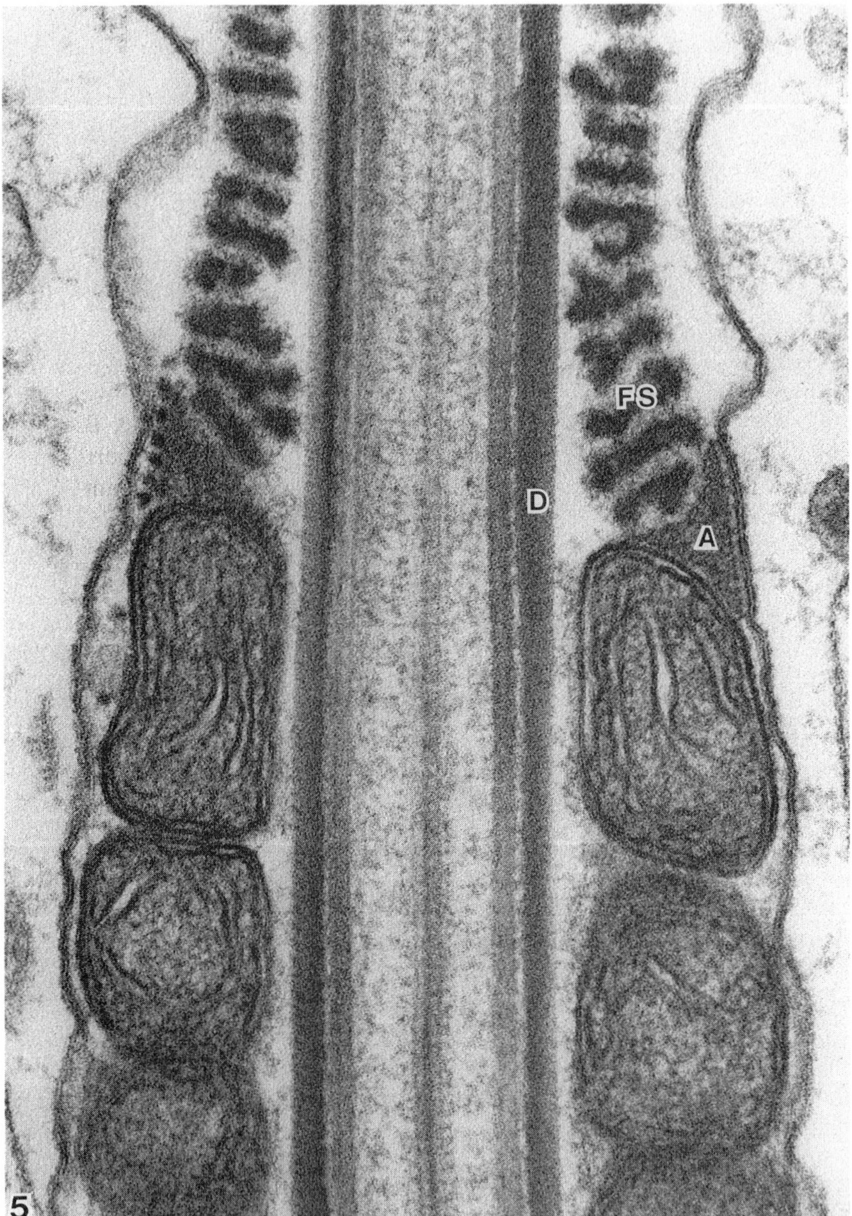

Fig. 5 Longitudinal section through the annulus (A) of a Rhesus monkey spermatozoon. Fibrous sheath (FS), dense fibers (D). × 130 000.

spermiogenesis when, together with the associated plasma membrane, it moves posteriorly until it encounters the tubular complex and the forming fibrous sheath (Fig. 6) (Phillips, 1974, 1980; Fawcett, 1975). Thus the molecules which will comprise the dense fibers and fibrous sheath must pass through the narrow neck region at the annulus in order to eventually form the definitive structures. Why the fibrous sheath forms only posterior to what will be the midpiece, and how the movement of the annulus is directed, are fascinating questions which are difficult to answer with our current technology.

The tubular complex

During spermiogenesis a complex array of tubules forms around the flagellum at the anterior end of the fibrous sheath (Fig. 6) (Rattner & Brinkley, 1970). This tubular complex disappears before spermiation. Immunocytochemical studies of recently developed monoclonal antibodies to fibrous sheath polypeptides should help to determine whether the tubular complex contains a pool of fibrous sheath proteins.

Satellite fibers

So-called satellite fibers occur in association with the flagellar dense fibers. In some species they are prominent structures, especially in the anterior midpiece (Fig. 7) (Fawcett & Phillips, 1970; Phillips & Bedford, 1985). Their function is unknown.

Possible contraceptive leads

It is clear from the above discussion that the mammalian sperm flagellum contains unique organelles. As more laboratories develop antibodies to flagellar components our knowledge of the uniqueness of the molecular composition of mammalian flagellar components will increase. A unique component suggests the possibility that an agent could specifically block the formation or function of this component and thus perhaps not have side effects on other tissues.

To have an effective immunologically based contraceptive to a flagellar component, the antibody would need to be directed to an immunogenic protein. Certainly some components of the fibrous sheath and dense fibers are present in relatively large amounts and appear to be immunogenic. The problem, of course, is that these are not surface proteins so they would not be recognized by an antibody in either the male or the female reproductive tract. There is evidence, however, that some sperm proteins may extend from inside the cell to the cell surface. The so-called zipper is an example

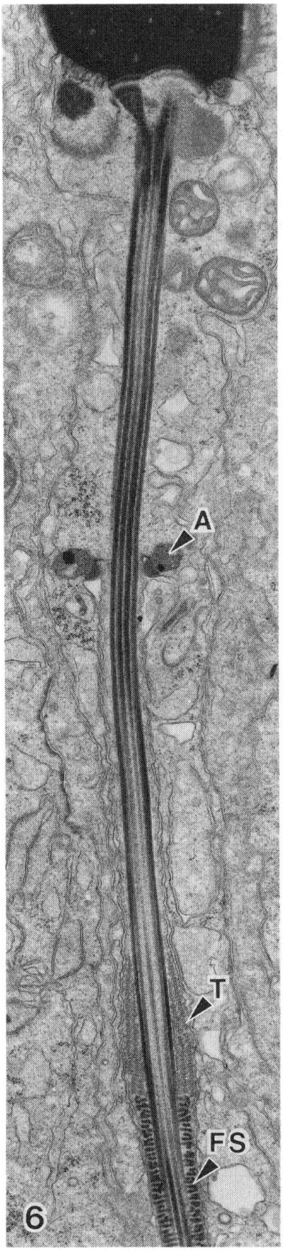

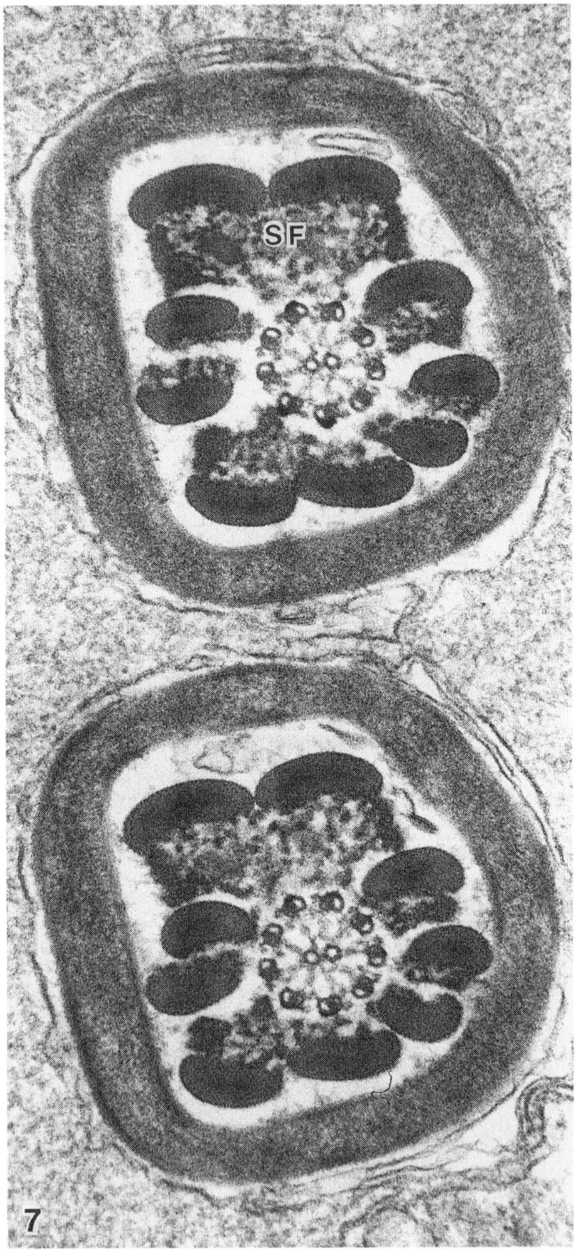

Fig. 6 In this monkey spermatid the annulus (A) has moved midway between the neck region and the tubular complex (T). Note that the plasmalemma is attached to the annulus and that the fibrous sheath (FS) is well formed. × 13 000.

of this. This structure, built of very regular rectangular subunits surrounding a central channel, is clearly seen in surface replicas of sperm principal pieces of several mammalian species (Fawcett, 1975; Phillips, 1977; Friend & Heuser, 1981). In thin section it can be seen that the dense material of the zipper extends from the outside of the plasmalemma through the plasmalemma and attaches to the fibrous sheath opposite to dense fiber number 1 (Fawcett, 1975; Friend & Heuser, 1981). Suzuki & Nagano (1980) described a surface structure which extends through the plasmalemma and attaches to flagellar tubules in the end piece of boar spermatozoa. Escalier (1984) has also described several structures which appear to extend from the plasmalemma to mitochondria, dense fibers, fibrous sheath and flagellar tubules. Producing an antibody to an intracellular flagellar polypeptide in the hope that it will bind to the sperm surface and interfere with sperm function is a risky option. However, leads for alternative contraceptives must be followed even when they are somewhat risky.

Acknowledgements

I am very grateful to Vanaja Zacharopoulos and Susan Warren for their first-rate technical assistance. The work in our laboratory was partially funded by a grant from the Andrew W. Mellon Foundation.

References

Baccetti, B., Pallini, V. & Burrini, A.G. (1973). The accessory fibers of the sperm tail I. Structure and chemical composition of the bull coarse fibers. **Journal of Submicroscopic Cytology** 5:237–256.

Baccetti, B., Pallini, V. & Burrini, A.G. (1976). The accessory fibers of the sperm tail. III High-sulfur and low-sulfur components in mammals and cephalopods. **Journal of Ultrastructure Research** 57:289–308.

Bedford, J.M. & Calvin, H.I. (1974). Changes in -S-S- linked structures of the sperm tail during epididymal maturation, with comparative observations in submammalian species. **Journal of Experimental Zoology** 187:181–204.

Bradley, F.M., Meth, B.M. & Bellvé, A.R. (1981). Structural proteins of the mouse spermatozoan tail: An electrophoretic analysis. **Biology of Reproduction** 24:691–701.

Brito, M., Figueroa, J., Vera, J.C., Cortes, P., Hott, R. & Burzio, L.O. (1986). Phosphoproteins are structural components of bull sperm outer dense fiber. **Gamete Research** 15:327–336.

Calvin, H.I. (1975). Keratinoid proteins in the heads and tails of mammalian spermatozoa. In: **Biology of the Male Gamete** (R.G. Duckett & P.A. Racey, eds.) Academic Press, London pp. 257–273.

Escalier, D. (1984). The cytoplasmic matrix of the human spermatozoon: Cross-filaments link the various cell components. **Biology of the Cell** 51:347–364.

Caption for Fig. 7 (cont.)

Fig. 7 Satellite fibers (SF) are well developed in these water shrew (*Neomys*) spermatozoa. × 87 000.

Fawcett, D.W. & Phillips, D.W. (1970). Recent observations on the ultrastructure and development of the mammalian spermatozoon. In: **Comparative Spermatology**, Academia de Lincei, Rome Italy; Academic Press, New York, pp. 13–29.

Fawcett, D.W. (1975). The mammalian spermatozoon. **Developmental Biology** 44:394–436.

Fenderson, B.A., Toshimori, K., Muller, C.H., Lane, T.F. & Eddy, E.M. (1987). Identification of a protein in the fibrous sheath of the sperm flagellum. **Developmental Biology**, in the press.

Friend, D.S. & Heuser, J.E. (1981). Orderly particle arrays on the mitochondrial outer membrane in rapidly-frozen sperm. **Anatomical Record** 199:159–175.

Irons, M.J. & Clermont, Y. (1982a). Formation of outer dense fibers during spermiogenesis in the rat. **Anatomical Record** 202:463–471.

Irons, M.J. & Clermont, Y. (1982b). Kinetics of fibrous sheath formation in the rat spermatid. **American Journal of Anatomy** 165:121–130.

O'Brien, D.A. & Bellvé, A.R. (1980). Protein constituents of the mouse spermatozoon. I. An electrophoretic characterization. **Developmental Biology** 75:386–404.

Olson, G.E., Hamilton, D.W. & Fawcett, D.W. (1976). Isolation and characterization of the fibrous sheath of rat epididymal spermatozoa. **Biology of Reproduction** 14:517– 530.

Olson, G.E. & Sammons, D.W. (1980). Structural chemistry of outer dense fibers of rat sperm. **Biology of Reproduction** 22:319–332.

Phillips, D.M. (1972). Comparative analysis of mammalian sperm motility. **Journal of Cell Biology** 53:561–573.

Phillips, D.M. (1974) **Spermiogenesis**. New York: Academic Press.

Phillips, D.M. & Olson, G.E. (1974). Mammalian sperm motility – structure in relation to function. In: **The Functional Anatomy of the Spermatozoon** (B. Afzelius, ed.) Pergamon Press, New York, pp. 117–126.

Phillips, D.M. (1975). Mammalian Sperm Structure. In: **Male Reproductive System**, Handbook of Physiology, Section 7: Endocrinology, Volume 5. (D.W. Hamilton & R.O. Greep, eds.) Waverly Press, Baltimore pp. 405–419.

Phillips, D.M. (1977). Mitochondrial disposition in mammalian spermatozoa. **Journal of Ultrastructure Research** 58:144–154.

Phillips, D.M. (1980). Observations on mammalian spermiogenesis using surface replicas. **Journal of Ultrastructure Research** 72:102–117.

Phillips, D.M. & Kalay, D. (1984). Observations on mechanisms of flagellar motility deduced from backwards swimming bull sperm. **Journal of Experimental Zoology** 231:109–116.

Phillips, D.M. & Bedford, J.M. (1985). Ultrastructure of spermatozoa of the musk shrew *Suncus murinus*. **Journal of Experimental Zoology** 235:119–126.

Rattner, J.B. & Brinkley, B.R. (1970). Ultrastructure of mammalian spermiogenesis. I. A tubular complex in developing sperm of the cottontop marmoset (*Sequinus oedipus*). **Journal of Ultrastructure Research** 32:316–322.

Sakai, Y., Koyama, Y., Fujimoto, H., Nakamoto, T. & Yamashina, S. (1986). Immunocytochemical study on fibrous sheath formation in mouse spermiogenesis using a monoclonal antibody. **Anatomical Record** 215:119–126.

Suzuki, F. & Nagano, T. (1980). Morphological relationship between the plasma membrane and the microtubules in the endpiece of the boar spermatozoon. **Journal of Electron Microscopy** 29:190–192.

Vera, J.C., Brito, M., Zuvic, T. & Burzio, L.O. (1984). Polypeptide composition of rat sperm outer dense fibers. **Journal of Biological Chemistry** 259(9):5970–5977.

Woolley, D.M. (1971). Striations in the peripheral fibers of rat and mouse spermatozoa. **Journal of Cell Biology** 49:935–939.

Centrioles, the cell cycle and microtubule formation

RYOKO KURIYAMA AND CATERINA SELLITTO

Department of Cell Biology and Neuroanatomy, 4-135 Jackson Hall, 321 Church St. SE, University of Minnesota, Minneapolis, Minnesota 55455, USA.

Abstract

The centriole has a characteristic structure composed of 9 triplet microtubules. It is located at the base of the sperm flagellum and at spindle poles, in a juxtanuclear position in interphase somatic cells. In order to understand the relationship between centriolar structure and its physiological function, we have investigated the morphological changes in centrioles during the cell cycle, and their capacity to initiate microtubule formation *in vitro*. The centriole cycle was documented in a graphic form which allowed a quantitative description of the different stages of the centriole cycles. Unlike axonemal microtubules, which grow directly from tubular elements of the centriolar wall in spermatozoa, microtubules in spindles as well as those in the interphase network in somatic cells were shown to originate from an amorphous cloud of pericentriolar material surrounding the centrioles. The capacity of pericentriolar material to initiate microtubule assembly was demonstrated *in vitro*. Two distinct modes of microtubule formation, either directly from the base plate as in the sperm centriole, or in an astral fashion from interphase and mitotic centrosomes in somatic cells, are dependent on the presence or absence of active pericentriolar material around the centrioles. We have also isolated monoclonal antibodies specific to the centriole-containing structures. A preliminary study on immunostaining patterns and immunoblots of some of the positive clones are also presented.

Key words

Centrioles, centriole cycle, centrosomes, pericentriolar material, microtubules, monoclonal antibodies, Chinese hamster ovary cells, spermatozoa.

Introduction

Centrioles are one of the most common organelles in animal cells. They are found not only at the bases of cilia and flagellae, but also at the mitotic poles and in juxtanuclear positions in somatic cells. Centrioles have a characteristic cylindrical structure composed of 9 triplet microtubules and are believed to be involved in the mechanism of microtubule organization in living cells (Fulton, 1971; Wheatley, 1982). Although they have been described in the literature since the late 19th century (see Wilson, 1928), our detailed knowledge on centriolar structure and function is still quite limited. Even very basic questions are still open: What functions do centrioles perform in cells? Why do they possess such a peculiar structure and what does the structure mean? What are their components?

In order to learn more about the centrioles we have studied their morphological changes during the cell cycle, and have identified a centriole-associated activity that initiates microtubule formation *in vitro*. An attempt has also been made to prepare specific antibodies to centriole-containing structures with the aim of elucidating the molecular nature of this organelle. Therefore, preliminary results to analyze the centriolar components with antibody probes will be presented here together with our observations on centriolar structure and its activity in nucleation of microtubules.

Centrioles in somatic and sperm cells

Centrioles in proliferating cultured mammalian cells are surrounded by an amorphous cloud of pericentriolar material (Murray *et al.*, 1965) and the combined subcellular structure, consisting of a pair of centrioles and the pericentriolar material, is referred to as the centrosome. Centrosomes are located at the mitotic poles of spindles and serve as an organizing center for spindle microtubules (Wilson, 1928). Throughout most of the cell cycle, however, the nucleus is intact and the centrosome occupies a juxtanuclear position, as has been confirmed by several techniques including electron as well as immunofluorescence microscopy (Murray *et al.*, 1965; Connolly & Kalnins, 1978).

During studies on interphase centrosomes in Chinese hamster ovary (CHO) cells, the centrosome and nucleus were found to be isolated together as a coherent complex (Kuriyama & Borisy, 1981). After extraction of isolated nuclei with detergent in a low ionic strength medium, the ultrastructure of the attached centriole could be discerned easily by electron microscopy of whole mount preparations. This has enabled us to examine a large number of centriolar configurations in a quite simple and accurate manner. Thus, an examination of the structural changes in centrioles throughout the cell cycle has become possible with the aim of collecting quantitatively significant data.

A method was developed to document the centriole cycle in a graphic form in which centriolar profiles were placed into 6 different categories according to their orientation and the length ratio between daughter and parent centrioles (Fig. 1). Category I represents profiles of two centrioles, the parent being full-sized and the daughter being almost full-sized, that

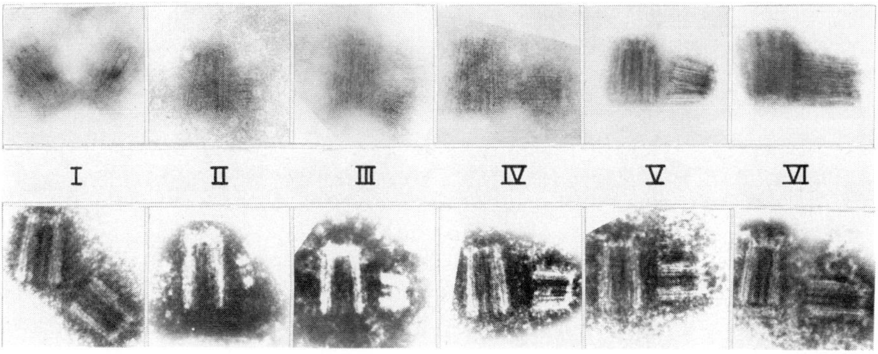

Fig. 1 Electron micrographs of whole-mount preparations of centrioles placed into 6 categories. Two types of centrioles attached to the nucleus (upper row) and freed from the nucleus (lower row) are shown. Category I represents centriole pairs, the parent is full-sized, and the daughter is almost full-sized, but they are not in an orthogonal orientation. Categories II to VI include parent and daughter centrioles oriented perpendicularly to each other. The ratios of daughter centriole to parent length are 0.0–0.2, 0.2–0.4, 0.4–0.6, 0.6–0.8 and 0.8–1.0 in categories II, III, IV, V and VI, respectively.

are not in an orthogonal configuration. Categories II to VI were generally observed as two pairs of centrioles. In each of these categories, parent and daughter centrioles were closely juxtaposed and in an orthogonal configuration. The ratios of daughter to parent centriole length were 0.0–0.2, 0.2–0.4, 0.4–0.6, 0.6–0.8, and 0.8–1.0, respectively. The morphological changes in the centriole cycle were characterized by 3 distinct events: nucleation, elongation and disorientation (Fig. 2). The proportion of centrioles in each category was then plotted as a frequency histogram.

Figure 3 shows a frequency histogram for categories of centriole pairs in exponentially growing CHO cells, which have a mean generation time of about 14 h. Category I is the most abundant, indicating that centrioles spend a greater proportion of their time in this phase than in any of the other phases. Categories II to VI are roughly equal in frequency, suggesting that centriolar elongation occurs in a more or less uniform manner.

The progress of centrioles was then examined in synchronous populations of cells. Centriole distribution in cells starting from S phase is shown in Fig. 4A and from M phase in Fig. 4B. After blocking the cells with 2–5 mM thymidine for 12-16 h, most of the cells are distributed through S phase and at the boundary between G1 and S. As shown in the panel (a) in Fig. 4A, centrioles in the blocked cells belong most abundantly to categories I to III. That is, categories containing centriole pairs before formation of the new daughter centrioles and the centriole pairs in which the daughter had a maximum of 40% of full length. When the cells were washed free of

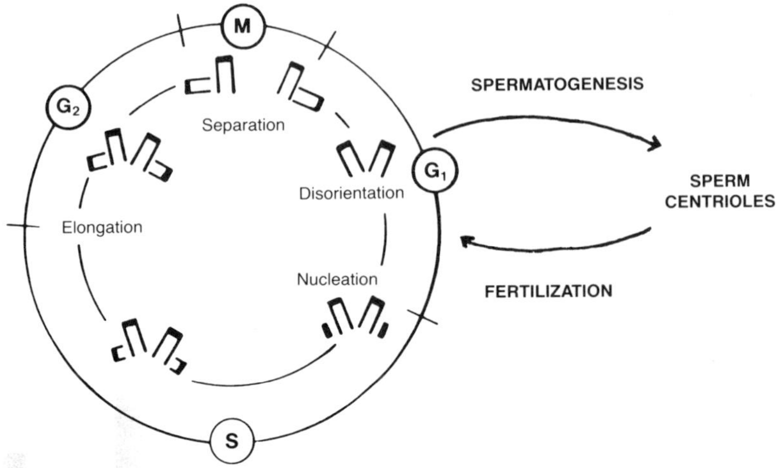

Fig. 2 Diagram of centriole cycle in somatic and sperm cells. The morphological changes of centrioles are characterized by 4 distinct events including disorientation, nucleation, elongation and separation.

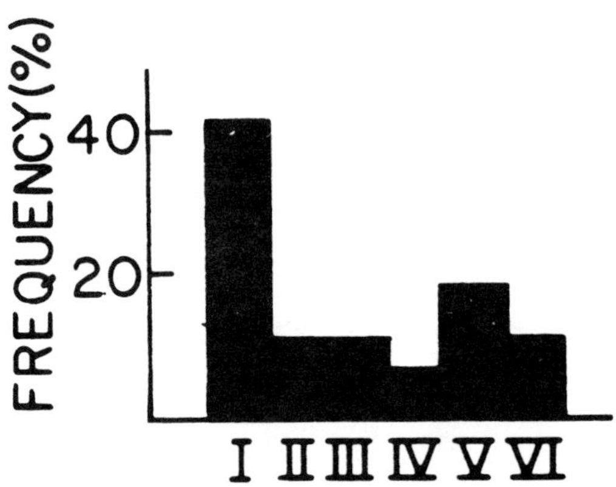

Fig. 3 Frequency histogram of centriole profiles in exponentially growing non-synchonized population of CHO cells. Frequency in percentage (ordinate) and category of centriole profiles (abscissa) are shown. 100–150 photographs of centrioles were evaluated to construct one histogram.

thymidine, they proceeded through their cycles and centriolar category I declined, indicating daughter centriole formation (b), and the peak in the frequency histogram shifted to later categories (b–c), indicating elongation of the daughter centrioles. The maximum mitotic index was obtained at 7 h after removal of thymidine. At this time, almost all of the cells had

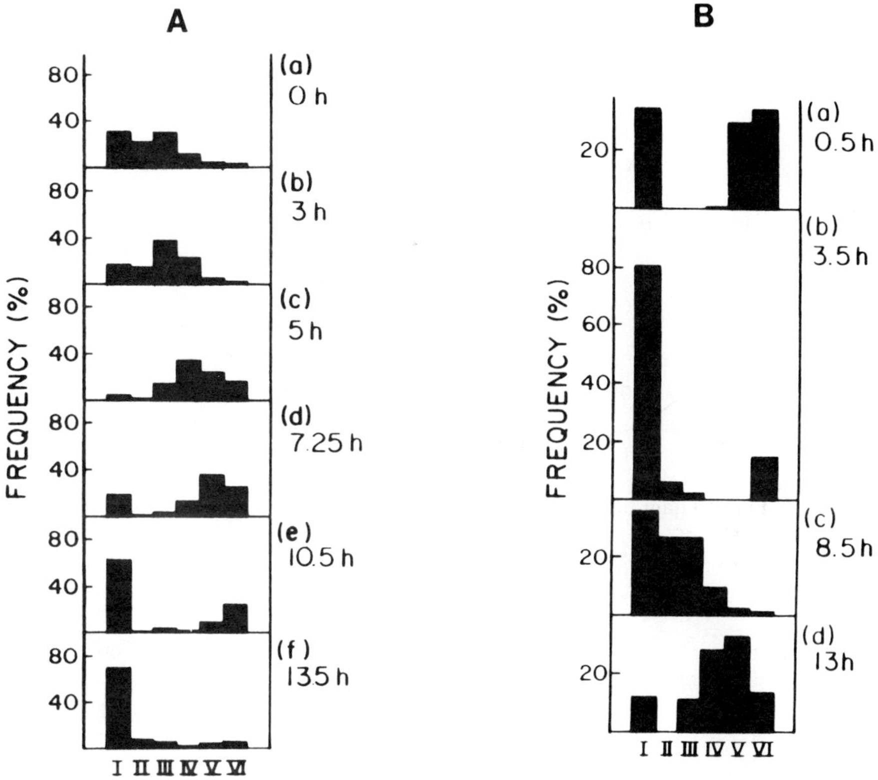

Fig. 4 Centriole cycle in CHO cells synchronized at S (A) and M (B) phase. (A) After treatment with 2 mM thymidine for 12 h, cells were washed and continued to culture with fresh medium for 0 (a), 3 (b), 5 (c), 7.25 (d), 10.5 (e) and 13.5 (f) h, respectively.
(B) Isolated mitotic cells were plated in culture dishes and continued in culture with fresh medium for 0.5 (a), 3.5 (b), 8.5 (c) and 13.0 (d) h.

long daughter centrioles whose length was 60–100% of full length, and at the same time category I began to increase (d). At 13.5 h, daughter centrioles, 20% of full length, began to appear again (f).

When cells were synchronized and started from M phase, essentially the same centriole cycle was observed (Fig. 4B). Mitotic cells contained centriole profiles primarily in categories V, VI but also in I (a). Panels (b) through (d) show the progress of disorientation, nucleation and elongation of daughter centrioles during the progress of the cell cycle.

This method provides a quantitative description of the events of the centriole cycle and permits us to study the effects of various treatments of cells on the centriole cycle. Results obtained thus far are summarized in Table 1 (Kuriyama & Borisy, 1981; Kuriyama, 1982; Kuriyama *et al.*, 1986).

Table 1. *Effect on centriole cycle*

Stimulus	Disorientation	Nucleation	Elongation
Enucleation	none	inhibition	none
Thymidine	none	none	none
Cycloheximide	none	inhibition	inhibition
Cytochalasin B	none	none	none
Colcemid	none	inhibition	inhibition
Taxol	none	inhibition	promotion

None of the treatments blocked centriolar disorientation, whereas the formation of new centrioles was specifically inhibited by protein synthesis inhibitor (Phillips & Rattner, 1976) and microtubule drugs (colcemid and taxol). Colcemid preferentially inhibited centriole elongation, but the microtubule-stabilizing reagent taxol did not; rather, it acted to accelerate the elongation of centrioles. Since centrioles contain microtubules as their main structural elements, and inhibition and promotion of microtubule assembly results from treatment with colcemid and taxol, respectively, the events of centriolar elongation may merely reflect the polymerization of centriolar microtubules. Results shown in Table 1 also indicate that cells that failed to duplicate DNA due to exposure to a high concentration of thymidine (Xeros, 1962), could still duplicate new centrioles (Rattner & Phillips, 1973; Kuriyama & Borisy, 1981). On the other hand, taxol-treated cells could not duplicate centrioles but did synthesize DNA as normally as control cells (Kuriyama *et al.*, 1986). Therefore, the events of DNA synthesis initiation and centriole nucleation, although coordinated in time, appear to be independent of each other.

Removal of the nucleus of the cell by enucleation with cytochalasin B and centrifugation did not prevent elongation and disorientation of centrioles, but nucleation of centrioles was specifically blocked, suggesting that formation of new centrioles is facilitated by a physical association between the centrosome and the nucleus. Some nuclear factor, not necessarily involved in signaling initiation of DNA synthesis, is suggested to be involved in centriole formation.

In addition to the juxtanuclear position at the spindle poles in somatic cells, centrioles are also located at the bases of sperm tails and participate in the formation of flagellar microtubules (Fawcett & Phillips, 1969). Sperm centrioles are originally derived from a pair of centrioles in spermatids. Spermatids, formed by completion of meiosis of spermatocytes, receive a pair of orthogonally arranged, almost full-sized, centrioles which are in association with the Golgi complex. Sooner or later the centrioles start to migrate from the Golgi region to the opposite side of the cell, resulting in the determination of the asymmetric shape of sperm cells (Fawcett, 1971).

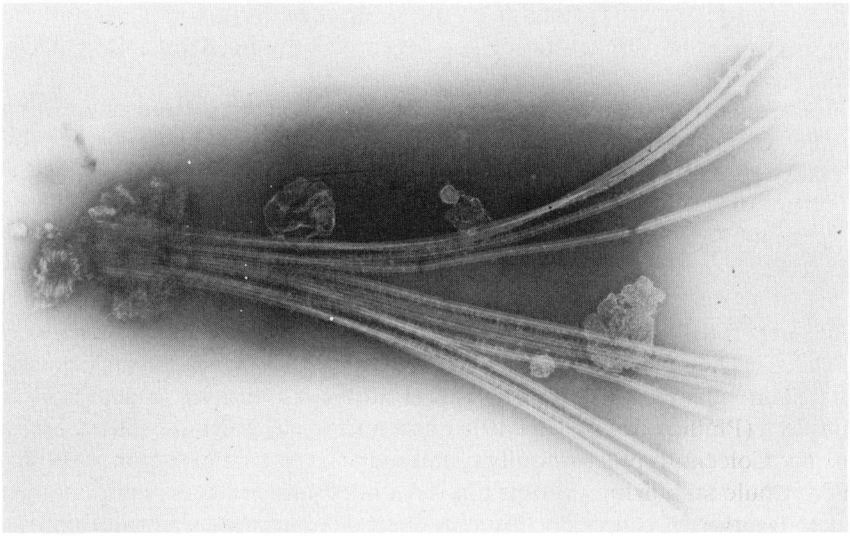

Fig. 5 Whole-mount electron micrograph of an isolated centriolar complex from starfish spermatozoa with associated short axonemal microtubules. A number of microtubules (9 + 2) are attached just at the end of the distal centriole. × 32 000.

Although there are some morphological variations in the centriolar region of spermatozoa among different animal species, 9 doublet microtubules directly rising from the base of the centrioles seems to be a common feature (Fulton, 1971). Such an example is shown in Fig. 5. A centriolar complex with short axonemal microtubules was isolated from starfish spermatozoa and observed by electron microscopy of whole-mount preparations. Up to '9 + 2' tubules adhering just beneath the distal end of the centrioles were counted. After introduction into mature eggs by fertilization, sperm-derived centrioles start their cycle to serve as division centers during development (Kuriyama & Kanatani, 1981). Morphological changes of centrioles in somatic and sperm cells are summarized in Fig. 2.

Centriolar function in initiation of microtubule assembly

There are distinct differences in the manner of microtubule organization by centrioles in zygotes, somatic cells and spermatozoa. The centriole at the base of sperm flagellae serves as a template-organizing center for flagellar outer fibers (Fulton, 1971). As demonstrated in Fig. 5, the 9-fold array of doublet microtubules grows directly from the tubular elements of the centriolar wall. *In vitro* experiments show that polymerization of microtubule proteins purified from porcine brain could be initiated by the structures at the ends of both proximal and distal centrioles. Very few, if any, microtubules are formed from the sides of each centriole, even in the presence of a high concentration of exogenous tubulin (Kuriyama & Kanatani, 1981).

There is, however, another mode of microtubule formation in addition to the template manner described above. Almost immediately after the entrance of a spermatozoon into a mature egg by fertilization, the identical structure of sperm centrioles starts to initiate the aster formation with egg microtubules (Wilson, 1928). However, all microtubules in asters and spindles can be seen to originate from material that surrounds the centriole, rather than directly from it. An amorphous cloud of pericentriolar material can be seen to surround the centrioles at the center of asters (Robbins *et al.*, 1968). The pericentriolar material has been shown to have the capacity to nucleate microtubules both *in vivo* (Rattner & Berns, 1976) and in vitro (Gould & Borisy, 1977). It is thus reasonable to conclude that the two distinct methods of microtubule formation (template from sperm centrioles and astral from interphase and mitotic centrosomes) depend only on whether or not active pericentriolar material is present surrounding the centrioles. Stored or newly synthesized pericentriolar material in eggs may be accumulated around sperm centrioles after fertilization and thereafter activated into microtubule formation.

Figure 6 demonstrates two different types of microtubule polymerization in vitro onto centrosomes isolated from mitotic CHO cells. As shown in (a), an isolated intact centrosome could nucleate an average number of

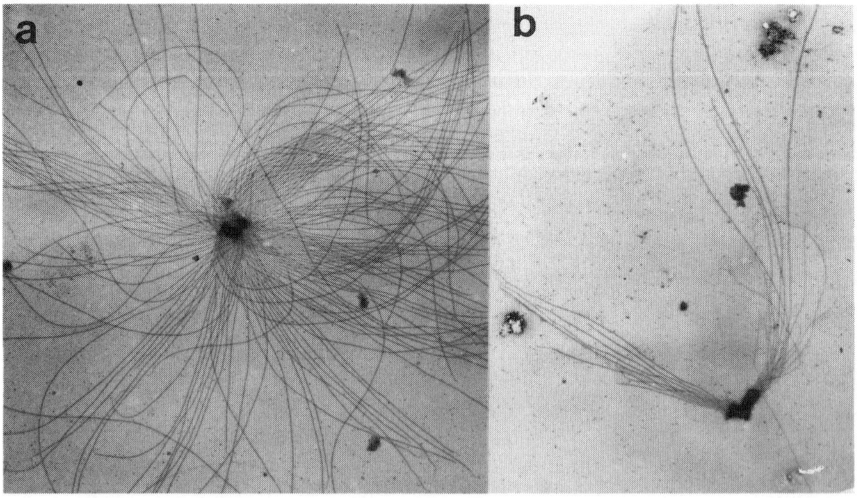

Fig. 6 Electron micrographs of whole-mount preparations of centrosome after incubation with brain microtubule protein *in vitro*. Centrosomes were isolated from mitotic CHO cells and treated without (a) or with (b) trypsin. × 5 000 approx.

200–250 microtubules associated with every side of the centriolar wall. Centrosomes treated with trypsin to digest pericentriolar material initiate microtubules only in a template manner directly from the centrioles; microtubules no longer emanate from all sides of the centrioles (Fig. 6b). When centrosomes are treated with a relatively high concentration of trypsin, centrioles are also damaged and their structure starts to disintegrate. Even the surviving, partially dissociated, centrioles, however, can still serve as sites for microtubule initiation *in vitro* (Kuriyama, 1984).

If centrioles do not directly participate in the initiation of microtubule formation, what is their function in the centrosomal structure? Centrioles and pericentriolar material appear to interact strongly with each other. Pericentriolar material has a tendency to form large aggregates in the absence of centrioles. When centrioles are introduced into cells, a cluster of pericentriolar material then dissociates to associate with the introduced centrioles, causing the formation of many microtubule-nucleating sites. That is, the total number of microtubule organizing sites is dependent upon the number of centrioles present. In order to proliferate properly, cells should include bipolar spindles at mitosis to ensure that their genetic material segregates equally (Stubblefield, 1968). It is the number of functioning centrioles that determines the exact number of mitotic poles. This might be a reason why the number of centrioles is regulated so strictly within cells.

Preparation of monoclonal anti-centrosomal antibodies

To understand the mechanism by which centriole-containing structures organize different arrays of microtubules, it is important to determine their macromolecular composition and to establish that such components are functional entities of the structure. Several approaches are possible for such studies. Treatment with proteases has indicated that the activity of centrosomes resides in protein components (Kuriyama, 1984). Although this approach is a simple and straightforward way to identify functional components of the centrosome, it does not reveal much about specific centrosomal components. The most direct approach would be to analyze purified centrosomal fractions biochemically (Mitchison & Kirschner, 1984). However, partly because of the small amount included in cells, and partly because of their unstable activity, routine biochemical fractionation appears to be inadequate to completely purify active centrosomes. That is, direct biochemical analyses have not yet provided much useful molecular information about the function and regulation of centrosomes.

As an alternative approach to this problem, it was decided to screen specific antibodies reactive to centriole-containing structures. Such an immunological approach seems to offer a better chance for identifying components of centriole-containing structures (Connolly & Kalnins, 1978; Lydersen & Pettijohn, 1980; Oliver *et al.*, 1981; Sherline & Mascardo, 1982;

Calarco-Gillan *et al.*, 1983; Nigg *et al.*, 1985; Salisbury *et al.*, 1986). In particular, preparation of monoclonal antibodies would be useful to determine the molecular composition of such heterogenous fractions as centrioles/centrosomes which are obviously difficult to further purify for production of specific polyclonal antibodies (Ring *et al.*, 1980; Davis *et al.*, 1983; Kuriyama & Borisy, 1985). Thus the monoclonal antibody technique provides useful reagents for identification, characterization and purification of antigenic components of centriole-containing structures.

Mitotic spindles from CHO cells were isolated in a medium containing taxol and were used for immunization in mice. After fusion with spleen cells, the supernatants of hybridomas were screened by indirect immunofluorescence staining of whole CHO cells. In other experiments, the mitotic apparatus from cleaving sea urchin embryos was prepared. Functional components of centrosomes were fractionated by treatment of spindles with high ionic strength medium, and 0.5 M NaCl eluates were used as immunogens for antibody preparation. A total of 2–3 dozen different hybridoma clones were obtained and some of the clones were subcloned by limiting dilution methods at least twice to obtain stable monoclonals. They were then grown in mass cultures of ascites fluid to produce large amounts of antibodies. Preliminary characterization of these antibodies is summarized in Table 2.

CHO1, derived from a mouse immunized with CHO spindles, reacted with interphase centrosomes as well as nuclei. The staining pattern was speckle-like and cell cycle-dependent (Fig. 7). Maximum staining of nuclei was observed in cells at late G2 stage. In mitotic cells, it was not the mitotic poles, but midbodies, that were stained specifically with the CHO1 antibody (arrow in Fig. 7b). A detailed distribution pattern of CHO1 antigen in spindles and midbodies is represented in Fig. 8. As in (f), CHO1 stained

Table 2. *Monoclonal anti-centrosomal antibodies*

Antibody	Immunoglobulin	Stained structures			
		Centrosomes		Nuclei	Other structures
		interphase	mitosis		
CHO1	IgM	+	–	+	Midbodies
CHO2	IgG	+	+	–	–
CHO3	IgM	+	+	+	Phosphoproteins
CHO4	IgM	+	+	–	Intermediate filaments
CHO5	IgM	+	+	–	Golgi apparatus
CHO6	IgM	+	+	–	Stress fibers
CHO7	IgM	+	+	–	–
SU8	IgM	–	+	+	–

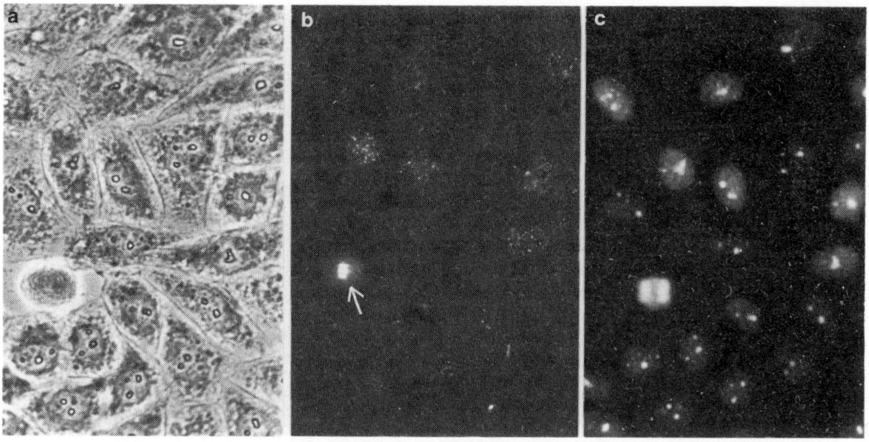

Fig. 7 Light micrographs of non-synchronized populations of CHO cells. Phase-contrast (a) and immunofluorescence (b,c). Cells were fixed with methanol and stained with CHO1 (b) and ethidium bromide (c). The staining of the midbody with CHO1 in a mitotic cell is indicated by an arrow. × 700.

faintly the total spindle area during metaphase. However, the antigen was eventually concentrated around the metaphase plate, resulting in the formation of discrete short lines in the interzonal region of the spindle (g). The bright fluorescent lines remained at the metaphase plate throughout the process of chromosome movement (h–i), then fluorescence became localized at the central region of the intracellular bridge between two daughter cells (j). Double fluorescence staining with tubulin antibody revealed that CHO1 antigen was specifically localized at the middle portion of the spindle and midbody, where microtubules were entrapped by the matrix components and blocked from binding to the anti-tubulin antibody. CHO1 recognized 105 and 95 kD polypeptides on immunoblots (Fig. 9).

Antibodies CHO2, CHO4, CHO5 and CHO6 not only stained centrosomes in cells at interphase but also at mitosis. CHO2 antigen was localized only at the centrosomal region (Fig. 10a), whereas CHO4, CHO5 and CHO6 could interact with other structures besides centrosomes. Those structures included cytoskeletal elements such as intermediate filaments (CHO4) (Fig. 10b), stress fibers (CHO6) (Fig. 10e) as well as the Golgi complex (CHO5) (Fig. 10c,d). Centrosomes appear to share either common molecular components or similar epitopes to these structures. CHO7 reacted with either two or four small dots in interphase and mitotic cells (Fig. 10f). We assume that CHO7 antigen is in the centriole rather than the centrosome.

CHO3 is bound to a phosphoprotein epitope on the centrosomal structure. Strong fluorescence was also seen in interphase nuclei (Fig. 10g). As

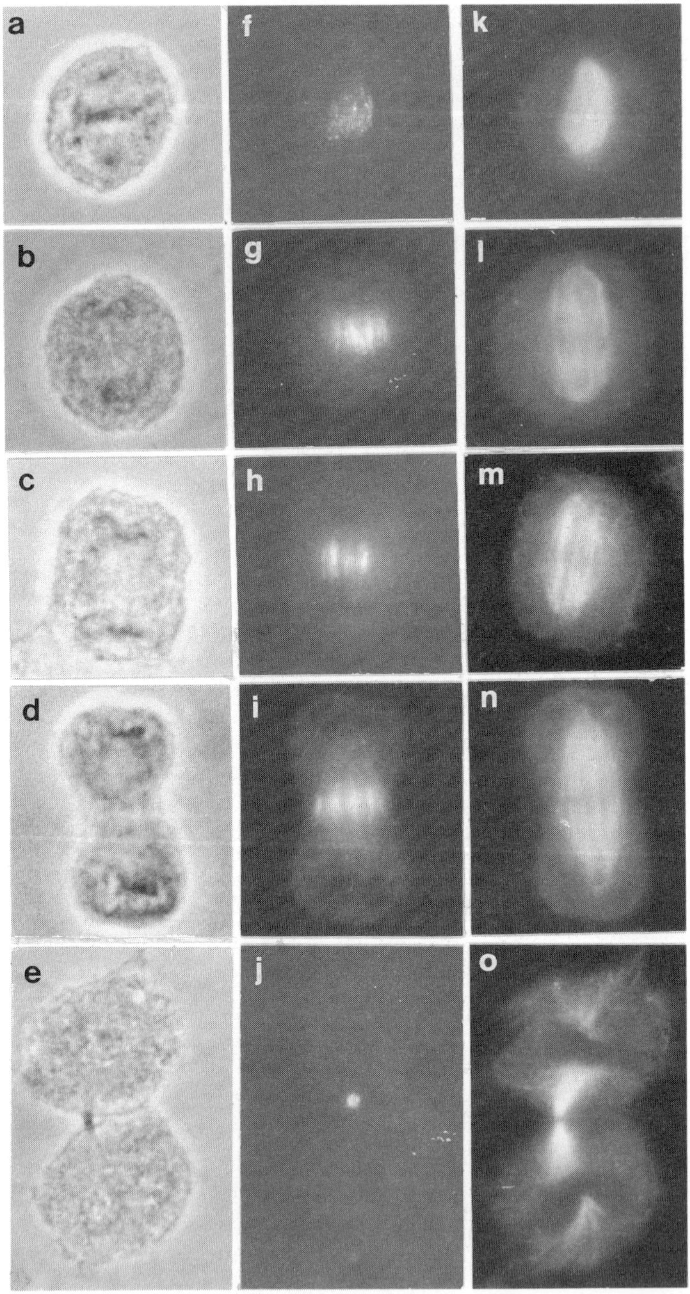

Fig 8 Light micrographs of non-synchronized populations of CHO cells. Phase-contrast (a–e) and immunofluorescence (f–o). Micrographs in the same line show the same cell in both phase and epifluorescence. Cells were double- stained with CHO1 (middle column) and anti-tubulin antibody (right column). a–c, e–h, j–m, o, × 1 200; d,i,n, × 1 550.

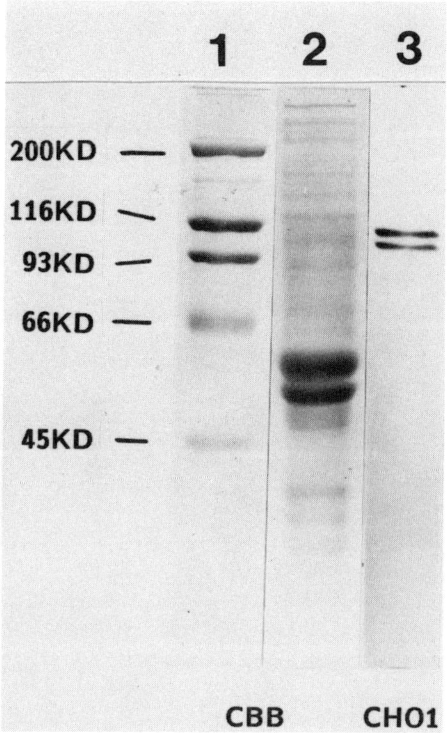

Fig. 9 Immunoelectrophoretic detection of antigen reactive with CHO1. CBB, gel stained with Coomassie Blue. CHO1, immunoreaction of nitrocellulose blot with CHO1 antibody. Lane 1, molecular weight markers. Lane 2 and 3, fraction of isolated spindles and midbodies.

cells entered mitosis, nuclear staining disappeared and immunoreactive material gradually concentrated around the mitotic poles (data not shown). Sea urchin eggs also include common antigens reactive to CHO3 and strong fluorescence could be detected at the polar region of the spindle in whole mitotic eggs (Fig. 10h). Figure 10i shows the isolated mitotic spindle from sea urchin eggs to localize antibody distribution in more detail. The photograph was taken after exposure to phase-contrast and epifluorescence optics simultaneously, indicating antibody specifically localized at the mitotic poles.

Although careful consideration is required to draw general conclusions from results of simple immunostaining patterns, further analysis with these antibodies could provide useful information on molecular components of centrioles and centrosomes.

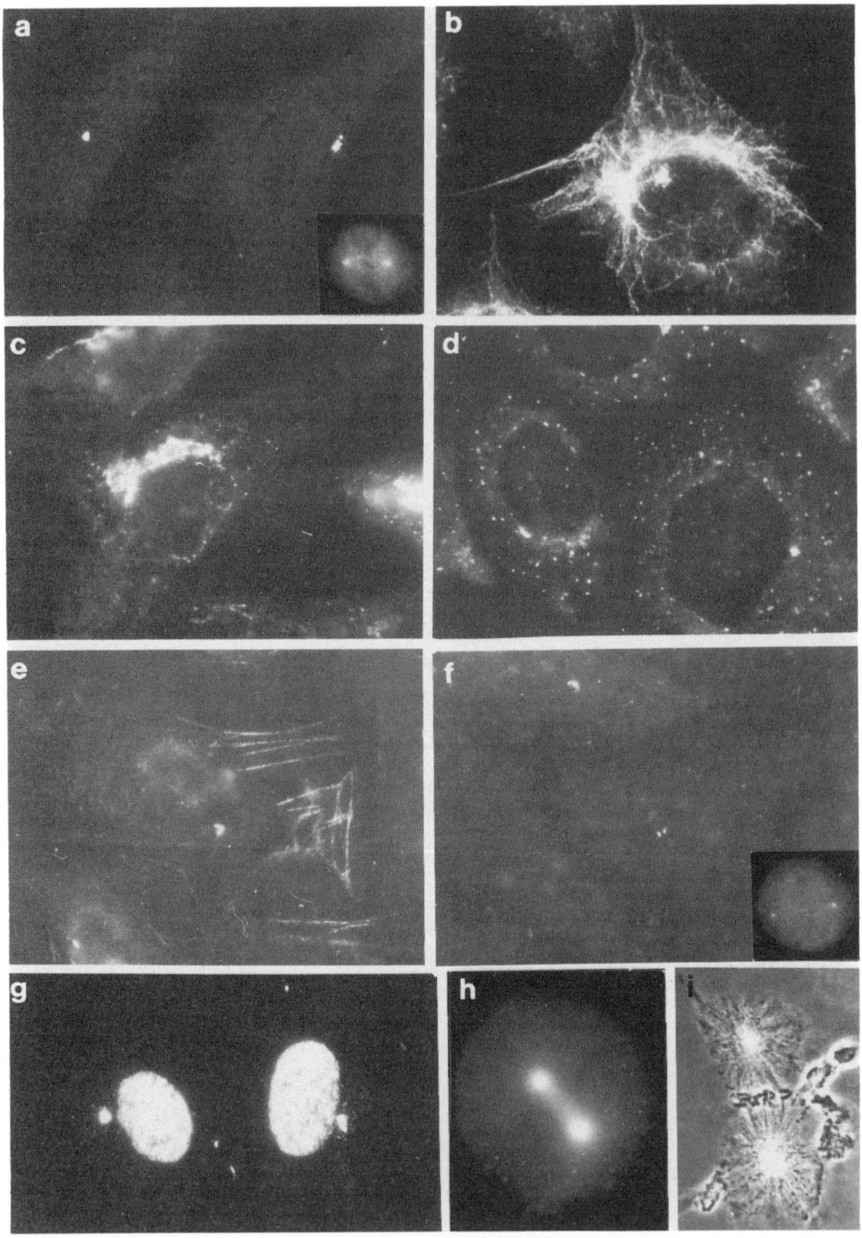

Fig. 10 Light micrographs of CHO cells (a–g), a mitotic sea urchin egg (h) and an isolated sea urchin spindle (i). Immunofluorescence (a–h) and double exposing with phase-contrast and epifluorescence optics (i). Cells were stained with CHO2 (a), CHO4 (b), CHO5 (c,d), CHO6 (e), CHO7

Acknowledgement

Some of the work presented here was done while one of the authors (R.K.) was in the laboratories of Dr G.G. Borisy in the University of Wisconsin, Madison and of Dr H. Kanatani in the National Institute for Basic Biology, Japan. Work reporting new data was supported by NSF grant DCB-8510151 to RK.

References

Calarco-Gillan, P.D., Siebert, M.C., Hubble, R., Mitchison, T. & Kirschner, M.W. (1983). Centrosome development in early mouse embryos as defined by autoantibody against pericentriolar material. **Cell** 35:621–629.

Connolly, J.A. & Kalnins, V.I. (1978). Visualization of centrioles and basal bodies by fluorescent staining with nonimmune rabbit sera. **Journal of Cell Biology** 79:526– 532.

Davis, F.M., Tsao, T.Y., Fowler, S.K. & Rao, P.N. (1983). Monoclonal antibodies to mitotic cells. **Proceeding of the National Academy of Sciences USA** 80:2926–2930.

Fawcett, D.W. (1971). Observations on cell differentiation and organelle continuity in spermatogenesis. In: Beatty, R.A. & Gluecksohn-Waelsch, S., eds. **The Genetics of the Spermatozoon**. Edinburgh, New York, pp. 37–68.

Fawcett, D.W. & Phillips, D.M. (1969). The fine structure and development of the neck region of the mammalian spermatozoon. **Anatomical Record** 165:153–184.

Fulton, C. (1971). Centrioles. In: Reinert, J. & Ursprung, H., eds. **Origin and Continuity of Cell Organelles**. New York, Heidelberg, Berlin, Springer, pp. 170–221.

Gould, R.R. & Borisy, G.G. (1977). The pericentriolar material in Chinese hamster ovary cells nucleates microtubule formation. **Journal of Cell Biology** 73:601–615.

Kuriyama, R. (1982). Effect of colcemid on the centriole cycle in Chinese hamster ovary cells. **Journal of Cell Science** 53:155–171.

Kuriyama, R. (1984). Activity and stability of centrosomes in Chinese hamster ovary cells in nucleation of microtubules *in vitro*. **Journal of Cell Science** 66:277–295.

Kuriyama, R. & Borisy, G.G. (1981). Centriole cycle in Chinese hamster ovary cells as determined by whole mount electron microscopy. **Journal of Cell Biology** 91:814– 821.

Kuriyama, R. & Borisy, G.G. (1985). Identification of molecular components of the centrosphere in the mitotic spindle of sea urchin eggs. **Journal of Cell Biology** 101:524– 530.

Kuriyama, R., Dasgupta, S. & Borisy, G.G. (1986). Independence of centriole formation and initiation of DNA synthesis in Chinese hamster ovary cells. **Cell Motility and the Cytoskeleton** 6:355–362.

Caption for Fig. 10 (cont.)

(f) and CHO3 (g–i). (c) and (d) show CHO cells stained with CHO5 before and after treatment with 0.12 μg/ml colcemid for 3.5–4 h. The insets in (a) and (f) show mitotic cells stained with CHO2 and CHO7, indicating two immunofluorescent dots at poles. a,a′, b, d , e , f′, × 1 000; c, f, × 1 150; g, × 1 500; h, × 400; i, × 800.

Kuriyama, R. & Kanatani, H. (1981). The centriolar complex isolated from starfish spermatozoa. **Journal of Cell Science** 49:33–49.
Lydersen, B.K. & Prittijohn, D.E. (1980). Human-specific nuclear protein that associates with the polar region of the mitotic apparatus: distribution in a human/hamster hybrid cell. **Cell** 22:489–499.
Mitchison, T. & Kirschner, M. (1984). Microtubule assembly nucleated by isolated centrosomes. **Nature** 312:232–237.
Murray, R.G., Murray, A.S. & Pizzo, A. (1965). The fine structure of mitosis in rat thymic lymphocytes. **Journal of Cell Biology** 26:601–619.
Nigg, E.A., Schäfer, G., Hilz, H. & Eppenberger, H.M. (1985). Cyclic-AMP-dependent protein kinase type II is associated with the Golgi complex and with centrosomes. **Cell** 41:1039–1051.
Oliver, J.M., Osborne, W.R.A., Pfeiffer, J.R., Child, F.M. & Berlin, R.D. (1981). Purine nucleoside phosphorylase is associated with centrioles and basal bodies. **Journal of Cell Biology** 91:837–847.
Phillips, S.G. & Rattner, J.B. (1976). Dependence of centriole formation on protein synthesis. **Journal of Cell Biology** 70:9–19.
Rattner, J.B. & Berns, M.W. (1976). Distribution of microtubules during centriole separation in rat kangaroo (*Potorus*) cells. **Cytobios** 15:37–43.
Rattner, J.B. & Phillips, S.G. (1973). Independence of centriole formation and DNA synthesis. **Journal of Cell Biology** 57:359–372.
Ring, D., Hubble, R., Caput, D. & Kirschner, M. (1980). Isolation of microtubule organizing centers from mouse neuroblastoma cells. In: DeBrabander, M. & DeMey, J., eds. **Microtubules and Microtubule Inhibitors**. Amsterdam, Elsevier, pp. 297–310.
Robbins, E.L., Jentzsch, G. & Micali, A. (1968). The centriole cycle in synchronized HeLa cells. **Journal of Cell Biology** 36:329–339.
Salisbury, J.L., Baron, A.T., Coling, D.E., Martindale, V.E. & Sanders, M.A. (1986). Calcium-modulated contractile proteins associated with the eucaryotic centrosome. **Cell Motility and the Cytoskeleton** 6:193–197.
Sherline, P. & Mascardo, R.N. (1982). Epidermal growth factor induced rapid centrosomal separation in HeLa and 3T3 cells. **Journal of Cell Biology** 93:507–512.
Stubblefield, E. (1968). Centriole replication in a mammalian cell. In: **The Proliferation and Spread of Neoplastic Cells**. Baltimore, Williams and Wilkins, pp. 175–193.
Wheatley, D.N. (1982). **The Centriole: a Central Enigma of Cell Biology**. Elsevier, Amsterdam, New York, Oxford, 1982.
Wilson, E.B. (1928). **The Cell in Development and Heredity**. 3rd edition, MacMillan, Inc., New York.
Xeros, N. (1962). Deoxyriboside control and synchronization of mitosis. **Nature** 194:682–683.

Interactions of drugs with microtubules

LESLIE WILSON

Department of Biological Sciences, University of California, Santa Barbara, CA 93106, USA.

Abstract

Microtubules participate in a large number of diverse cellular processes, including several that are critical to normal spermiogenesis. The assembly and disassembly of microtubules is finely regulated in cells, and it is clear that assembly/disassembly dynamics are integrally linked with the functions of microtubules in the development and maintenance of cell shape and in various kinds of intracellular movement. A large number of chemical substances interact with tubulin, the building block unit of the microtubule, and with the microtubule itself, either at the ends or along the microtubule surface, and modify the dynamics of tubulin addition and loss at microtubule ends. Perturbation of assembly dynamics results in impaired microtubule function. The purpose of this review is to describe the mechanism of action of several drugs whose interactions with tubulin and microtubules are becoming well understood, and to suggest possible directions for the development of drugs that might interfere selectively with microtubules during spermiogenesis.

Key Words

Microtubules, tubulin, drug mechanisms, spermiogenesis, microtubule-associated proteins (MAPs), colchicine, vinblastin, taxol.

Introduction

Microtubule assembly dynamics

Microtubules (MTs) are important components of the cytoskeleton of eukaryotic cells. During the past 20 years much has been learned about the occurrence, organization, and the functions of MTs (reviewed in Roberts & Hyams, 1979; Dustin, 1984). Microtubules participate in the development and maintenance of cell shape and, therefore, perform a cytoskeletal role in cells. They also participate in many kinds of movement, as for example, chromosome movement during mitosis and meiosis and the beating of sperm flagella.

MTs in cells can be distinguished on the basis of differences in their stabilities. Such differences must reflect the different functions of MTs. Most cytoplasmic MTs are dynamic, and assemble and disassemble rapidly in response to specific requirements of the cell. Some are totally stable once formed, such as the outer doublet and central pair MTs of sperm flagellae. The molecular reasons for these differences are not understood at present, but they must reflect the presence of chemically distinct tubulins, and/or the presence, absence, or modification of specific factors that regulate MT polymerization and function.

There is now considerable knowledge regarding the structure and chemistry of tubulin, the basic building block of the MT. We are also beginning to learn about the molecules that bind to MT surfaces, that mediate the interactions of MTs with other cytoskeletal elements and various cytoplasmic components, and that serve to determine and control MT polymerization and function. Tubulin is a dimeric protein composed of two different 50 kD molecular weight polypeptide subunits, α and β. The tubulin dimer associates noncovalently to form long strands of protofilaments, and 13 protofilaments associate laterally in the shape of a hollow tube 25 nm in diameter to form the basic MT structure. Although there are chemical differences among tubulins even within the same species (e.g., Wilson *et al.*, 1984; see the chapter by Hecht in this volume), tubulins are generally highly conserved in evolution and the ability to polymerize into the MT structure is especially well conserved. MTs are also coated on their external surfaces with accessory proteins known as MT-associated proteins, or MAPs. There appear to be a large number of MAPs, that differ substantially in number and diversity from cell type to cell type. MAPs modulate the polymerization properties of tubulin, and are undoubtedly responsible for the functional and structural diversity of MTs. For example, when highly purified tubulin derived from stable outer doublet MTs is repolymerized in the test tube, single MTs are formed that are labile and dynamic. Thus, the extreme stability and doublet structure of sperm flagellar outer doublet MTs must be determined in significant part by MAPs specific to sperm flagellae.

Much has been learned in recent years about the mechanism of MT polymerization, primarily from studies carried out *in vitro* using MT protein preparations purified from vertebrate brain, but also with tubulin with or without MAPs from other cell sources, such as sea urchin sperm tail outer doublet MTs (reviewed in Roberts & Hyams, 1979; Margolis & Wilson, 1981; Kirschner & Mitchison, 1986). Early studies revealed that MT polymerization exhibited the properties of a nucleated helical polymerization of the kind described for actin (Oosawa & Asakura, 1975). Nucleation is the formation of a short MT fragment, and growth and shortening of the MT occurs by addition and loss of tubulin at the MT ends. The nucleated helical polymerization model predicted the existence of a simple equilibrium between the ends of the MT and the subunits in solution. The subunit concentration in equilibrium with a MT end is the critical concentration. Early studies established that growth at the two ends occurred at different rates, thereby demonstrating that the kinetics of tubulin addition at the two ends were different, and that MTs possess an intrinsic growth polarity (reviewed in Roberts & Hyams, 1979). But kinetic differences do not necessarily indicate that thermodynamic differences exist, and early theories for polymerization considered that the critical subunit concentrations for growth at the two ends were identical.

However, in recent years it has become clear that the dynamics of assembly and disassembly at MT ends are much more complicated and interesting than initially thought. For example, using a MAP-rich preparation of bovine brain MTs, Margolis & Wilson (1977) discovered that, at apparent equilibrium *in vitro*, net growth occurs at one MT end (called the net assembly end) and a precisely balanced net shortening occurs at the opposite end (the net disassembly end), a behavior called 'treadmilling'. These results have been confirmed with a number of other MT protein preparations, including a MAP-free preparation of MTs made from the outer doublet tubulin of sea urchin sperm tail (Farrell *et al.*, 1979). More recently, another interesting behavior of microtubules was discovered (see Kirschner & Mitchison, 1986) called 'dynamic instability'. It was observed that under certain conditions, a small number of MTs in a suspension at polymer mass steady state depolymerize completely, while the remaining MTs grow at both ends due to the constant addition of subunits to the soluble pool. Thus, both shortening and lengthening MTs can coexist in the same solution. MT assembly dynamics are strongly influenced by MAPs (Farrell *et al.*, 1987), and the roles of individual MAPs in controlling the various MT dynamic behaviors remain to be determined. The potential uses of treadmilling and dynamic instability behaviors in determining MT organization and function in cells have been discussed extensively (e.g., see Margolis & Wilson, 1981; Kirschner & Mitchison, 1986).

The mechanistic basis for the complex assembly dynamics is not understood, but it is clear that it involves the hydrolysis of GTP. A current and attractive model is that tubulin with bound GTP adds to the ends of elongating polymers and the GTP becomes hydrolyzed subsequently. Thus the ends contain a short 'cap' of GTP tubulin and the remaining core of the MT is comprised of tubulin with GDP bound at the nucleotide site. Differential utilization of the energy from GTP hydrolysis at one end as compared with the opposite end creates a thermodynamic difference between the two ends that can give rise to treadmilling. Loss and regain of a GTP cap at MT ends could be responsible for transitions between growing and shortening phases and, thus, could be responsible for dynamic instability.

Drug mechanisms

A large number of chemical agents perturb MT function in cells (see Roberts & Hyams, 1979, Dustin, 1984, for reviews). Among the most extensively studied and best understood are colchicine, vinblastine and taxol (Fig. 1). All three compounds affect the kinetics of tubulin addition and loss at MT ends, but exert their effects through different mechanisms. Recently we have been analyzing the effects of colchicine, vinblastine, and taxol on the dynamics of tubulin addition and loss at MT ends *in vitro* using purified MAP-rich bovine brain MTs and a double-isotope labelling procedure with carbon-14 and tritium-labeled guanine nucleotides. The double-label procedure permits us to determine the effects of a drug at both MT ends simultaneously. The method, which depends upon using assembly conditions in which net tubulin addition occurs at one end and loss at the opposite end (i.e., treadmilling), is diagrammatically represented in Fig. 2 and described in detail elsewhere (Wilson & Farrell, 1986).

Colchicine

The binding reaction between colchicine and tubulins from a variety of sources has been studied extensively (for an early review, see Wilson & Bryan, 1974). There is one colchicine binding site per molecule of tubulin dimer, and the binding reaction exhibits a number of unusual properties. For example, the rate of binding is slow, and binding appears to be accompanied by conformational changes both in the tubulin and the colchicine (Detrich *et al.*, 1981). Early work established that the drug must first bind to soluble tubulin in order to inhibit polymerization (Margolis & Wilson, 1977). Early studies in cells and *in vitro* also indicated that colchicine poisons MT polymerization by a substoichiometric mechanism, with inhibition of polymerization occurring when only a small proportion of the tubulin free in solution is bound by the drug (Taylor, 1965; Olmsted & Borisy, 1973; Margolis & Wilson, 1977).

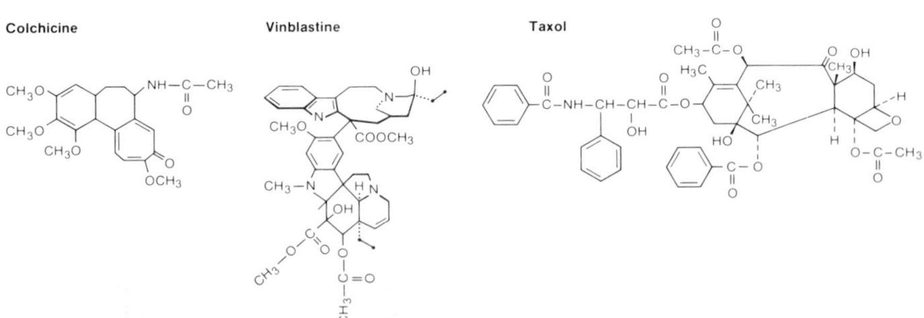

Fig. 1 Structural formulae of colchicine, vinblastine, and taxol.

The molecular mechanism by which colchicine inhibits polymerization at microtubule ends is still not fully understood, despite considerable effort in many laboratories (discussed in Wilson & Farrell, 1986). In fact, the precise mechanism seems to differ when studied with different experimental conditions and various compositions of MT protein. Most of our studies have been carried out using a preparation of MAP-rich MTprotein purified from bovine brain. When colchicine is added to a suspension of these MTs already at steady state, the TC complexes that form in solution produce a 'kinetic cap' at the A ends, with the rate constant for tubulin dissociation reduced as well as the rate constant for tubulin association. Under these conditions, small numbers of TC complexes bind or become incorporated at the A ends of the MTs, and a single TC-complex bound per assembly end inhibits the rate of net tubulin addition at that end by 50% (Margolis *et al.*, 1980).

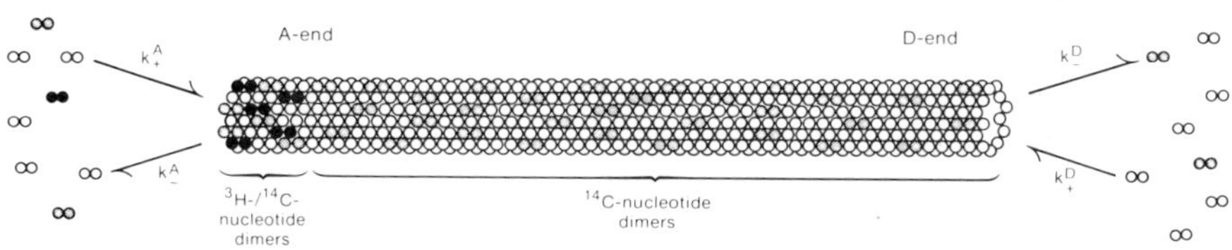

Fig. 2 A schematic diagram of microtubules (MTs) differentially labelled with carbon-14 and tritium at opposite ends. Subscripts '+' and '−' refer to association rate constants and dissociation rate constants, respectively. Superscripts refer to operationally defined assembly (A) and disassembly (D) ends. Double circles represent tubulin dimers. Open circles represent unlabeled dimers; hatched circles represent dimers labeled with [^{14}C]guanine nucleotide; filled circles represent dimers labeled with [^{3}H]guanine nucleotide. MTs are constructed so that they are labeled only with [^{14}C] at the D ends, and a mixture of [^{14}C] and [^{3}H] at the A ends. Labeled GTP binds exchangeably to soluble tubulin, and the labeled nucleotide becomes non-exchangeable upon incorporation into the MT. When MTs are polymerized to steady-state in the presence of [^{14}C]GTP and a GTP regenerating system, they become uniformly labelled with [^{14}C]guanine nucleotide. Upon pulsing with [^{3}H]GTP, the [^{3}H]guanine nucleotide becomes incorporated at the operationally-defined A ends and serves as the marker for the steady-state rate of A end net tubulin addition, while the [^{14}C]guanine continues to be added at A ends and to be lost from D-ends in an unaltered fashion. The rate of steady-state D end tubulin loss is determined by monitoring [^{14}C]guanine nucleotide loss after addition of 25-fold excess unlabeled GTP. The labeled MTs are collected on glass fiber filters for analysis of label incorporation. Apparent off rate constants can be obtained by analysis of isotope loss after dilution-induced disassembly. Mean MT lengths are obtained by analysis of negative stain electron micrographs of MT suspensions.

A typical double-label experiment in which the effects of colchicine added at steady state were examined on tubulin addition and loss at opposite ends is shown in Fig. 3A in its entirety, and the pulse portion is shown in expanded form in Fig. 3B. In this experiment, MTs were polymerized to steady state in the presence of [^{14}C]GTP. The suspension was split into two portions, and 0.2 μM colchicine was added to one portion at 45 min (after the MTs had reached steady state). This was followed at 75 min by a pulse of [^{3}H]GTP, that permitted determination of the rate of net tubulin addition to the MT A ends in the presence and absence of colchicine. This was then followed by a chase with excess unlabeled GTP, which permitted determination of the net rate of tubulin loss at the disassembly ends. Under the conditions used, the rate of net tubulin addition at the assembly ends was 1.1/s (66 molecules of tubulin/min/MT), and the rate of net tubulin loss at the disassembly ends was closely similar, at 1.14/s (68 molecules of tubulin/

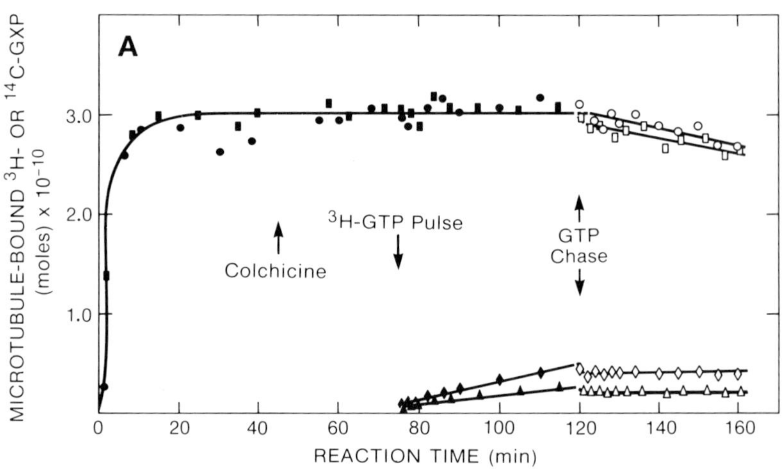

Fig. 3 (A) Effect of colchicine on the incorporation of tubulin at A ends and loss of tubulin at D ends of steady-state MTs. Two identical portions of a bovine brain MT protein solution, 2.96 mg/ml total protein, were polymerized to steady state at 30°C in reassembly buffer containing [^{14}C]GTP as described elsewhere (Wilson & Farrell, 1986). At 45 min (arrow), 0.2 μM colchicine was added to one portion, and incubation was continued ([^{14}C]guanine nucleotide data: circles, untreated; squares, colchicine-treated). A [^{3}H]GTP pulse was added to both MT suspensions at 75 min (arrow) ([^{3}H]guanine nucleotide pulse data: untreated suspension, filled diamonds; colchicine-treated, filled triangles). A 25-fold excess unlabeled GTP chase was initiated at 119 min (arrow), and was continued for an additional 45 min (open symbols denote chase conditions). Data on the ordinate reflect the quantity of bound nucleotide per filter disc (40 μl of MT suspension containing 0.049 mg of tubulin in MTs was applied to each disc).

min/MT). Colchicine inhibited the growth rate at the A ends by 49% under the conditions used, to 0.57/s (34 molecules of tubulin/min/MT), but exerted little effect at the disassembly ends (colchicine inhibited the disassembly end loss rate by 18% (i.e., 0.93/s, or 56 molecules of tubulin/min/MT). Thus, the rate of growth at A ends is reduced, while disassembly continues initially at D ends at approximately the same rate as untreated MT D ends.

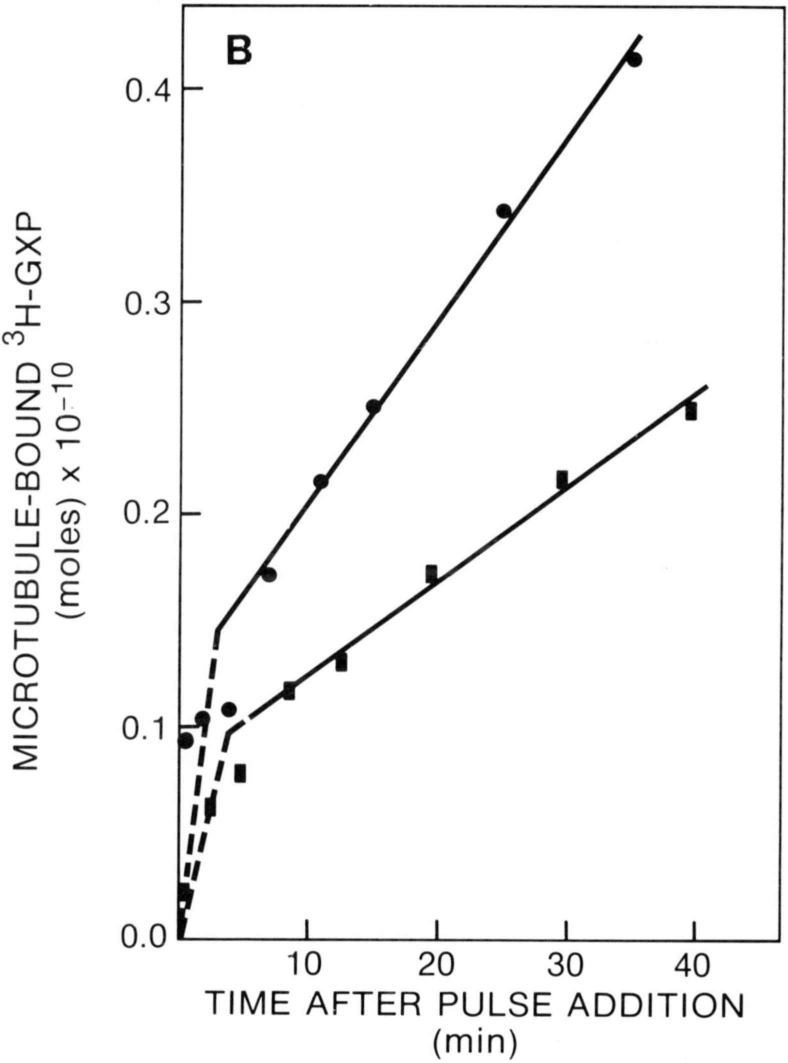

(B) Incorporation of ^{3}H-guanine nucleotide into untreated and colchicine-treated MTs at plateau. Data represent the pulse portion of the experiment shown in (A) with expanded scales. Circles, control MTs; squares, colchicine-treated MTs.

The interaction of TC complexes with MT ends can also inhibit the dissociation rate for tubulin loss, resulting in the formation of a 'kinetic cap'. The apparent dissociation rate constants for tubulin loss at each end can be determined by dilution-induced disassembly of double label MTs. For example, at the colchicine concentration used in the experiment described in Fig. 3, the dissociation rate constant at the assembly end was reduced by 78% from 96/s to 21/s, while the dissociation rate constant at the disassembly end was unaffected (384/s in the control, 366/s in the presence of 0.2 μM colchicine).

A kinetic cap can be produced at D ends also under certain conditions. When bovine brain MT protein is polymerized in the presence of TC complexes, the TC complexes become incorporated into the polymer at ratios determined by the ratio of TC complex to free tubulin in solution (Sternlicht & Ringel, 1979; Farrell & Wilson, 1984). By dilution-induced disassembly of double-labeled bovine brain MTs composed of various ratios of TC complexes and tubulin, it was found that the dissociation rate constants at both A- and D-ends are reduced in the TC-complex containing polymers as compared with untreated MTs, but that the A-ends of the copolymers were substantially more sensitive than the D-ends at all TC complex to free colchicine concentrations tested (Farrell & Wilson, 1984). The preferential kinetic capping of MT A ends by colchicine leads to a dramatic enhancement of the kinetic differences between A- and D-ends. Such MTs would exhibit extremely biased polar growth in cells, and it is interesting to consider the possibility that colchicine-like molecules exist in cells and regulate MT growth polarity by a similar differential kinetic stabilization of MT ends.

The results just described both with TC-complex-containing bovine brain MTs and with MTs treated with substoichiometric colchicine concentrations differ in some respects from results of other studies on the mechanism of action of colchicine (Bergen & Borisy, 1983; Deery & Weisenberg, 1981). For example, Bergen & Borisy examined the rates of MT elongation of brain tubulin onto flagellar axonemal seeds in the presence and absence of TC complexes, and found that the drug reduced the rate constants for tubulin addition equally at the two MT ends without affecting the rate constants for tubulin loss. An explanation for the differences in the colchicine poisoning mechanisms that have been described is that the experimental conditions and microtubule protein compositions significantly influence the way the drug inhibits polymerization. This raises the possibility that mechanistic differences might exist in the way colchicine inhibits assembly of various classes of MTs.

Vinblastine

The vinca alkaloid, vinblastine (Fig. 1), is a potent inhibitor of cell growth, and appears to exert its antiproliferative effects predominantly by disrupting MTs. The binding reaction between vinblastine and tubulin has been studied in detail in a number of laboratories. Binding of vinblastine to tubulin occurs rapidly, is rapidly reversible and relatively independent of temperature. There are two specific vinblastine binding sites per tubulin dimer, but there is still some uncertainty about the magnitude of the binding affinities for the sites (reviewed in Jordan *et al.*, 1986).

Interestingly, vinblastine exerts a number of mechanistically distinct actions on tubulin and on MTs, depending upon the conditions and vinblastine concentrations used. At low concentrations, below approximately 1 μM both in cells and *in vitro*, vinblastine inhibits the polymerization of MT protein into MTs. No unusual polymorphic effects of vinblastine on tubulin or on MTs have been observed below 1 μM vinblastine. At intermediate concentrations, between 10 and 100 μM, vinblastine has the unusual property of causing cytoplasmic tubulin to rearrange to form birefringent uniaxial crystals composed of tubulin complexed with vinblastine (reviewed in Dustin, 1984). Also at intermediate concentrations >2–4 μM *in vitro*, addition of vinblastine to preformed brain MTs causes peeling of protofilament arrays from both MT ends as shown in Fig. 4. At high concentrations, >1 mM, vinblastine precipitates tubulin nonspecifically.

Like colchicine, vinblastine is capable of inhibiting MT polymerization *in vitro* at concentrations that are considerably below the concentration of free tubulin. For example, 50% inhibition of net tubulin addition to steady-state bovine brain MTs *in vitro* occurs at 0.14 μM vinblastine, at a ratio of vinblastine-bound tubulin to uncomplexed tubulin of approximately 2 : 100 (see Jordan *et al.*, 1986). The binding of vinblastine to steady-state bovine brain MTs has been measured over a wide concentration range. Interestingly, two affinity classes of sites exist. At low concentrations, a high affinity and low capacity class of sites can be titrated (16–17 sites per MT; dissociation equilibrium constant, 1.9 μM). The most likely location of these sites is at one or both ends of the MT. At 0.14 μM, the vinblastine concentration that produces half-maximal inhibition of tubulin addition at the net assembly ends of the MTs, approximately 1.2 molecules of vinblastine were bound per MT.

The mechanism of inhibition of MT polymerization by vinblastine differs in several significant respects from that of colchicine. Vinblastine binds rapidly and reversibly to the high affinity binding sites at MT ends in the absence of soluble tubulin. Thus, the poisoning reaction is not mediated by binding to MT ends of drug-modified tubulin molecules. The kinetically rapid binding of vinblastine to a limited number of high affinity binding sites at the net assembly end of a MT appears to be all that is required to effectively reduce the rate of further tubulin addition. At concentrations that produce substoichiometric poisoning, there is little or no effect of vinblastine at the disassembly ends. Therefore, vinblastine affects the MT ends differentially. It is also incapable of becoming incorporated along with tubulin into the lattice structure of the polymer.

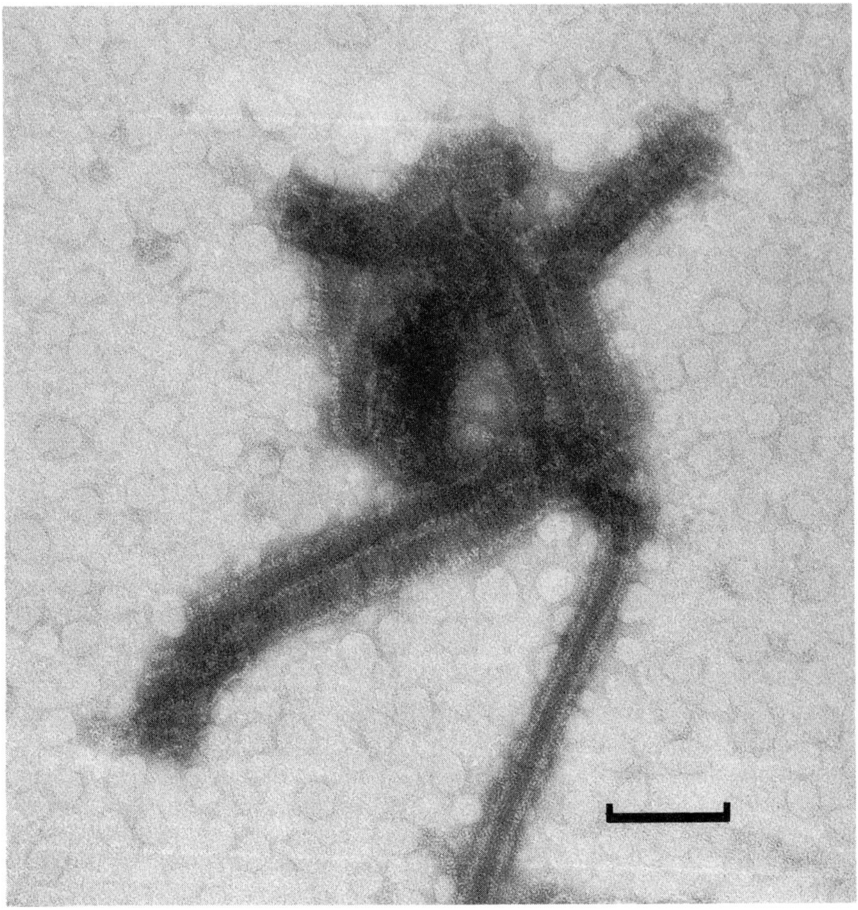

Fig. 4 Peeling of protofilament arrays at an MTend by vinblastine. Shown is a steady-state bovine brain MT *in vitro* that had begun to depolymerize in the form of spirals at the ends after 4 min of incubation with 190 μM vinblastine (see Jordan *et al.*, 1986, for additional details). Bar = 100 nm. Electron micrograph by Dr Mary Ann Jordan.

It seems unlikely that the interaction of vinblastine with the high affinity class of sites is responsible for the direct depolymerization of MTs and the protofilament peeling reaction that occurs at intermediate vinblastine concentrations. Low affinity binding to a large number of sites located on the MT surface is detected in the same vinblastine concentration range that results in protofilament peeling, and binding and the degree of peeling increase in parallel with one another (reviewed in Jordan *et al.*, 1986). A likely possibility is that the low affinity binding of vinblastine to MTsurfaces represents the binding of vinblastine with its specific binding sites on tubulin, but the sites have been modified in such a way as to reduce the affinity for vinblastine considerably. Thus, the vinblastine binding sites of tubulin can display different affinities depending upon their location in the MT surface lattice. The low affinity sites may be sterically blocked, or alternatively, the tubulin may undergo a conformational change during assembly that reduces the affinity for vinblastine.

Taxol

The effects of taxol on MT polymerization differ markedly from those of colchicine and vinblastine. Taxol (Fig. 1) is a macrocyclic alkaloid from the western yew, *Taxus brevifolia*, which stabilizes MTs in cells and *in vitro* (Schiff *et al.*, 1979; Schiff & Horwitz, 1980). Taxol *in vitro* stabilizes MTs against the depolymerizing action of dilution, cold, calcium ions, and of a number of MT-disrupting drugs. Initiation of MT polymerization is enhanced by taxol and the critical subunit concentration is reduced.

In a recent study, we investigated the effects of taxol on the exchange of tubulin at the opposite ends of bovine brain MTs at steady-state, and on the apparent off rate constants at the two MT ends by dilution-induced disassembly using the double isotope procedure described earlier (Wilson *et al.*, 1985). It was found that addition of taxol to steady-state MTs produced a concentration-dependent decrease in the net uptake of tubulin at assembly ends, with half-maximal inhibition occurring between 5 and 7 μM drug. At steady-state in the absence of taxol, the rate of tubulin addition at A ends (0.73/s) was balanced by a very similar loss rate at D ends (0.76/s). After addition of 6.5 μM taxol to the steady-state MTs, the growth rate at A ends as determined with the tritium-label incorporation data was reduced by 48%, and the loss rate at D ends was reduced by 47%. These data indicate that the flux of tubulin from the A ends of the microtubules to the D ends was inhibited by the drug.

Estimates of the molecular dissociation rate constants at microtubule A- and D-ends were obtained by dilution of untreated and taxol-treated double-labeled MTs. Taxol reduced the off rate constants at two microtubule ends to exactly the same extent, while exerting little or no effect on the apparent association rate constants. Combined with results obtained by Parness & Horwitz (1981) on the binding of taxol to MTs, the data indicate that the drug binds stoichiometrically with tubulin along the surface of the MT and stabilizes inter-tubulin bonding uniformly.

Possibilities for drug intervention in microtubule formation and function during spermiogenesis

The foregoing discussion was intended to provide a mechanistic overview of how three of the well-known tubulin-binding drugs act. It is clear that

MT polymerization is a highly complex process, and just considering the tubulin backbone itself, there are many potential ways to interfere with polymerization.

It is conceivable that drugs could be identified that inhibit MT polymerization by interacting selectively with sperm flagellar tubulin. Evidence obtained by isoelectric focusing of highly purified outer doublet tubulin of sea urchin sperm flagellae and of egg tubulin from the same species, *Strongylocentrotus purpuratus*, indicates that there are chemical differences between them (Fig. 5). It is interesting that there are also significant differences in the colchicine binding activity of these same tubulins (Wilson *et al.*, 1984). Van't Hoff plots for the binding of colchicine to purified native tubulins from sperm flagellar outer doublet MTs and from unfertilized eggs of the sea urchin are shown in Fig. 6 from which the thermodynamic parameters governing the binding of colchicine to tubulin were derived. The data demonstrate that significant chemical differences in the tubulins

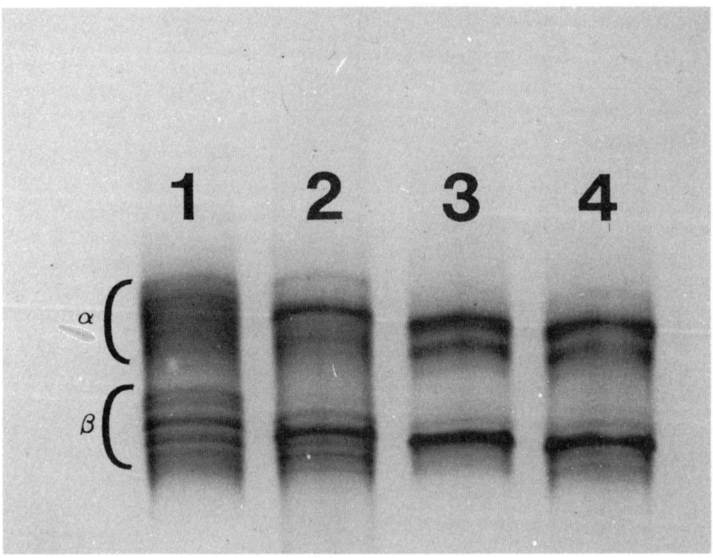

Fig. 5 Isoelectric focusing of highly purified sperm flagella outer doublet and egg tubulins from the sea urchin *Strongylocentrotus purpuratus*, and comparison with tubulin from bovine brain. Tubulins were prepared as described by Wilson *et al.* (1984), and Wilson & Farrell (1986), and subjected to isoelectric focusing as described by Sullivan & Wilson (1984). Lane 1, purified bovine brain tubulin; lane 2, a mixture of purified outer doublet and egg tubulins; lane 3, purified egg tubulin; and lane 4, purified outer doublet tubulin. The putative positions of the alpha and beta peptides are indicated. (Data of Mr Herbert Miller.)

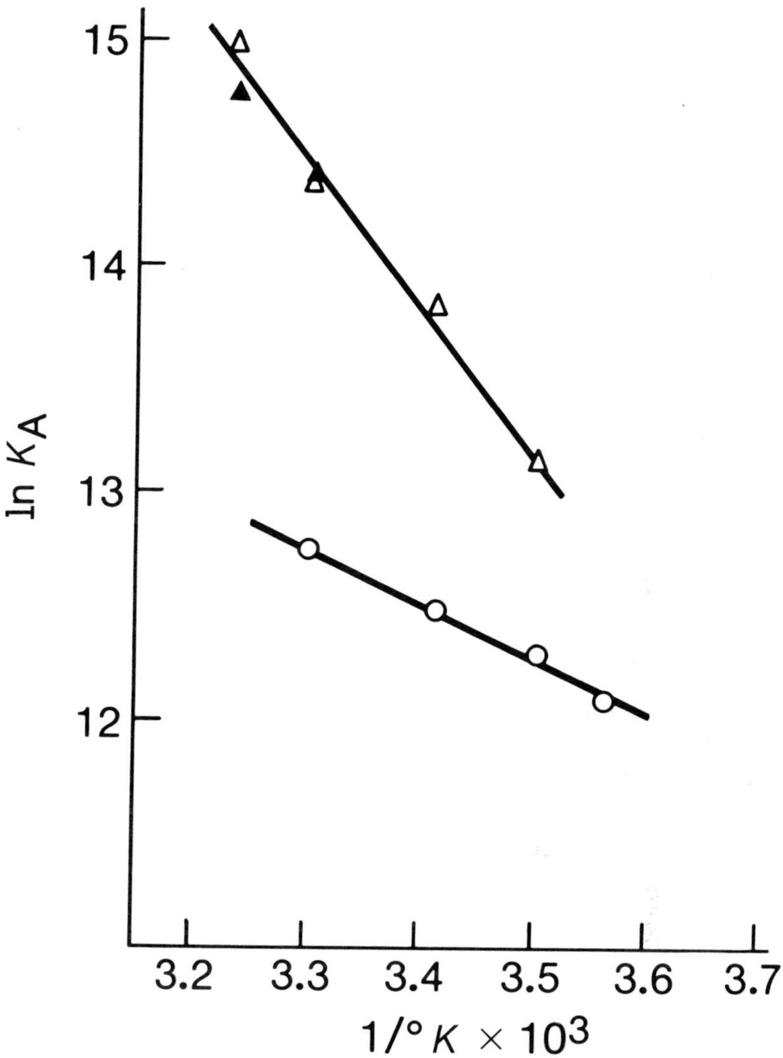

Fig. 6 Van't Hoff analysis of colchicine binding to sea urchin sperm tail outer doublet and egg tubulins. The binding of tritium-labeled colchicine to purified tubulin obtained from unfertilized eggs and sperm tail outer doublet MTs of the sea urchin *Strongylocentrotus purpuratus*, was ascertained at a variety of temperatures by Scatchard analysis, and the data plotted in the form of Van't Hoff plots (see Wilson *et al.*, 1984, for experimental details). Open and closed triangles, outer doublet tubulin; circles, purified egg tubulin. Thermodynamic parameters for egg tubulin were: $\triangle H_o$, 4.6 kcal/mol; $\triangle S_o$, 40.4 entropy units; and $\triangle G_{13°C}$, −6.9 kcal/mol. Thermodynamic parameters for outer doublet tubulin were: $\triangle H_o$, 13.6 kcal/mol; $\triangle S_o$, 73.4 entropy units; and $\triangle G_{13°C}$ −7.4 kcal/mol.

from two functionally distinct MT systems of this species exist, and that the differences are expressed at the level of the colchicine binding site on the native tubulin dimer. The affinity of colchicine for the sperm flagellar outer doublet tubulin was significantly greater than that of the egg tubulin at all temperatures, suggesting that flagellar tubulin might be a more sensitive target for colchicine than egg tubulin.

However, the tubulins are very strongly conserved proteins, and while the differences in the colchicine binding activity of sperm flagellar and egg tubulins are appreciable, the differences are not sufficient to permit a useful selectivity. Even a small extent of inhibition of MT polymerization in a non-sperm cell by a potential male antifertility drug would make the drug too dangerous to use. There is the remote possibility that a tubulin-binding drug with a near absolute selectivity for flagellar tubulin could be found, but this approach does not appear to be practical.

A much more attractive possibility for selective intervention with MT function during spermiogenesis might involve a search for compounds that impair the function of sperm-specific MAPs. It should be remembered that the drugs that we presently know about are only those that interact with the tubulin backbone of the MT. It is reasonable to assume, since MT function requires large numbers of associated proteins, that chemical agents ultimately may be found that interfere with the function of the associated proteins. It is clear that the ability of sperm tail outer doublet tubulin to form the doublet MT structure, and to attain the remarkable degree of stability that the doublet MTs display, are due to the MAPs (e.g., see Farrell & Wilson, 1978). These proteins probably are functioning to create the stable doublet structure during formation of the flagellum. Conceivably, one or several of the proteins functions exclusively during flagellar formation, and do not participate in MT function in any other type of cell (e.g., are not involved in the formation of cilia). Unfortunately, there is virtually no knowledge regarding any flagellar MAPs. It seems reasonable to suggest that a search for flagellae-specific MAPs could be enormously fruitful. Once such proteins are identified and their functions understood, it might be possible to design drugs to interfere with their function. With this approach, one might be able to create a drug that inhibits flagellar assembly during spermiogenesis without exerting any effect on formation or function of other classes of MTs.

Acknowledgements

Original data mentioned in this review was supported by USPHS grant NS13560 from the National Institute of Neurological and Communicative Disorders and Stroke, and by American Cancer Society Grant CH-381.

References

Bergen, L.G. & Borisy, G. (1983). Tubulin-colchicine complex inhibits microtubule elongation at both plus and minus ends. **Journal of Biological Chemistry** 258:4190–4194.

Deery, W.J.& Weisenberg, R.C. (1981). Kinetic and steady-state analysis of microtubules in the presence of colchicine. **Biochemistry** 20:2316–2324.

Detrich III, H.W., Williams, Jr, R.C., Macdonald, T.L., Wilson, L. & Puett, D. (1981). Changes in the circular dichroic spectrum of colchicine associated with its binding to tubulin. **Biochemistry** 21:5999–6005.

Dustin, P. (1984). **Microtubules**. Springer-Verlag, New York.

Farrell, K.W. & Wilson, L. (1978). Microtubule reassembly *in vitro* of *S. purpuratus* sperm tail outer doublet tubulin. **Journal of Molecular Biology** 121:393–410.

Farrell, K.W., Kassis, J.A. & Wilson, L. (1979). Outer doublet tubulin reassembly: evidence for opposite end assembly-disassembly at steady state and a disassembly end equilibrium. **Biochemistry** 12:2642–2647.

Farrell, K.W. & Wilson, L. (1984). Tubulin-colchicine complexes differentially poison opposite microtubule ends. **Biochemistry** 23:3741–3748.

Farrell, K.W., Jordan, M.A., Miller, H.F. & Wilson, L. (1987). Phase dynamics at microtubule ends: the coexistence of microtubule length changes and treadmilling. **Journal of Cell Biology** 104:1035–1046.

Jordan, M.A., Margolis, R.L., Himes, R.H. & Wilson, L. (1986). Identification of a distinct class of vinblastine binding sites on microtubules. **Journal of Molecular Biology** 187:61–73.

Kirschner, M. & Mitchison, T. (1986). Beyond self-assembly: from microtubules to morphogenesis. **Cell** 45:329–342.

Margolis, R.L. & Wilson, L. (1977). Addition of colchicine-tubulin complex to microtubule ends: the mechanism of substoichiometric colchicine poisoning. **Proceedings of the National Academy of Sciences USA**. 74:3466–3470.

Margolis, R.L. & Wilson, L. (1981). Microtubule treadmills-possible molecular-machinery. **Nature** 293:705–711.

Margolis, R.L., Rauch, C.T. & Wilson, L. (1980). Mechanism of colchicine-dimer addition to microtubule ends. Implications for the microtubule polymerization mechanism. **Biochemistry** 19:5550–5557.

Olmsted, J.B. & Borisy, G. (1973) Characterization of microtubule assembly in porcine brain extracts by viscometry. **Biochemistry** 12:4282–4289.

Oosawa, F. & Asakura, S. (1975). **Thermodynamics of the Polymerization of Protein**. Academic Press, London.

Parness, J. & Horwitz, S.B. (1981). Taxol binds to polymerized tubulin *in vitro*. **Journal of Cell Biology** 91:479–487.

Roberts, K. & Hyams, K.S. (1979). **Microtubules**. Academic press, Inc. New York.

Schiff, P.B., Fant, J. & Horwitz, S.B. (1979). Promotion of microtubule assembly *in vitro* by taxol. **Nature** 277:665–667.

Schiff, P.B. & Horwitz, S.B. (1980). Taxol stabilizes microtubules in mouse fibroblast cells. **Proceedings of the National Academy of Sciences USA** 77:1561–1565.

Sternlicht, H. & Ringel, I. (1979). Colchicine inhibition of microtubule assembly via copolymer formation. **Journal of Biological Chemistry** 254:10540–10550.

Sullivan, K.F. & Wilson, L. (1984). Developmental and biochemical analysis of chick brain tubulin heterogeneity. **Journal of Neurochemistry** 42:1363–1371.

Taylor, E. (1965). The mechanism of colchicine inhibition of mitosis. I. Kinetics of

inhibition and the binding of ^{3}H-colchicine. **Journal of Cell Biology** 25:145–160.
Wilson, L. & Bryan, J. (1974). Biochemical and pharmacological properties of microtubules. **Advances in Cellular and Molecular Biology** 3:21–78.
Wilson, L., Miller, H.P., Pfeffer, T.A., Sullivan, K.F. & Detrich III, H.W. (1984). Colchicine binding activity distinguishes sea urchin egg and outer doublet tubulins. **Journal of Cell Biology** 99:47–41.
Wilson, L., Miller, H.P., Farrell, K.W., Snyder, K.S., Thompson, W.C. & Purich, D.L. (1985). Taxol stabilization of microtubules *in vitro*: dynamics of tubulin addition and loss at opposite microtubule ends. **Biochemistry** 24:5254–5262.
Wilson, L. & Farrell, K.W. (1986). Kinetics and steady state dynamics of tubulin addition and loss at opposite microtubule ends: the mechanism of action of colchicine. **Annals of the New York Academy of Sciences** 466:690–708.

Tektin filaments in the structure and assembly of sperm flagellar microtubules

R. W. LINCK AND M. J. GOGGIN

Department of Cell Biology and Neuroanatomy, University of Minnesota, Minneapolis, Minnesota 55455, USA.

Abstract

Microtubules from marine invertebrate sperm can be fractionated to yield filaments 2–6 nm in diameter composed of a novel set of proteins (tektins) that resemble mammalian intermediate filment proteins. This paper reviews the work on the isolation and characterization of these proteins and on their structural organization within the microtubule. The paper also discusses the possible function of these proteins and the possibility of their being important targets for the control of male fertility in humans.

Keywords

Flagellum, tubulin, tektins, sea urchin, axoneme, microtubules, antibodies.

Introduction

Microtubules form the structural and functional basis for sperm flagellae and flagellar motility, and are therefore likely to be important targets for the regulation of male fertility (Fawcett & Bedford, 1979; Gibbons, 1981; Dustin, 1984). In addition, microtubules play essential roles in the development of sperm: e.g., mitosis, meiosis, centriolar activities, the manchette of spermatids, and the cytoskeleton and motile events within the Sertoli cells (cf., Bardin & Sherins, 1982). Until recently, knowledge of flagellar microtubule structure, composition and function has been mainly limited to the tubulin proteins that polymerize to form the walls of microtubules and to the dyneins, the ATPase transducers that generate forces for sperm tail propulsion (Gibbons, 1981; Dustin, 1984; Soifer, 1986). Within the last few years a new set of filamentous proteins, tektins, have been discovered to be important components of invertebrate flagellar and ciliary microtubules. This chapter will review the evidence for the existence of these novel proteins, their structural organization within flagellar microtubules, and their possible functions in sperm.

Briefly, microtubules are hollow cylindrical organelles, 25–30 nm in diameter, usually composed of 13 longitudinally arranged 'protofilaments' (Dustin, 1984). The major proteins forming the microtubule wall are α- and β-tubulin, M_r = ~50 kD (Stephens, 1970; Soifer, 1986). In solution, purified tubulin exists mainly in the form of a 6S heterodimer composed of the globular α- and β-subunits (Bryan & Wilson, 1971; Luduena *et al.*, 1977; Soifer, 1986). In native microtubules, the dimers are polymerized in a polar fashion to form most or all of the longitudinally arranged protofilaments (Amos, 1979; Amos & Klug, 1974). Soluble tubulin from a variety of sources can be polymerized in vitro to yield microtubule-like structures (Soifer, 1986, Weisenberg, 1972). From mammlian brain tissue the repolymerized tubulin contains a set of high-molecular-weight microtubule-associated proteins (HMW-MAPs) that form lateral projections on the surface of microtubules in vivo (Sloboda *et al.*, 1975; Soifer, 1986). Another set of MAPs, referred to as tau, with molecular weights between 55 kD and 62 kD, have also been reported (Cleveland *et al.*, 1979; Soifer, 1986), but their structural association with the microtubule is not understood. Both the HMW MAPs and tau have been shown to influence the assembly of tubulin in vitro, but neither have been shown to be present in sperm flagellar microtubules.

Sea urchin sperm are a model system for the study of microtubules. Flagella have a clearly defined function (motility) and can be isolated intact and in a functionally active state (Gibbons & Gibbons, 1972; Gibbons, 1981). In most species flagellae (and cilia) are composed of the evolutionarily conserved '9 + 2' arrangement of microtubules, known as the axoneme (see Fig. 1). The 9 outer doublet microtubules are continuous with the 9 triplet microtubules of the centriole-like basal bodies (Dustin, 1978; Gibbons, 1981). The doublet tubule is composed of A- and B-subfibers; the A-fiber is a complete microtubule with 13 protofilaments, to which are attached several accessory structures. These include the inner and outer arms, radial spokes, nexin fibers that circumferentially connect the 9 doublet tubules, and the B-subfiber (Gibbons, 1981). The inner and outer arms are composed of a complex set of dynein ATPase proteins (Gibbons, 1981; Johnson, 1985). The various axonemal components are attached along the A-tubule axis at regular intervals that are multiples of the underlying microtubule subunit repeat, 4 and 8 nm for the tubulin monomer and dimer respectively (Amos & Klug, 1974; Amos *et al.*, 1976).

Flagellar microtubules and their associated components can be selectively fractionated and, to some extent, reconstituted in vitro (Stephens, 1970; Kuriyama, 1976; Linck, 1973, 1976; Stephens & Edds, 1976; Binder & Rosenbaum, 1978; Farrel & Wilson, 1978; Gibbons, 1981; Linck & Langevin, 1981). Thus, it is possible to study the individual protein components in the structure, biochemistry and function of flagellar microtubules. Importantly, flagellar microtubules can be fractionated by extraction with sarkosyl detergent (sodium dodecyl sarcosinate) into ribbons of 3 protofilaments (Meza *et al.*, 1972; Witman *et al.*, 1972a, b). Linck (1976) reported that these sarkosyl-resistant protofilament ribbons were composed of residual

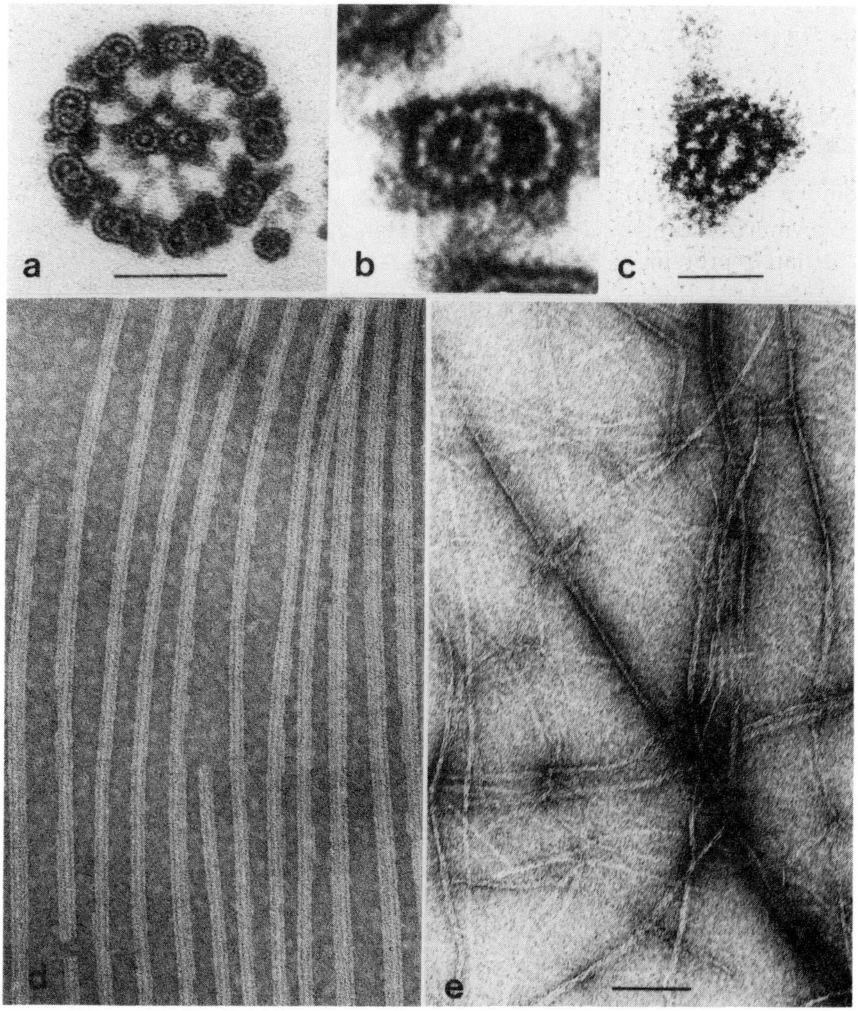

Fig. 1 Electron micrographs showing the products of the fractionation of sea urchin sperm flagella (*S. purpuratus*): a, axonemes; b, doublet microtubules; c, singlet A-microtubules; d, protofilament ribbons; and e, tektin filaments. A-microtubules were obtained by thermal fractionation of doublet tubules; protofilament ribbons were prepared by extraction of doublet tubules with 0.5% sarkosyl detergent; and tektin filaments were obtained by extraction of doublet tubules with 0.5% sarkosyl, 2 M urea. Tannic acid fixation, a–c; negative stain, d and e. Magnification bars = 0.1 μm for a, 25 nm for b and c, and 100 nm for d and e.

tubulin and a set of what appeared to be non-tubulin proteins. These non-tubulin proteins are the subject of this chapter.

Fractionation of flagellar microtubules into tektin filaments

For review, the ultrastructures of flagellar axonemes and doublet microtubules are shown in Fig. 1. Doublet microtubules are prepared by low ionic strength dialysis of flagellar axonemes (Gibbons, 1981; Linck & Langevin, 1981) and can be fractionated in several ways, as illustrated. Thermal fractionation or gentle heating at 40°C for 6 min (Stephens, 1970; Linck & Langevin, 1981) solubilizes the B-tubule fraction, yielding singlet A-microtubules (Fig. 1c). The A-tubule is characterized by having 13 protofilaments and an 'adlumenal' component associated with the inside of the A-tubule wall across from the inner B-tubule junction. Doublet microtubules or the thermally-derived A-tubules can be further fractionated by extraction with 0.5% sarkosyl, yielding ribbons of protofilaments (Fig. 1d). The number of protofilaments remaining in these sarkosyl-insoluble ribbons is species dependent: in *Strongylocentrotus purpuratus* the number ranges from 3 to 5, with an average of 4 protofilaments; in *Strongylocentrotus droebachiensis* the number is uniformly 3 protofilaments. If urea is added to the sarkosyl solution, the number of protofilaments appearing in the ribbons is reduced. At a concentration of 0.5% sarkosyl–2M urea, the only structures remaining are filaments (Fig. 1e) of extended length with diameters of 2-6 nm, which themselves appear to be composed of 2 nm subfibrils. Such filaments can be obtained by sarkosyl-urea extraction of either doublet microtubules, thermal A-tubules or sarkosyl-ribbons.

The polypeptide composition of the products of the fractionation of *S. purpuratus* doublet microtubules by sarkosyl–urea are illustrated in Fig. 2. In this experiment aliquots of doublet microtubules were added to control buffer, to a 0.5% sarkosyl solvent, or to 0.5% sarkosyl containing varying amounts of urea. Lane a represents the polypeptide composition of doublet microtubules that were treated with a control buffer in which the tubules are stable (0.1 M KCl, 5 mM $MgCl_2$, 10 mM Tris, pH 8). Lane b shows the composition of the ribbons of 3–4 protofilaments (Fig. 1b) obtained by 0.5% sarkosyl extraction; note the preparation consists principally of residual α- and β-tubulin, a pair of polypeptides with molecular masses of 77 kD and 83 kD, and a set of polypeptides between 47 kD and 55 kD. Lanes c and d show that the addition of urea to the sarkosyl solvent achieves the removal of the 77 kD and 83 kD polypeptides and the tubulins, so that at 2 M urea only three principal bands remain with molecular masses of 47 kD, 51 kD and 55 kD. This latter material corresponds to the filamentous fraction shown in Fig. 1e. At a higher urea concentration (4 M) only the 51 kD and 55 kD polypeptides are retained in the fraction composed of filaments 2–4 nm in diameter, which have an α-helical content of 71%, as measured by circular dichroism (Linck & Langevin, 1982).

Characterization of the tektins

The major polypeptides (47 kD, 51 kD and 55 kD) of the sarkosyl–urea insoluble filaments have been collectively named tektins, and they appear

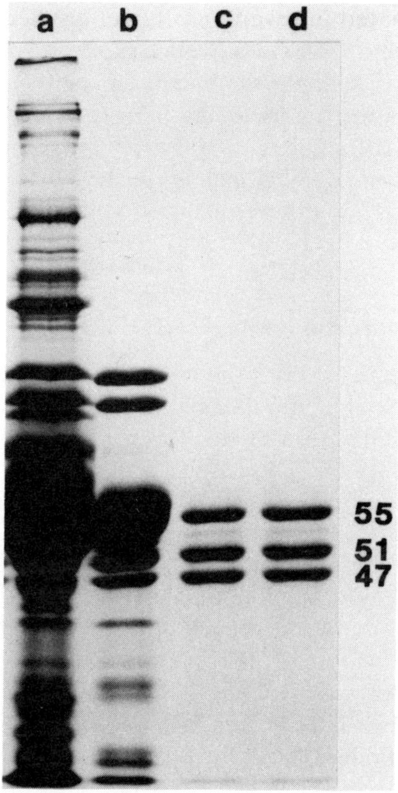

Fig. 2 SDS-PAGE analysis of the fractionation of *S. purpuratus* flagellar doublet microtubules. a, doublet microtubules showing the major tubulin components and the lesser amounts of associated proteins; b, pellet from extraction with 0.5% sarkosyl; c, pellet from extraction with 0.5% sarkosyl, 2 M urea; d, pellet from extraction with 0.5% sarkosyl, 2.5 M urea. The three major bands remaining after the sarkosyl–urea extractions are the three tektin polypeptides described in this paper, with molecular masses indicated in Daltons × 10^{-3}; this material corresponds to that shown in Fig. 1e.

to form structures called tektin filaments (Linck, 1982; Linck & Langevin, 1982; Linck *et al.*, 1985). The three main tektins have been individually purified by SDS-PAGE and HPLC and biochemically characterized, as presented in detail elsewhere (Linck & Langevin, 1982; Linck *et al.*, 1985; Linck & Stephens, 1987). Briefly, amino acid analysis and tryptic peptide mapping suggest that the three tektins so far characterized may share a significant degree of homology, yet are distinctly different gene products, and that they are strikingly different from the tubulins (Linck & Stephens, 1987). These results explain the earlier findings that antibodies to tubulin do not recognize the tektins and vice versa (Linck *et al.*, 1985). Each of the three major tektins runs as a single major peak on HPLC (Linck & Stephens, 1987); by 2-D isoelectric focussing-SDS-PAGE each tektin appears as a single major spot, although secondary spots suggest isoelectric heterogeneity (Linck & Langevin, 1982; Linck *et al.*, 1987). The results of HPLC and amino acid analysis, together with their solubility properties, suggest that the tektins are a relatively hydrophobic class of proteins.

Characterization of antitektin antibodies

Antisera have been prepared against both the individually purified tektins and against whole tektin filaments composed of all three tektins. The preparation and complete characterization of these antitektins are presented elsewhere (Linck *et al.*, 1985, 1987). The results of immunoblotting with the individually prepared antitektins are briefly described here (see Fig. 3). For both *S. purpuratus* and *L. pictus* each antibody is primarily specific for its respective tektin, i.e., the anti-55 kD tektin (anti-55) is primarily specific for the 55 kD polypeptide band, the anti-51 for the 51 kD band, and the anti-47 for the 47 kD band. However, some cross-reaction does occur between a given antitektin and the other tektins. The rabbit antitektin antibodies have been shown to cross-react with specific polypeptides in a variety of species, as shown elsewhere (Linck *et al.*, 1985, 1987), including sea urchin embryonic cilia and molluscan gill cilia. However, in some cases a given antitektin cross-reacts with one or more polypeptides with apparent molecular weights somewhat different from the original tektin antigen. Cross-reaction with mammalian sperm has not yet been observed.

Localization of tektins in flagellar microtubules

By immunofluorescence microscopy (Fig. 4) the antitektins stain axonemes of sea urchin sperm and embryonic cilia (Linck *et al.*, 1985) that have been fixed with −20°C methanol. Depending on the antibody, there also appears to be specific staining of the basal body and a region corresponding to the sperm head envelope; for these details, the reader is referred to Linck *et*

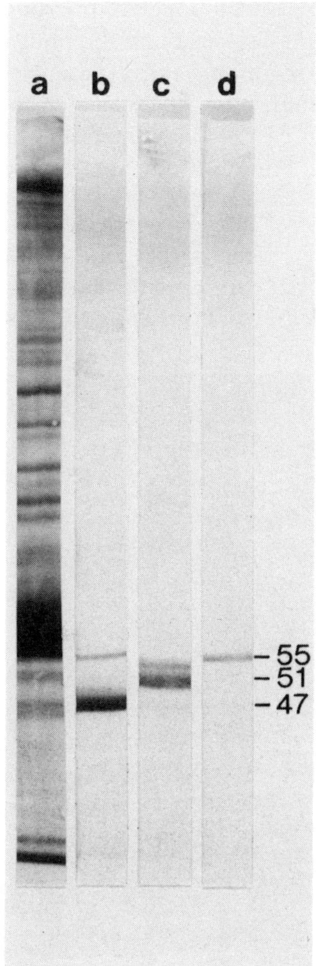

Fig. 3 SDS-PAGE immuno blot analysis of antitektin (antibodies) against flagellar axonemes. A curtain of *S. purpuratus* flagellar axonemes was electrophoresed and transferred to nitrocellulose. Strips were then stained as follows: a, with Amido Black for total protein; b, with anti-47 kD tektin; c, with anti-51 kD tektin; d, with anti-55 kD tektin. Note the specificities of each antitektin, particularly of the anti-47 kD tektin for the 55-kD tektin.

al., (1987). In sperm tails which have been splayed apart, all 9 doublet tubules stain with each purified antitektin (Steffen & Linck, 1988).

The localization of the tektins at the electron microscope level is not as staightforward as the immunofluorescence microscopy. By indirect immuno-

EM staining procedures, intact flagellar microtubules do not stain with antitektins along their lengths, although the ends of protofilament ribbons in these preparations do show antitektin staining (Linck *et al.*, 1985). However, if flagellar microtubules are first attached to EM grids and extracted in place with sarkosyl–urea, 2–4 nm filaments are then apparent, which then stain with the antitektin antibodies.

Identity of the tektin proteins

The three tektins that have been partially characterized are unique proteins; they share a significant degree of homology with each other but are distinctly

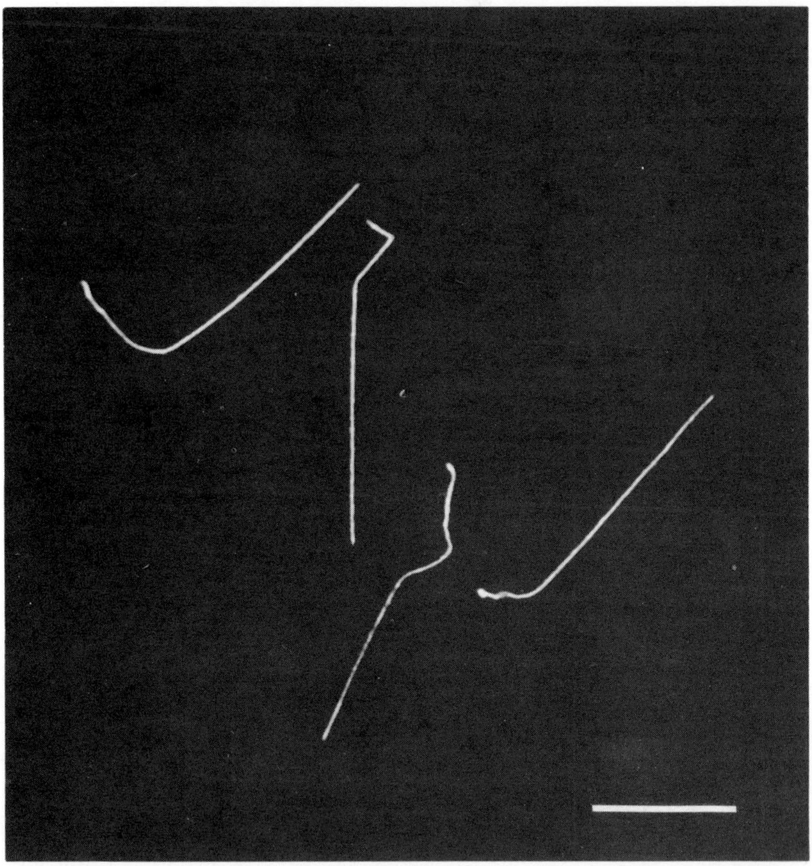

Fig. 4 Immunofluorescence micrograph of *S. purpuratus* sperm stained with antisera to the 55 kD tektin. The anti-55 kD tektin stains only the flagella. The anti-51 kD and anti-47 kD tektins show staining of the basal bodies and a region corresponding to the sperm head envelope (see Linck, *et al.*, 1987). Magnification bar = 20 μm.

different gene products (Linck & Stephens, 1987). Amino acid analysis demonstrates that the three tektins are entirely different from both α- and β-tubulin, but are strikingly similar to each other; where differences do exist, one tektin deviates slightly from a trait common to the other two. Interestingly, the tektins also show a significant similarity to intermediate filament proteins (desmin, vimentin and keratin), although the tektins have more cysteines and tryptophans than the IF proteins, which characteristically have a single tryptophan and, except for keratins, a single cyteine (cf. Weber & Geisler, 1985). Tryptic peptide mapping has demonstrated that the tektins share a significant degree of homology based on the coincidence of peptides; between a given pair of tektins the percent homology is estimated between 63 and 73%, a feature that is also typical between different IF proteins. By immunological analysis (Fig. 3) the tektins are distinguishable, but with polyclonal antibodies the tektins appear to share some homology with each other (Linck *et al.*, 1987). Thus, the tektins represent a unique class of proteins that are IF-like but, in the case of sperm, are associated with microtubules.

Organization of tektins in flagellar microtubules

It seems reasonably clear that the 3 tektins are major components of the A-microtubule of flagellar doublet microtubules, and further that they are localized as extended filaments in the sarkosyl–insoluble protofilament ribbon fraction of the A-tubule. The tektins cofractionate with thermally-derived A-tubules and with the sarkosyl-insoluble ribbons (cf. Linck, 1976; Linck & Langevin, 1981). In neither the A-tubule nor sarkosyl-ribbon fractions are 2–6 nm filaments observed by negative staining; such filaments are only observed following the solubilization of the tubulin. Antitektin antibodies stain the individual doublet microtubules after methanol fixation for immunofluorescence microscopy (Linck *et al.*, 1985; Amos *et al.*, 1986; Steffen & Linck, 1988). The antitektins stain the ends of protofilament ribbons, particularly the thin (2–4 nm) fibrils that protrude from the ends of the ribbons (Linck *et al.*, 1985). Finally, the antitektins stain the extended tektin filaments that are exposed when doublet microtubules are extracted with sarkosyl-urea (Linck *et al.*, 1985). Thus the major tektins would seem to exist as integral and continuous components of the sarkosyl protofilament ribbons; however, this structural model has not been directly proven.

The precise location(s) of the sarkosyl protofilament ribbons and tektin filaments have not been determined. Linck (1976) provided quantitative data suggesting that there may be two classes of ribbons, and that one of them probably originates from the protofilaments of the A-tubule nearest the inner B-tubule attachment site. Such a location is particularly interesting, since this ribbon segment of the A-tubule makes contact with the B-tubule, it is associated with the 'adlumenal component' of the A-tubule, and depending on how far the ribbon compartment extends circumferentially, it might include the attachment sites for the row of radial spokes (Linck, 1976; Linck & Langevin, 1982). The tektins themselves may form thin fibrils between tubulin protofilaments on the inside or outside surface of the tubule wall, or the tektins may actually form unique protofilaments of the microtubule wall (Linck *et al.*, 1985). Whatever the case, by immuno-EM the antigenic sites of the tektins are masked in the tubule or in the ribbons, such that they do not label with antibodies, unless the microtubule structure is partially disrupted.

Possible function of tektins in flagellar microtubule assembly

There are numerous possible functions of the tektins in flagellar microtubules. For example, tektins could be important in the assembly, stability, length determination, and 3-dimensional architecture of flagellar microtubules. The evidence for their role is limited but intriguing.

Although the precise arrangement of the tektins in the microtubule is not known, the best available evidence suggests that they are organized as filaments running parallel with the microtubule protofilaments (Linck *et al.*, 1985). Considering this model, it is particularly interesting that the tektin filaments have been reported to have high (71%) α-helical content as measured by circular dichroism (Linck & Langevin, 1982), that they yield strong α-type X-ray patterns (Beese, 1984), and that they show evidence for an axial periodicity of 48 nm (Amos *et al.*, 1986), which is in turn a characteristic repeat of the tubulin subunit lattice (Linck & Langevin, 1982; Amos *et al.*, 1986). Proposed models (Linck, 1982; Linck & Langevin, 1982; Amos *et al.*, 1986) have indicated that a fundamental period of 48 nm (or 96 nm) would allow the tektin filament to make lateral bonds with tubulin subunits in a regular way along the tektin filament–tubulin protofilament axis. If such a model were correct, the tektin–tubulin interactions could be important in microtubule assembly and could also help to stabilize initially or finally the flagellar microtubule. It is relevant to point out that the flagellar A-tubules are the most stable class of microtubules (Stephens, 1970).

One of the polypeptide components ('Component 20') of the sarkosyl-insoluble ribbons was reported to have a molecular weight of ~ 55 kD (Linck, 1976). Stephens (1977) subsequently showed in experiments on ciliary regeneration in sea urchin blastulae embryos, that the ~ 55 kD component of the sarkosyl protofilament ribbons was synthesized de novo in a limited amount sufficient for one round of ciliary growth (i.e., ~ 6 μm of ciliary length); on each subsequent regeneration, an additional amount of this component was again synthesized. Based on the characterization of the 55 kD polypeptide as a protofilament ribbon component (Linck, 1976) and on his de novo quantal synthesis data, Stephens proposed that the 55 kD polypeptide could be a length-determining component of the microtubule. Stephens' hypothesis is particularly interesting now that it is known that the tektins are present in sea urchin blastula cilia (Amos *et al.*, 1985; Linck *et al.*, 1987) and that they seem to be associated as filaments that extend the length of flagellar microtubules.

A final possible function of the tektins in microtubules is to specify 3-dimensional information for assembly of the various axonemal accessory structures (Amos *et al.*, 1976, 1985; Linck *et al.*, 1987). Again, this concept is based on our model wherein tektin filaments extend the length of the microtubule and are associated regularly with tubulin subunits. In such a model, the actual positions of the tektin filaments around the microtubule wall and their axial periodicities could specify the site of assembly of the B-subfiber and the complex attachment sites of the dynein arms (24 nm axial repeats), radial spokes (24, 36 and 48 nm), and the nexin filament linkages (96 nm).

Tektins as targets for the control of male fertility

From the above discussion it seems reasonable to speculate that the tektins may be central to many aspects of sperm flagellar microtubule assembly and function and, therefore, could be important targets for the control of male fertility. Pathological conditions or pharmacological drugs that might alter the gene expression or synthesis of these proteins might be expected in turn to alter the assembly or motility of sperm flagella. For example, in human Kartagener's syndrome and various other mutant forms of ciliary dyskinesis in humans and primitive organisms (i.e., *Chlamydomonas* or *Drosophila*), flagellar axonemes lack dynein arms or radial spokes or show disoriented arrangements of the basic 9 + 2 axoneme pattern (Eliasson *et al.*, 1977; Huang *et al.*, 1981; Raff & Fuller, 1984; Sturgess *et al.*, 1986). The molecular basis for these alterations is not known, but it is possible in some of these cases that the basis for the axoneme defect is an aberrant tektin which determines the placement of a particular axoneme component. Nothing is known about pharmacological agents that might affect tektin synthesis or assembly, although possible clues emerge from the intermediate filament-like nature of the tektins. Several chemical agents have been reported to alter IF protein synthesis or assembly in other cell systems, for example, acrylamide (Eckert & Yaegle, 1986) and β,β'-iminodiproprionitrile (Papasozomenos, 1986). While these substances are toxic, carcinogenic or neuropathological, the observations nonetheless suggest that drugs might be found that are nontoxic and that have desired, specific effects on inhibiting male fertility by interfering with the expression or assembly of flagellar microtubule proteins, including tektins. If tektins are truly a fundamental component of flagellar microtubules, then screening for the effects of drugs on flagellar development and motility in protozoa and invertebrates might provide a rapid and economical way to develop nontoxic antifertility agents that could then be tested in mammalian systems.

Acknowledgements

One of us (R.W.L.) is pleased to acknowledge collaborations on parts of this work with Drs L. A. Amos, W. B. Amos and R. E. Stephens. Funds for this research were provided by U.S.P.H.S. grant GM 35648 from the National Institute of General Medical Sciences.

References

Amos, L.A. (1979). Structure of microtubules. In: Roberts, K. & Hyams, J.S., eds. **Microtubules**. New York, Academic Press Inc. pp. 1–64.

Amos, L.A. & Klug, A. (1974). Arrangement of subunits in flagellar microtubules. **Journal of Cell Science** 14:523–549.

Amos, L.A., Linck, R.W. & Klug, A. (1976). Molecular structure of flagellar microtubules. In: Goldman, R., Pollard, T. & Rosenbaum, J., eds. Cell Motility. New York, Cold Spring Harbor Laboratory. pp 847–867.

Amos, W.B., Amos, L.A. & Linck, R.W. (1985). Proteins closely similar to flagellar tektins are detected in cilia but not in cytoplasmic microtubules. **Cell Motility** 5:239–249.

Amos, W.B., Amos, L.A. & Linck, R.W.(1986). Studies of tektin filaments from flagellar microtubules by immuno-electron microscopy. **Journal of Cell Science**. Suppl. 5:55–68.

Bardin, C.W. & Sherins, R.J., eds. (1982). Cell biology of the testis. **Annals of the New York Academy of Sciences**, Vol. 383.

Beese, L.(1984). Microtubule structure: an analysis by X-ray diffraction. Ph.D. Thesis. Brandeis University, Waltham, MA.

Binder, L.I. & Rosenbaum, J.L. (1978). The in vitro assembly of flagellar outer doublet tubulin. **Journal of Cell Biology** 79:500–515.

Bryan, J. & Wilson, L. (1971). Are cytoplasmic microtubules heteropolymers? **Proceedings of the National Academy Sciences** U.S.A. 68:1762–1766.

Cleveland, D.W., Spiegelman, B.M. & Kirschner, M.W. (1979). Conservation of microtubule associated proteins. **Journal of Biological Chemistry** 254:12670–12678.

Dustin, P. (1984). Microtubules. Springer-Verlag, Heidelberg.

Eckert, B.S. & Yaegle, P.L. (1986). NMR investigation of acrylamide-induced changes in intermediate filament organization. **Journal of Cell Biology** 103:422a.

Eliasson, R., Mossberg, B., Camner, P. & Afzelius, B.A. (1977). The immotile-cilia syndrome. A congenital ciliary abnormality as an etiologic factor in chronic airway infections and male sterility. **New England Journal of Medicine** 297:1–6.

Farrel, K.W. & Wilson, L. (1978). Microtubule reassembly in vitro of *Strongylocentrotus purpuratus* sperm tail outer doublet tubulin. **Journal of Molecular Biology** 121:393–410.

Fawcett, D.W. & Bedford, J.M., eds. (1979). **The Spermatozoon. Maturation, Motility, Surface Properties and Comparative Aspects**. Urban & Schwarzenberg, Baltimore.

Gibbons, B.H. & Gibbons, I.R. (1972). Flagellar movement and adenosine triphosphatase activity in sea urchin sperm extracted with Triton X-100. **Journal of Cell Biology** 54:75–97.

Gibbons, I.R. (1981). Cilia and flagella of eukaryotes. **Journal of Cell Biology** 91:107s–124s.

Huang, B., Piperno, G., Ramanis, A. & Luck, D. (1981). Radial spokes of Chlamydomonas flagella: genetic analysis of assembly and function. **Journal of Cell Biology** 88:80–88.

Johnson, K.A. (1985). Pathway of the microtubule-dynein ATPase and the structure of dynein: A comparison with actomyosin. **Annual Reviews of Biophysics and Biophysical Chemistry** 14:161–188.

Kuriyama, R. (1976). In vitro polymerization of flagellar and ciliary outer fiber tubulin into microtubules. **Journal of Biochemistry (Tokyo)** 80:153–165.

Laemmli, U.K. (1970). Cleavage of structural proteins during the assembly of the head of bacteriophage T4. **Nature, Lond**. 227:680–685.

Linck, R.W. (1973). Chemical and structural differences between cilia and flagella from the lamellibranch mollusc, Aequipecten irradians. **Journal of Cell Science** 12:951–981.

Linck, R.W. (1976). Flagellar doublet microtubules: fractionation of minor components and α-tubulin from specific regions of the A-tubule. **Journal of Cell Science** 20:405–439.

Linck, R.W. (1982). The structure of microtubules. **Annals of the New York Academy of Sciences** 383:98–121.

Linck, R.W., Amos, L.A. & Amos, W.B. (1985). Localization of tektin filaments in microtubules of sea urchin sperm flagella by immunoelectron microscopy. **Journal of Cell Biology** 100:126–135.

Linck, R.W., Goggin, M.J., Norrander, J.M. & Steffin, W. (1987). Characterization of antibodies as probes for structural and biochemical studies of tektins from ciliary and flagellar microtubules. **Journal of Cell Science** 88:453–466.

Linck, R.W. & Langevin, G.L. (1981). Reassembly of flagellar B(β) tubulin into singlet microtubules: consequences for cytoplasmic microtubule structure and assembly. **Journal of Cell Biology** 89:323–337.

Linck, R.W. & Langevin, G.L. (1982). Structure and chemical composition of insoluble filamentous components of sperm flagellar microtubules. **Journal of Cell Science** 58:1–22.

Linck, R.W. & Stephens, R.E. (1987). Biochemical characterization of tektins from sperm flagellar doublet microtubules. **Journal of Cell Biology** 104:in press.

Luduena, R.F., Shooter, E.M. & Wilson, L. (1977). Structure of the tubulin dimer. **Journal of Biological Chemistry** 252:7006–7014.

Meza, I., Huang, B. & Bryan, J. (1972). Chemical heterogeneity of protofilaments forming the outer doublets from sea urchin flagella. **Experimental Cell Research** 74:535–540.

Papasozomenos, S.Ch. (1986). Reorganization of axonal cytoskeleton following β,β'-iminodipropionitrile (IDPN) intoxication. In: T.W. Clarkson, P.R. Sager, & T.L.M. Syverson, eds. **The Cytoskeleton: A Target for Toxic Agents**. New York, Plenum. pp. 67–82.

Raff, E.C. & Fuller, M.T. (1984). Genetic analysis of microtubule function in Drosophila. In: **Molecular Biology of the Cytoskeleton**. G. G. Borisy, D.W. Cleveland & D.B. Murphy, eds. Cold Spring Harbor Laboratory, New York. p. 293–304.

Sloboda, R.D., Rudolph. S.A., Rosenbaum, J.L. & Greengard, P. (1975). **Proceedings of the National Academy of Sciences U.S.A.** 72:177–181.

Soifer, D., ed. (1986). Dynamic aspects of microtubule biology. **Annals of the New York Academy of Sciences** 466:950–978.

Steffen, W. & Linck, R.W. (1988). Evidence for tektins in centrioles and axonemal microtubules. **Proceedings of the National Academy of Sciences U.S.A.** 85:2643–2647.

Stephens, R.E. (1970). Thermal fractionation of outer doublet microtubules into A-and B-subfiber components: A- and B-tubulin. **Journal of Molecular Biology** 47:353–363.

Stephens, R.E. (1977). Differential protien synthesis and utilization during cilia formation in sea urchin embryos. **Developmental Biology** 61:311–329.

Stephens, R.E. & Edds, K.T. (1976). Microtubules: structure, chemistry, and function. **Physiological Reviews** 56:709–777.

Sturgess, J.M., Thompson, M.W., Czegledy-Nagy, E. & Turner, J. (1986). Genetic aspects of immotile cilia syndrome. **American Journal of Medical Genetics** 25:149–160.

Weber, K. & Geisler, N. (1985). Intermediate filaments: structural conservation and divergence. **Annals of the New York Academy of Sciences** 455:126–143.

Weisenberg, R.C. (1972). Microtubule formation *in vitro* in solutions containing low calcium concentrations. **Science** 177:1104–1105.

Witman, G.B., Carlson, N.K., Berliner, J. & Rosenbaum, J. (1972a). *Chlamydomonas* flagella. I. Isolation and electrophoretic analysis of microtubules, matrix, membranes, and mastigonemes. **Journal of Cell Biology** 54:507–539.

Witman, G.B., Carlson, K. & Rosenbaum, J. (1972b). *Chlamydomonas* flagella. II. The distribution of tubulins 1 and 2 in the outer doublet microtubules. **Journal of Cell Biology** 54:540–555.

Synthesis and assembly of flagellar proteins during flagellar regeneration in *Chlamydomonas*

JOEL L. ROSENBAUM

Department of Biology, Yale University, New Haven, CT 06520, USA.

Abstract

Chlamydomonas sp. can be deflagellated by chemical and mechanical means and the flagellae will regenerate. This has been used as a model system to study the regulation of genes encoding flagellar constituents, particularly tubulin genes. Control of tubulin genes and assembly into microtubules is complex and may be regulated by tubulin itself. Acetylation of alpha-1-tubulin occurs at the time of tubulin assembly in flagellae and is reversed when flagellae are degraded. Mutants have proved to be useful in studying flagellar assembly and disassembly in *Chlamydomonas*. Future work will involve transfection of immotile mutants with well-characterized wild type genes to study the roles of specific genes and gene sets in flagellar function.

Keywords

Flagellum, tubulin, *Chlamydomomas*, alpha-tubulin, assembly, disassembly.

Introduction

When the flagellae of the bi-flagellate alga *Chlamydomonas* are detached by chemical or mechanical means, two new 12 μm flagellae regenerate within about 70 min. If only one of the two flagellae is detached, the intact flagellum shortens while the other elongates; when both reach the same intermediate length they both grow out together (see Lefebvre & Rosenbaum, 1986). The flagellae, which are composed of over 150 different polypeptides including the alpha and beta tubulins, can be isolated at different stages in their regeneration and fractionated into their component parts. The flagellae have been shown to assemble new subunits at their distal tips even though the distal ends of the A-fiber of the outer doublet and the central tubules have 'caps' on their ends which attach the distal ends to the tip of the flagellar membrane.

Other studies have shown that particles attached to the flagellar membrane are moved rapidly up and down the length of the flagellum at speeds approaching 10 μm/s. The movement requires energy and Ca^{2+} and is probably related to the structures which can be observed to attach the membrane to the outer portion of the outer doublet microtubules of the axonemes – a locus which has been shown in EM histochemical studies to have ATPase activity. The function of this membrane/axoneme motility is unknown.

Use of mutants to study flagellar development

Studies on flagellar development and motility in *Chlamydomonas* are aided by the availability of many mutants defective in either motility or development (see Lefebvre & Rosenbaum, 1986). A large number of motility mutants have been obtained and the structural defects have been determined; there are also many developmental mutants including cells with extra-long, extra-short or no flagellae. These mutants have been key to determining the molecular mechanisms underlying flagellar movement and development, and *Chlamydomonas* is one of the few systems available in which genetics can be utilized in such studies of the eukaryotic flagellum.

Expression of tubulin genes

Detachment of the flagellum rapidly induces the synthesis of tubulin and other flagellar mRNAs which reach their peak within 15–30 min after flagellar removal, and then quickly recede to the constitutive levels present prior to deflagellation (Baker *et al.*, 1986). The tubulin mRNAs can accumulate to 30–40% of their pre-deflagellation levels within minutes of deflagellation. This new synthesis of flagellar mRNAs does not require protein synthesis, so all the factors necessary for up-regulation of the flagellar genes are already present prior to deflagellation. However, the degradation of the accumulated flagellar mRNAs to their constitutive levels does require protein synthesis, and some of the evidence suggests that the proteins required for decreasing the amount of tubulin mRNAs may be the tubulins themselves (Baker *et al.*, 1986; Ben Ze'ev *et al.*, 1979; Cleveland *et al.*, 1981, 1983). Current studies are aimed not only at determining the nature of those DNA flanking sequences which interact with proteins to up-regulate the tubulin genes following deflagellation, but also at the mechanism by which the accumulated flagellar mRNAs are specifically degraded following flagellar regrowth. Recent results suggest that the localization of the induced mRNAs – free, or on polysomes – may be important in determining whether or not they are degraded.

Chlamydomonas has two genes for alpha-tubulin and two for beta-tubulin (Silflow & Rosenbaum, 1981; Brunke *et al.*, 1982). The two genes for beta-tubulin, although making mRNAs which are slightly different, code for identical beta-tubulins. The two alpha-tubulin genes code for alpha-tubulins differing in two amino acids – one of which is charged. When the flagellae are detached, all four genes transcribe tubulin mRNAs. However, even though two alpha-tubulin mRNAs are synthesized, only one, alpha-1, making alpha-1 tubulin is translated. The other mRNA is not used, *in vivo*, or *in vitro*, when the total mRNA population is translated in the reticulocyte system. However, both mRNAs can be shown to be present on *Chlamydomonas* polysomes. Thus, there is a translational selection of alpha-tubulin mRNAs, probably at the level of chain elongation on termination.

Acetylation of tubulin

Once alpha-tubulin is synthesized on the polysomes, it is transported to the tip of the flagellum as alpha-1 tubulin. At this point, either just before or during tubulin assembly, the alpha-1 tubulin is acetylated on the epsilon amino group of one of its lysines (Greer *et al.*, 1985; Brunke *et al.*, 1982; L'Hernault & Rosenbaum, 1983, 1985b; Maruta *et al.*, 1986). Work by others (e.g., Piperno & Fuller, 1985) has shown that the alpha-tubulin of all flagellae, regardless of where they are from, are acetylated. When the flagellae are induced to shorten by experimental means, the acetyl group is removed (L'Hernault & Rosenbaum, 1985a). We have isolated and characterized the tubulin acetyl transferase from *Chlamydomonas* flagellae (Maruta *et al.*, 1986). The function of the acetylation of alpha-tubulin is as yet unknown. In addition to the microtubules of basal bodies, a small subset of cytoplasmic microtubules, usually emanating from the basal bodies, are also acetylated.

Transformation of paralyzed mutants

We are endeavoring to transform paralyzed (non-motile) *Chlamydomonas* flagellar mutants back to the wild-type (motile) state by use of wild-type genes. These studies are directed toward rescue of paralyzed 'radial link' mutants by the use of well-defined radial link genes (Williams *et al.*, 1986). Similar studies are being carried out with dynein genes and paralyzed dynein mutants. Ultimately, this transformation system will be essential in defining the molecular mechanisms underlying both flagellar growth and motility.

References

Baker, E.J., Keller, L.R., Schloss, J.A. & Rosenbaum, J.L. (1986). Protein synthesis is required for rapid degradation of tubulin mRNA and other deflagellation-induced RNAs in *Chlamydomonas reinhardi*. **Molecular and Cellular Biology** 6:54–61.

Ben Ze'ev, A., Farmer, S.R. & Penmans, O. (1979). Mechanisms of regulating tubulin synthesis in cultured mamalian cells. **Cell** 17:319–325.

Brunke, K.J., Collis, P.S. & Weeks, D.P. (1982). Post-translational modification of tubulin dependent on organelle assembly. **Nature** 297:516–518.

Cleveland, D.W., Lopata, M.A., Sherline, P. & Kirschner, M.W. (1981). Unpolymerized tubulin modulates the level of tubulin mRNAs. **Cell** 25:537–546.

Cleveland, D.W., Pittenger, M.F. & Feramisco, J.R. (1983). Elevation of tubulin levels by microinjection suppresses new tubulin synthesis. **Nature** 305:738–740.
Greer, K., Maruta, H., L'Hernault, S.W. & Rosenbaum, J.L. (1985). Alpha-tubulin acetylase activity in isolated *Chlamydomonas* flagella. **Journal of Cell Biology** 101:2081–2084.
L'Hernault, S.W. & Rosenbaum, J.L. (1983). *Chlamydomonas* alpha-tubulin is post-translationally modified in the flagella during flagellar assembly. **Journal of Cell Biology** 97:258–263.
L'Hernault, S.W. & Rosenbaum, J.L. (1985a). Reversal of the post-translational modification on *Chlamydomonas* alpha-tubulin occurs during flagellar resorption. **Journal of Cell Biology** 100:457–462.
L'Hernault, S.W. & Rosenbaum, J.L. (1985b). *Chlamydomonas* alpha-tubulin is post-translationally modified by acetylation on the epsilon-amino acid group of a lysine. **Biochemistry** 24:473–478.
Lefebvre, P.A. & Rosenbaum, J. (1986). Regulation of the synthesis and assembly of ciliary and flagellar proteins during regeneration. **Annual Reviews of Cell Biology** 2:517–546.
Maruta, M., Greer, K. & Rosenbaum, J.L. (1986). The acetylation of alpha-tubulin and its relationship to the assembly and disassembly of microtubules. **Journal of Cell Biology** 103:571–579.
Piperno, G. & Fuller, M.T. (1985). Monoclonal antibodies specific for an acetylated form of alpha-tubulin recognize the antigen in cilia and flagella from a variety of organisms. **Journal of Cell Biology** 101:2085–2094.
Silflow, C.D. & Rosenbaum, J.L. (1981). Multiple α- and β-tubulin genes in *Chlamydomonas* and regulation of tubulin mRNA levels after deflagellation. **Cell** 24:81–88.
Williams, B.D., Mitchell, D.R. & Rosenbaum, J.L. (1986). Molecular cloning and expression of flagellar radial spoke and dynein genes of *Chlamydomonas*. **Journal of Cell Biology** 103:1–11.

IV Cellular and molecular events in spermiogenesis

Cell interactions in the seminiferous tubule: an overview of the roles of Sertoli cell–peritubular cell interactions in testicular functions

IRVING B. FRITZ AND PIERRE S. TUNG

Banting and Best Department of Medical Research, University of Toronto, Toronto, M5G 1L6 Canada.

Abstract

The nature of the interactions between Sertoli cells and peritubular cells is reviewed. During co-culture, the two populations of cells act co-operatively, eventually generating a tubule-like structure. The restructuring events are in part dependent on direct cell contacts, upon an intact cytoskeletal network, and upon interactions with specific extracellular matrix (ECM) components, particularly laminin. Other types of interactions reviewed include paracrine influences, and those involved in the regulation of proteolysis. Peritubular cells secrete protein(s), under androgen modulation, which enhance the production of androgen binding protein and transferrin by Sertoli cells in primary culture. In addition, peritubular cells secrete a specific inhibitor of plasminogen activator activity which may serve to limit the proteolysis elicited by plasminogen activator produced by Sertoli cells.

The possible relevances of these *in vitro* observations to tubulogenesis, to the maintenance of the cytoarchitecture of the seminiferous tubule at various stages of spermatogenesis, and to the role of the basement membrane in these processes are discussed. The viewpoint is presented that specific stages of spermatogenesis are dependent upon unique arrangements within the tubule generated by continuing sets of complex interactions between Sertoli cells and peritubular cells, and upon interactions of these cells with ECM components in the basement membrane. The model is developed that plasmalemmal interactions between closely apposing surfaces of Sertoli cells and peritubular cells initiate basement membrane formation in association with cytoskeletal elements in a highly organized array.

These observations and viewpoints are discussed within the context that the endocrine control of spermatogenesis is mediated *via* direct actions of hormones on somatic cells of the seminiferous tubule. The unique chemical and physical microenvironment created within the seminiferous tubule by Sertoli cells, peritubular cells and their interactions in response to various signals is postulated to be essential for optimal germinal cell development.

Key words

Spermatogenesis, Sertoli cells, peritubular myoid cells, seminiferous tubule, basement membrane formation, testis cell interactions, testis proteases and antiproteases, plasminogen activator, hormonal controls of spermatogenesis, extracellular matrix, laminin, Sertoli cell and peritubular cell cytoskeleton.

Introduction

We have reviewed previously some of the properties of somatic cells in the seminiferous tubule, and have developed the concept that the hormonal controls of spermatogenesis are mediated *via* direct actions on Sertoli cells and peritubular cells (Fritz, 1978). In further examination of possible mechanisms by which gonadal somatic cells generate the microenvironment within the seminiferous tubule required for germinal cell development, investigations of the nature of interactions between Sertoli cells and peritubular cells have continued. These two cell types, which comprise the major somatic cells in the seminiferous tubules of mammals, interact during co-culture to favor mutual survival and maintenance of optimal functions (for recent reviews, see Fritz, 1984; Fritz & Tung, 1986; Fritz *et al.*, 1986).

Several kinds of co-operativity and interactions have been detected:

(1) A morphogenetic cascade takes place when Sertoli cells are plated on top of confluent layer(s) of peritubular cells, resulting in the formation of a tubule-like structure. The key events during this remodelling consist of the following:

(a) Sertoli cells spread to form a monolayer of plaque-like colonies upon the free surfaces of underlying peritubular cells;
(b) Sertoli cells reaggregate, under the influence of invading ribbons of peritubular cells, become compacted, and then form macroscopically visible mounds;
(c) the two cell types act co-operatively to generate a highly organized basal lamina and associated lamellae which separate Sertoli cells from peritubular cells; and
(d) nodules form which are morphologically similar to germ cell-depleted seminiferous tubules (Tung & Fritz, 1980, 1987; Fritz & Tung, 1986).

(2) During co-culture, peritubular cells and Sertoli cells secrete various ECM components which are required for the subsequent formation of an organized basement membrane structure. If the two cell populations are not in direct contact with each other during co-culture, an organized lamellar structure fails to form (Tung & Fritz, 1987). In monoculture, Sertoli cells synthesize and release collagen

type IV, laminin and unique proteoglycans into the medium, but not fibronectin or collagen type I. In contrast, peritubular cells in monoculture secrete collagens types I and IV, fibronectin, and other classes of proteoglycans into the medium, but not laminin. Basement membrane formation occurs only during co-culture. Most of the ECM components released during monoculture remain in soluble form (Tung *et al.*, 1984; Skinner *et al.*, 1985; Skinner & Fritz, 1985c).

(3) Survival and optimal functions of peritubular cells and Sertoli cells are limited in monocultures maintained in serum-free medium. In contrast, during co-culture in otherwise simple medium, Sertoli cells continue to secrete androgen binding protein (ABP) at relatively high levels for weeks, and both cell types survive for months (Tung & Fritz, 1980; Hutson & Stocco, 1981; Cameron & Markwald, 1981).

(4) Primary cultures of peritubular cells, under androgenic modulation, synthesize and release a protein into the medium which greatly enhances the formation of ABP and transferrin by Sertoli cells in culture. This protein has been called 'P-Mod-S' to designate the presence of a component secreted by peritubular cells which modulates the functional activities of Sertoli cells (Skinner & Fritz, 1985a, b, 1986). Other identified components secreted by peritubular cells in culture include a plasminogen activator inhibitor (Hettle *et al.*, 1986b); a mitogenic factor which cross-reacts with antibodies against IGF-1 (Skinner & Fritz, 1986); and a lactalbumin-like protein (Skinner & Fritz, 1986).

(5) Perturbations of the cytoskeleton, elicited by agents which disperse the microfilament network, profoundly influence interactions between Sertoli cells and peritubular cells in co-culture. When added at varying times after plating Sertoli cells upon peritubular cells, dibutyryl cyclic AMP (dbcAMP) inhibits each successive stage of the morphogenetic cascade described above (Tung & Fritz, 1987). Addition either of the phosphodiesterase inhibitor methylisobutyl xanthine (MIX), or of cytochalasin D, elicits the same effects as dbcAMP, namely blockage of Sertoli cell migration and peritubular cell migration; formation of mounds of reaggregated Sertoli cells; and formation of basement membrane in a tubule-like structure. Data are consistent with the conclusion that agents which elevate cellular cAMP levels (dbcAMP or MIX), or which directly disrupt microfilaments (cytochalasin D), prevent each stage of remodelling which normally takes place when Sertoli cells are plated on top of peritubular cells (Tung & Fritz, 1987).

Observations reviewed demonstrate that Sertoli cells interact with peritubular cells during co-culture in complex manners involving direct cell contacts; contacts mediated *via* ECM components; and *via* diffusible molecules secreted by each cell type. This overview will consider the hypothesis that Sertoli cells and peritubular cells form a functional unit required for the maintenance of germinal cell development, and will evaluate the possible *in vivo* relevance of interactions described above. Emphasis is placed on work from the authors' laboratory. A broader coverage of the controls of spermatogenesis can be found elsewhere (Fritz, 1984; Fritz & Tung, 1986). General aspects of cell-matrix interactions, and to the role of cell adhesion during development, have been reviewed by Ekblom *et al.* (1986).

Perturbations of seminiferous tubule basement membrane structures are correlated with disturbances of spermatogenesis

A thickening of the basement membrane, together with disorganization of ECM structures, is associated with many pathological conditions of the testis, including several forms of testicular involution (Gravis, 1978; for review, see Fritz & Tung, 1986). In testes of testicular feminized mice (^{tfm}X/Y), the basement membrane is greatly thickened and convoluted (Fig. lB). This structure should be compared with the relatively thin basement membrane which normally separates peritubular cells from Sertoli cells in testes from wild-type sibling mice (Fig. lA). Spermatogenesis is impaired in tfm mutants, with few if any detectable spermatids. However, tfm testes are cryptorchid, and this condition alone can result in distortion and thickening of the tubule boundary layers (Kerr *et al.*, 1979). Thus, the abnormal basement membrane structure evident in tubules of tfm testes may not result primarily from a disturbance of spermatogenesis, but could be a consequence of the elevated temperature in the abdominal testis. From data reviewed above, co-operativity between peritubular cells and Sertoli cells is known to be required for basement membrane formation *in vitro* (Skinner *et al.*, 1985; Tung & Fritz, 1987). We therefore postulate that the thickening of the basement membrane evident in testes of tfm mice (Fig. 1A, B) demonstrates a disturbance of Sertoli cell–peritubular cell interactions in these tubules.

The organization of the basal lamellae may be directly perturbed experimentally by intratesticular injections of antibodies against specific ECM components or against whole testis basement membrane preparations. In the latter case, this procedure results in multifocal defoliation of germinal cells, and changes in the structure of peritubular cells (Lustig *et al*,, 1982; Denduchis *et al.*, 1985). We have extended these experiments by injecting monospecific antibodies against laminin and collagen type IV, respectively, into testes of adult rats. The anti-laminin treatment increased the permeability of the blood–testis barrier, as measured by the appearance of immunocytochemically detectable rat immunoglobulin in the adluminal

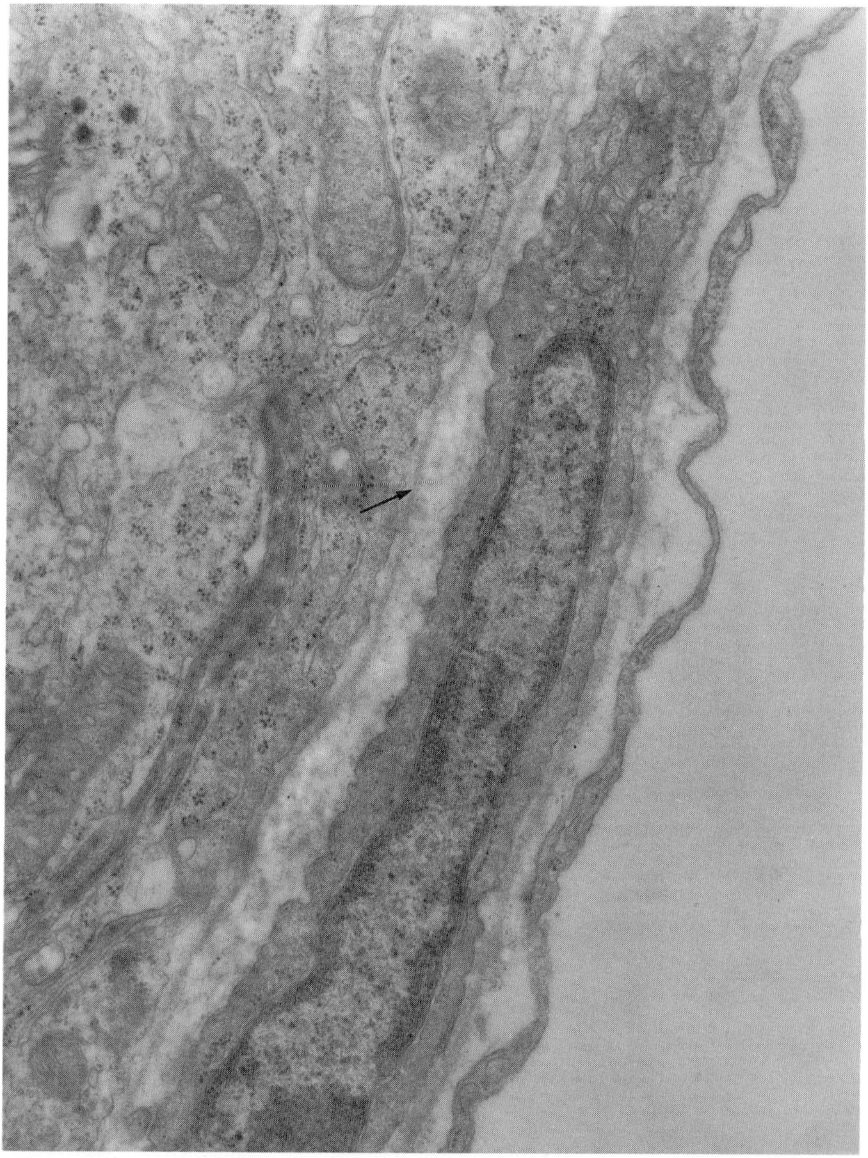

Fig. 1 Transmission electron micrographs of sections of testes from adult testicular feminized mice or ta/y wild-type littermates. Note that the basement membrane (arrows) is thick and convoluted in tfm/y testis tubule (panel B), while that in wild-type litter mate testis is thinner and more even (panel A). Testes were fixed, prepared and sectioned by conventional procedures, and photographs were taken with a Hitachi HU 11E transmission electron microscope. A,B × 26 000 approx.

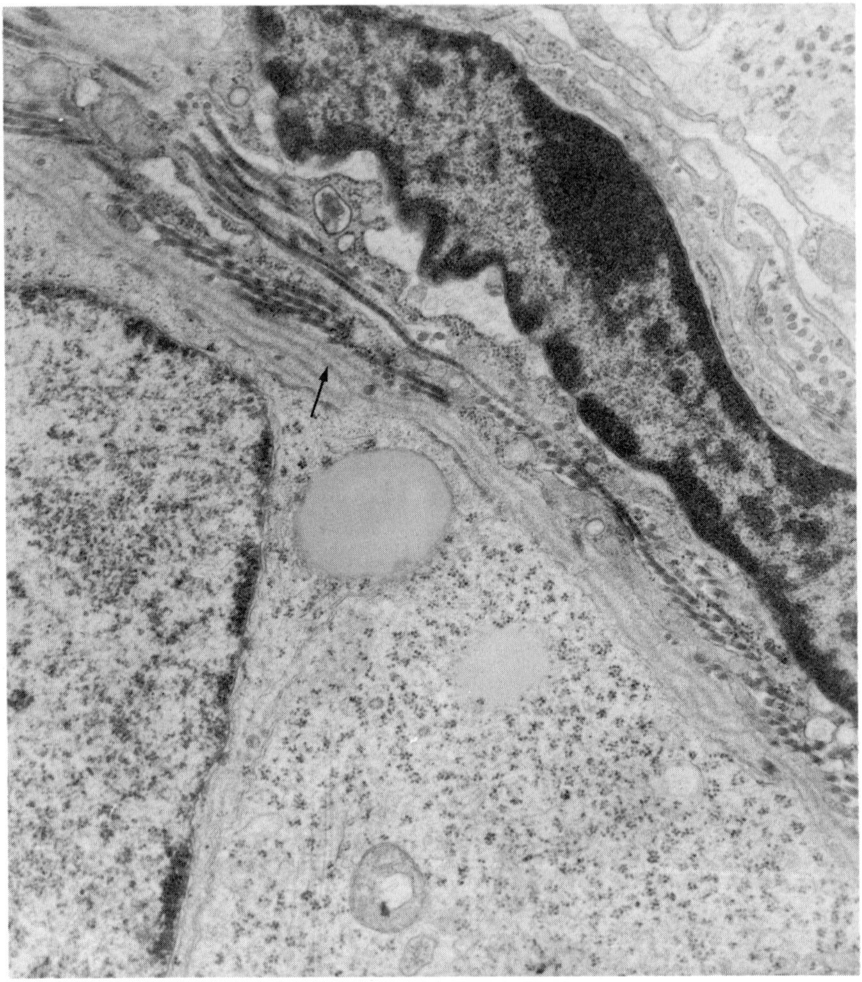

compartment, and it also elicited defoliation of germ cells. In contrast, treatment with the collagen IV antibody did not result in detectable changes (P.S. Tung & I.B Fritz, unpublished observations).

Antibodies against laminin block the migration of Sertoli cells plated on glass, plastic, collagen type I, or on Matrigel. In contrast, addition of antibodies against fibronectin do not block the migration of Sertoli cells on any of these substrates (Table 1). However, addition of anti-fibronectin antibody does prevent migration of Sertoli cells upon the free surfaces of peritubular cells plated on type I collagen, a process called 'cable formation' (Tung & Fritz, 1986a). We interpret these observations to indicate that fibronectin synthesized by peritubular cells, and present on their surfaces, permits Sertoli cells to form attachments, mediated presumably by specific receptors to fibronectin on Sertoli cell plasma membranes. Since Sertoli

Table 1. *Effects of antibodies against ECM components on the migration of purified populations of Sertoli cells plated on different substrata.*

	Substratum	Antibody added to basal medium	Migration pattern of Sertoli cells*
(1)	Uncoated plastic or glass	None, or anti-fibronectin	Plaque-like monolayer formation ~24 h
(2)	Plastic or glass coated with collagen type I, or Matrigel	None, or anti-fibronectin	Cells remain in aggregates for 4–6 days; then spread randomly to form a monolayer
(3)	Plastic or glass coated with collagen type I or Matrigel	Anti-laminin	Migration of cells is inhibited; cells remain in aggregates for longer periods before beginning to spread randomly
(4)	Layers of peritubular cells in collagen type I	None	Plaque-like monolayer formation followed by reaggregation, compaction and tubulogenesis
(5)	Layers of peritubular cells on collagen type I	Anti-fibronectin	Inhibition of migration, reaggregation, compaction and tubulogenesis

*For details, consult original publications (Tung & Fritz, 1980, 1984, 1986a, 1987).

cells do not synthesize fibronectin, antibodies against this ECM component would not be expected to influence the migration of Sertoli cells in monoculture on substrates which do not contain fibronectin (Table 1).

Under conditions cited, peritubular cells have the capacity to direct the migration of Sertoli cells along specific pathways. Thus, Sertoli cells attach preferentially to the surfaces of fibronectin-coated peritubular cells. They do not migrate on type I collagen until laminin or collagen type IV, synthesized by Sertoli cells, are deposited on the substrate (Tung & Fritz, 1986a and Table 1). We believe that this property may be of potential importance in guiding the movements of Sertoli cell precursors during tubulogenesis. If peritubular cell precursors form the initial bridge joining segments of seminiferous cords, it appears plausible that Sertoli cell precursors could migrate on top of this bridge to complete tubule formation, just as they do during cable formation (Tung & Fritz, 1986a; Fritz & Tung, 1986).

Sertoli cell and peritubular cell interactions are mediated in part *via* diffusible molecules

Peritubular cells in culture, under androgenic modulation, secrete components which increase the production of ABP and transferrin by primary cultures of Sertoli cells (Skinner & Fritz, 1985a, b, 1986) (Table 2). These effects are similar to the enhancement of ABP and transferrin production elicited by FSH and testosterone. In contrast, peritubular cell conditioned medium does not increase plasminogen activator secretion by Sertoli cells, even though FSH addition does. Peritubular cells in culture do release a specific inhibitor of plasminogen activator (Hettle *et al.*, 1986b). This inhibitor is cross-reactive with antibodies against the plasminogen activator inhibitor synthesized by bovine vascular endothelial cells (van Mourik *et al.*, 1984), kindly provided by Dr David Loskutoff; and also with antibodies against the plasminogen activator inhibitor synthesized by human fibrosarcoma cells (Andreason *et al.*, 1986), kindly provided by Dr Keld Dano (Hettle, J.O., Balekjian, E. & Fritz, I.B., unpublished observations).

Sertoli cells in culture secrete predominantly urokinase-type plasminogen activator under basal conditions, and predominantly tissue-type plasminogen activator when stimulated by FSH or by dbcAMP (Hettle *et al.*, 1986a). The same phenomena are observed in seminiferous tubules in organ culture when segments at defined stages are incubated in the presence or absence of dbcAMP. FSH addition also stimulates the formation of plasminogen activator of the tissue-type by cells in segments from all stages of the cycle of the seminiferous epithelium, but in cells from adult testes, it is necessary to have an inhibitor of phosphodiesterase present in order to detect the stimulation (Hettle *et al.*, 1986a). In confirmation of previous reports (Lacroix *et al.*, 1981; Fritz & Karmally, 1983; Vihko *et al.*, 1984), the highest rates of plasminogen activator secretion are obtained in segments at stages VII and VIII of the cycle of the seminiferous epithelium. Although addition of FSH resulted in increased production by Sertoli cells in all segments,

Table 2. *Effects of peritubular cell conditioned medium on the secretion of various products by Sertoli cells in culture.*

Additions to basal medium	Levels of protein released		Plasminogen activator activity
	ABP (ng/μg DNA/72 h)	Transferrin (ng/μg DNA/72 h)	(units/μg DNA/72 h)
None	4.5 ± 0.5	58 ± 10	60 ± 50
FIRT*	16.0 ± 4.0	180 ± 20	210 ± 20
PcMEM*	14.5 ± 3.5	200 ± 15	35 ± 5.0

*'FIRT' represents the addition of follicle stimulating hormone, insulin, retinol and testosterone, as described in the original publications (Skinner & Fritz, 1985a; Fritz *et al.*, 1986), while 'PcMEM' represents the partially purified 50–100 kD fraction prepared from 200 ×-concentrated serum-free conditioned medium obtained from secondary cultures of peritubular cells (1.5 μg protein/ml) (Skinner & Fritz, 1985a, 1986; Fritz *et al.*, 1986). PcMEM has subsequently been shown to contain a 55 kD protein which specifically inhibits plasminogen activator (Hettle *et al.*, 1986b).

the highest levels were secreted by cells in stages VII and VIII (Fig. 2; Hettle *et al.*, 1986a).

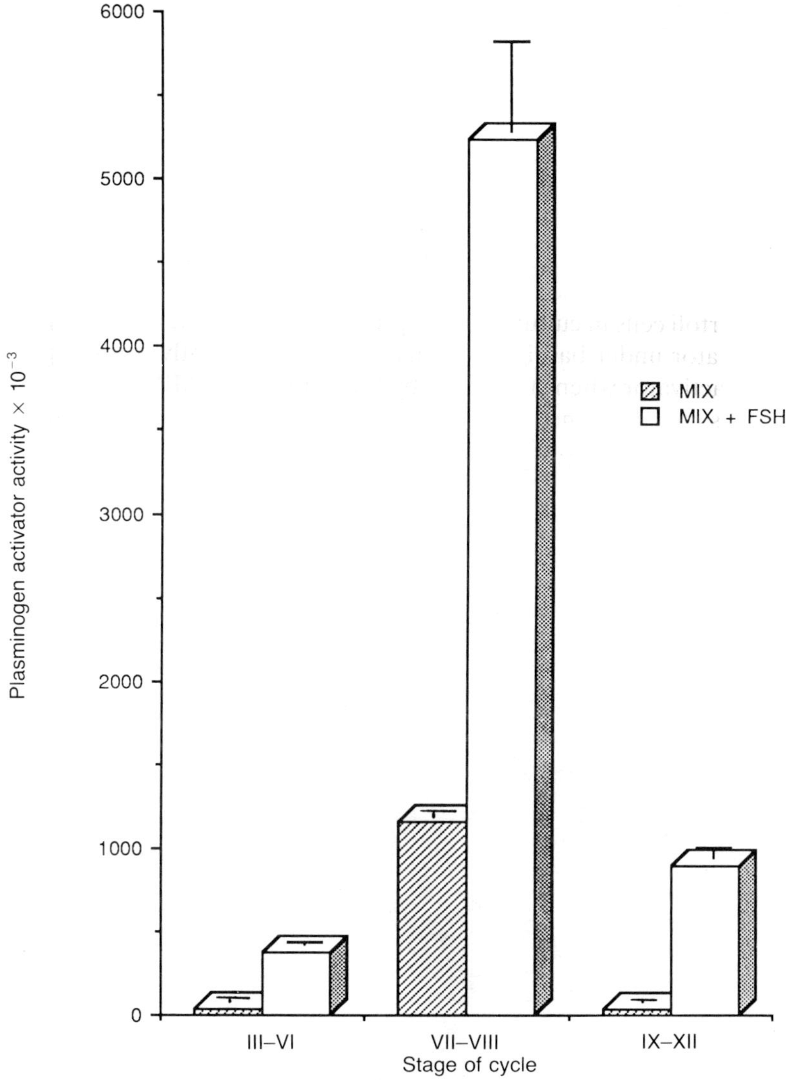

Fig. 2 Stimulation by FSH of plasminogen activator activities secreted by adult rat tubule segments at different stages of the cycle of the seminiferous epithelium. Values represent the mean (plus or minus standard deviation) from 4 determinations. Tubule segments designated were cultured for 24 h in Eagle's minimal essential medium containing 50 μM methylisobutylxanthine (MIX), in the presence or absence of ovine FSH (NIH-S13). Details of incubation and assay procedures are described in the original publication (Hettle *et al.*, 1986a).

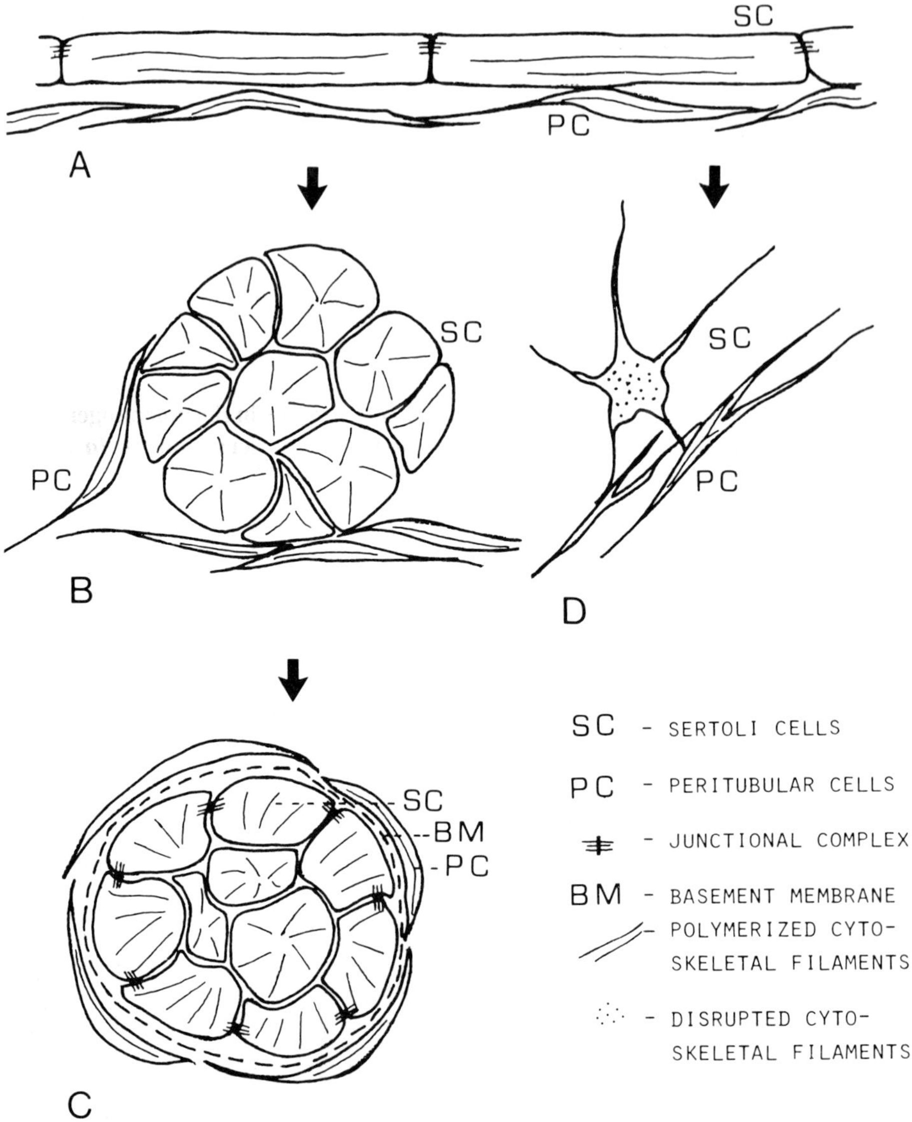

Fig. 3 Schematic representation of morphogenetic cascade during co-culture, after plating of Sertoli cells (SC) on top of pre-existing layers of peritubular cells (PC). Cells depicted refer to those prepared from testes of normal 20-day-old rats, and then maintained in culture under conditions previously described (Tung & Fritz, 1980, 1987). In panels A, B, and C, cells were cultured in the absence of one of the following; dbcAMP, MIX or cytochalasin D, as described in detail elsewhere (Tung & Fritz,1987). Panel A: Sertoli cells spread to form plaque-like colonies shortly after being plated on top of peritubular cells. The Sertoli cells, which have a

We have previously postulated that the plasminogen activator activity produced by Sertoli cells may be implicated in the restructuring in the seminiferous tubule which takes place as clones of preleptotene spermatocytes are separated at stage VII from the basement membrane by the invading cytoplasmic extensions of neighboring Sertoli cells (Fritz & Karmally, 1983; Fritz & Tung, 1986; Hettle *et al.*, 1986a). To evaluate this hypothesis further, the possibility is being explored that the protease inhibitor produced by peritubular cells may be involved in limiting proteolysis to discrete regions of the tubule during the restructuring processes associated with the translocation of clones of spermatocytes across the seminiferous tubule barrier.

The role of cytoskeletal and ECM components in the morphogenetic cascade: possible relationships of *in vitro* observations to *in vivo* behaviour

Models to interpret the interactions which take place between peritubular cells and Sertoli cells during co-culture (Fig. 3), and the possible relevance of these interactions to conditions prevailing *in vivo* are proposed. We

Caption for Fig. 3 (cont.)

flat shape, form tight junctional complexes, oriented at sites apical to their sites of attachment to underlying peritubular cells. The contacts between Sertoli cells and peritubular cells do not closely approximate each other at this early stage of co-culture.
Panel B: Sertoli cells subsequently become less polarized and more rounded. Their plasma membranes more closely approximate the surfaces of adjacent peritubular cells, and contact becomes increasingly intimate as the Sertoli cells become compacted during the reaggregation process. Migrating peritubular cells which are not in intimate contact with Sertoli cells have a fibroblast-like appearance, whereas peritubular cells in intimate contact with neighboring Sertoli cells assume a cobblestone arrangement characteristic of epithelial-type cells in culture. Peritubular cells in close contact with Sertoli cells also have a much lower rate of mitosis (Tung & Fritz, 1986a, c).
Panel C: After a period of about 2 weeks in co-culture, the close plasmalemmal interactions between depolarized Sertoli cells and peritubular cells result in the formation of a basement membrane (BM), which in turn facilitates histotypic expression of both cell types in a manner characteristic of the histotype observed *in situ* (Tung & Fritz, 1984, 1986a, b).
Panel D: The addition of dbcAMP, MIX, or cytochalasin D to the medium during different stages of co-culture prevents close plasmalemmal interactions between Sertoli cells and peritubular cells; blocks basement membrane formation; and precludes normal histotypic expression by either cell type (Tung & Fritz, 1987).

postulate that Sertoli cells plated onto the surfaces of peritubular cells are initially polarized, and remain so during their migration to form a monolayer. When the Sertoli cells reaggregate, however, they become non-polarized. This is manifested by random arrangements of microfilaments in the cytoskeleton. The non-polarized state permits close apposition between plasmalemmal surfaces of Sertoli cells and those of adjacent peritubular cells. Only at these interfaces can plasmalemmal interactions result in the organized deposition of soluble ECM components to form a basement membrane. Although specific ECM components synthesized by Sertoli cells and peritubular cells are present in concentrations sufficient to permit basement membrane formation, organized deposition does not occur except at sites of intimate contact between Sertoli cell plasmalemmal surfaces and peritubular cell plasmalemmal surfaces (Tung & Fritz, 1987). At these points of contact, plasma membranes undergo characteristic structural changes. The resulting plasmalemmal specializations are postulated to be correlated with the beginning of repolarization of the cytoskeletal network.

As described earlier, dbcAMP, MIX, or cytochalasin D addition inhibits each stage of the morphogenetic cascade during co-culture, presumably by eliciting dispersal of the microfilament network (Fig. 3). This cytoskeletal disruption is postulated to prevent migration of Sertoli cells and of peritubular cells; compaction of Sertoli cells; and normal plasmalemmal interactions between Sertoli cells and peritubular cells. Speculative interpretations of the blockage of restructuring elicited by the presence of dbcAMP, MIX, or cytochalasin D have been advanced (Tung & Fritz, 1987). Inhibition of early stages of the morphogenetic cascade by any reagents which cause disruption of the microfilament network of the cytoskeleton would concomitantly result in the prevention of tubulogenesis and organized basement membrane formation during later stages of co-culture. Similar effects may be predicted in co-cultures in which antibodies against laminin or fibronectin are added to block normal associations of cell receptors with these ECM components during the restructuring processes (Fig. 3). Antibodies against laminin and fibronectin block the morphogenetic cascade (Tung & Fritz, unpublished observations) (Table 1).

The plasmalemmal interactions described are believed to be crucial not only for Sertoli cell – peritubular cell interactions during co-culture, but also for interactions *in vivo*. If comparable types of interactions do in fact prevail *in vivo*, they would play an important role in basement membrane formation during tubulogenesis, and they would also be required to establish, as well as maintain, the histotypic expression characteristic of each cell type (Tung & Fritz, 1986a, b). During *in vivo* tubulogenesis, for example, Sertoli cell precursors are postulated to be relatively less polarized, compared to the state of Sertoli cells during the reaggregation process in the early stages of the *in vitro* morphogenetic cascade. When intimate associations with neighboring peritubular cells become established, repolarization can commence, in association with basement membrane formation. If Sertoli cells migrate onto the free surfaces of peritubular cells during *in vivo* tubulogenesis in a manner comparable to that observed during cable formation *in vitro* (Tung & Fritz, 1986b), it may be predicted that associated plasmalemmal interactions would occur concomitantly, thereby leading to basement membrane formation in a manner analogous to that depicted in our model (Fig. 3).

Taketo *et al.* (1984) reported that testis cord formation in explants of fetal testes is blocked by cyclic AMP derivatives. This inhibition may be related to the absence of close plasmalemmal interactions caused by dispersal of the microfilament network. If this postulate is valid, the ensuing disrupted cytoskeleton would lead to a failure of appropriate Sertoli cell – peritubular cell interactions required for basement membrane formation. The model, based on the behaviour of Sertoli cells and peritubular cells during the morphogenetic cascade in co-culture (Fig. 3), is useful in interpreting the prevention by dbcAMP of basement membrane formation, and hence of tubule cord formation. The postulate that dbcAMP prevents the normal morphogenetic cascade during co-culture or during testis cord formation by causing disintegration of the microfilament network is well correlated with changes in Sertoli cell shape and structure elicited by FSH, MIX or by dbcAMP (Tung *et al.*, 1975; Spruill *et al.*, 1981). Changes in Sertoli cell morphology caused by agents which elevate cellular levels of cyclic AMP have been shown to result primarily from a disintegration of the microfilament network, and to a lesser extent, from a redistribution of the microtubule assembly (Dedman *et al.*, 1979; Welsh *et al.*, 1980).

Conclusion

The postulate that plasmalemmal interactions between closely apposing surfaces of Sertoli cells and peritubular cells, in conjunction with specific organized cytoskeletal components, are responsible for initiating tubulogenesis and basement membrane formation dictates a need to gain deeper insight into mechanisms governing the complex sets of interactions, mediated both directly and indirectly, between Sertoli cells and peritubular cells, both *in vitro* and *in vivo*. Germinal cell development in mammals is dependent upon the intricate and delicate microenvironment contained within the extraordinarily complex cytoarchitecture of the seminiferous tubule. This microenvironment is created and maintained, in large part, by Sertoli cells acting in conjunction with peritubular cells, which respond appropriately to various hormonal and environmental signals during the modulation of spermatogenesis. The focal disturbances of spermatogenesis, which result from basement membrane disruption by antibodies against basement membrane or specific ECM components, pose interesting problems for future inquiry.

Acknowledgements

This work has been supported by grants (MT 3292) from the Medical Research Council of Canada, to whom we express our deep gratitude. Over the years a procession of marvellous students and postdoctoral fellows have contributed greatly to the projects described. We especially thank Martial Lacroix and Alan Hettle for their contributions to our understanding of the plasminogen activator problem; to Fannie Smith for her work on this problem and also for transmission electron micrographs of the tfm testis sections; and to Michael Skinner for his contributions to our understanding of paracrine factors implicated in Sertoli cell-peritubular cell interactions. We also thank E. Balekjian, K. Burdzy, B. Calvieri, D. McCabe, and K. Wong for expert technical assistance. We owe a special debt of gratitude to H. Kleinman and J. Sodek for their generous gifts of antibodies against ECM components used in this study.

References

Andreasen, P.A., Christensen, T.H., Huang, J.Y., Nielson, L.S., Wilson, E.L. & Danø, K. (1986). Hormonal regulation of extracellular plasminogen activators and M_r approximately 54 000 plasminogen activator inhibitor in human neoplastic cell lines, studied with monoclonal antibodies. **Molecular Cellular Endocrinology** 45:137–147.

Cameron, D.F. & Markwald, R.R. (1981). Structural response of adult rat Sertoli cells to peritubular fibroblasts in vitro. **American Journal of Anatomy** 160:343–358.

Dedman, J.R., Brinkley, B.R. & Means, A.R. (1979) Regulation of microfilaments and microtubules by calcium and cyclic AMP. **Advances in Nucleotide Research** 11:131– 174.

Denduchis, B., Satz, M.L., Sztein, M.B., Puig, R.R., Doncel, G. & Lustig, L. (1985). Multifocal damage of the testis induced in rats by passive transfer of antibodies prepared against non-collagenous fraction of basement membrane. **Journal of Reproduction & Immunology** 7:59–75.

Ekblom, P., Vestweber, D. & Kemler, R. (1986). Cell-matrix interactions and cell adhesion during development. **Annual Review of Cell Biology** 2:27–47.

Fritz, I.B. (1978). Sites of action of androgens and follicle stimulating hormone on cells of the seminiferous tubule. In: **Biochemical Actions of Hormones** (G. Litwack, ed.), Academic Press. New York; Vol. V. pp. 249–281.

Fritz, I.B. (1984). Past, present and future of molecular and cellular endocrinology of the testis. In: **Third European Testis Workshop** (J. Saez, M.G. Forest, A. Dazord & J. Bertrand, eds.) **Les Editions de l'INSERM** 123:15–54.

Fritz, I.B. & Karmally, K. (1983). Hormonal influences on formation of plasminogen activator by cultured testis tubule segments at defined stages of the cycle of the seminiferous epithelium. **Canadian Journal of Biochemistry & Cell Biology** 61:553–560.

Fritz, I.B., Skinner, M.K. & Tung, P.S. (1986). The nature of somatic cell interactions in the seminiferous tubule. In: Serono Symposia, Review No. 11 **Development and Function of the Reproductive Organs** (A. Eshkol, ed.). N.Y. Raven Press, pp. 185–202.

Fritz, I.B. & Tung, P.S. (1986). Role of interactions between peritubular cells and Sertoli cells in mammalian testicular functions. In: **Gametogenesis and the Early Embryo** (J. Gall, ed.). New York, Alan Liss, pp. 151–173.

Gravis, C.J. (1978). Testicular involution after optic enucleation: Ultrastructure and alkaline phosphatase cytochemistry of the peritubular tissue. **American Journal of Anatomy** 151:213–226.

Hettle, J.A., Waller, E.K. & Fritz, I.B. (1986a). Hormonal stimulation alters the type of plasminogen activator produced by Sertoli cells. **Biology of Reproduction** 34:895–904.

Hettle, J.D., Balekjian, E. & Fritz, I.B. (1986b). Testicular peritubular cells in culture secrete specific inhibitor(s) of plasminogen activators. **Federation Proceedings** 45:1918.

Hutson, J. C. & Stocco D. M. (1981) Peritubular cell influence on the efficiency of androgen binding protein secretion by Sertoli cells in culture. **Endocrinology** 108:1362–1368.

Kerr, J.B., Rich, K.A. & de Kretser, D.M. (1979). Effects of experimental cryptorchidism on the ultrastructure and function of the Sertoli cell and peritubular tissue of the rat testis. **Biology of Reproduction** 21:823–838.

Lacroix, M., Parvinen, M. & Fritz, I.B. (1981). Localization of testicular plasminogen activator in discrete portions (stages VII and VIII) of the seminiferous tubule. **Biology of Reproduction** 25:143–146.

Lustig, L., Satz, M.L., Sztein, M.B. & Denduchis, B. (1982). Antigens of the basement membrane of the seminiferous tubules induce autoimmunity in Wistar rats. **Journal of Reproduction & Immunology** 4:79–90.

Skinner, M. & Fritz, I.B. (1985a). Testicular peritubular cells secrete a protein under androgen control that modulates Sertoli cell functions. **Proceedings of the National Academy of Sciences, USA** 82:114–118.

Skinner, M.K. & Fritz, I.B. (1985b). Androgen stimulation of Sertoli cell function is enhanced by peritubular cells. **Molecular and Cellular Endocrinology** 40:115–122.

Skinner, M. & Fritz, I.B. (1985c). Structural characterization of proteoglycans produced by testicular peritubular cells and Sertoli cells. **Journal of Biological Chemistry** 260:11874–11883.

Skinner, M.K. & Fritz, I.B. (1986). Identification of a non-mitogenic paracrine factor involved in interactions between testicular peritubular cells and Sertoli cells. **Molecular and Cellular Endocrinology** 44:85–97.

Skinner, M., Tung, P.S. & Fritz, I.B. (1985). Cooperativity between Sertoli cells and testicular peritubular cells in the production and deposition of extracellular matrix components. **Journal of Cell Biology** 100:1941–1947.

Spruill, W.A., White, M.G., Steiner, A.L., Tres, L.L. & Kierszenbaum, A.L. (1981). Temporal sequence of cell shape changes in cultured rat Sertoli cells after experimental elevation of intracellular cAMP. **Experimental Cell Research** 131:131–148.

Taketo, T., Thau, R.B., Adeyemo, O. & Koide, S.S. (1984). Influences of adenosine 3′-5′ cyclic monophosphate analogues on testicular organization of fetal mouse gonads *in vitro*. **Biology of Reproduction** 30:189–198.

Tung, P.S., Dorrington, J.H. & Fritz, I.B. (1975) Structural changes induced by follicle stimulating hormone or dibuturyl cyclic AMP on presumptive Sertoli cells in culture. **Proceedings of the National Academy of Sciences, USA** 72:1838–1842.

Tung, P.S. & Fritz, I.B. (1980). Interactions of Sertoli cells with myoid cells *in vitro*. **Biology of Reproduction** 23:207–217.

Tung, P.S. & Fritz, I.B. (1984). Extracellular matrix promotes rat Sertoli cell histotype expression *in vitro*. **Biology of Reproduction** 30:213–229.

Tung P.S. & Fritz, I.B. (1986a). Extracellular matrix components and testicular peritubular cells influence the rate and pattern of Sertoli cell migration *in vitro*. **Developmental Biology** 113:119–134.
Tung, P.S. & Fritz, I.B. (1986b). Cell-substratum and cell-cell interactions promote testicular peritubular myoid cell histotypic expression *in vitro*. **Developmental Biology** 115:155–170.
Tung, P.S. & Fritz, I.B. (1987). Morphogenetic restructuring and formation of basement membranes by Sertoli cells and testis peritubular cells in co-culture: inhibition of the morphogenetic cascade by cyclic AMP derivatives and by blocking direct cell contact. **Developmental Biology** 120:139–153.
Tung, P.S., Skinner, M.K. & Fritz, I.B. (1984). Fibronectin synthesis is marker for peritubular cell contaminants in Sertoli cell-enriched cultures. **Biology of Reproduction** 30:199–211.
van Mourik, J.A., van Lawrence, D.A. & Loskutoff, D.J. (1984). Purification of an inhibitor of plasminogen activator (antiactivator) synthesized by endothelial cells. **Journal of Biological Chemistry** 259:14914–14921.
Vihko, K.K., Suominen, J.J.O. & Parvinen, M. (1984). Cellular regulation of plasminogen activator secretion during spermatogenesis. **Biology of Reproduction** 31:383–389.
Welsh, M.J., VanSickle, M. & Means, A.R. (1980). Possible involvement of cyclic AMP, calcium and the cytoskeleton in control of protein secretion by Sertoli cells. In: **Testicular Development, Structure and Function** (A. Steinberger & E. Steinberger, eds.) New York, Raven Press. pp. 89–103.

Energy metabolism of spermatids: a review

J.A. GROOTEGOED AND P.J. DEN BOER

Department of Biochemistry II, Medical Faculty, Erasmus University Rotterdam, The Netherlands.

Abstract

The metabolism of energy-yielding substrates by spermatids from mammalian species, in particular the rat, is reviewed in relation to the metabolic activities of Sertoli cells. The production of ATP in the germ cells seems to be dependent on a faultless performance of their mitochondria, as indicated by anti-spermatogenic effects of inhibitors of mitochondrial metabolism. This is further supported by results obtained using isolated spermatids, which show that the ATP content of the germ cells is maintained only in the presence of pyruvate and lactate, but not through metabolism of glucose. The properties of mitochondria and glycolytic enzymes in spermatids and also in spermatozoa are described. In view of these properties, it cannot be excluded that the energy metabolism of spermatids and spermatozoa might be a target for anti-fertility agents. However, there are as yet no clear indications that defined compounds can inhibit ATP production in the germ cells in a highly cell-specific manner.

Key words

Spermatids, Sertoli cells, energy, ATP, mitochondria, glycolysis.

Introduction

All living cells transform and use energy. The cellular adenosine triphosphate (ATP) system links energy-yielding catabolism with energy-requiring cell processes. The biochemical pathways and processes involved in energy metabolism are highly conserved during evolution, but the relative importance of the various metabolic pathways is cell type-dependent. Moreover, ATP is expended on widely different processes in various cell types.

Data from the literature indicate that the flow of energy in spermatids from mammalian species involves a peculiar pattern of utilization of energy-yielding metabolic pathways. The aim of the present chapter is to review information from the literature, to try to clarify the topic. The emphasis is on carbohydrate metabolism, because the metabolism of glucose by the testis, and by spermatids in particular, has quite unique features that could be of interest for a better understanding of the effects of various factors on spermatogenesis.

Testicular carbohydrate metabolism

The early studies on the metabolism of carbohydrates by total testicular tissue have been reviewed by Free (1970). At that time, it was very clear that carbohydrates are an exceedingly important energy source for the testis in mammals. It had been found that the incorporation of radioactive lysine into testicular proteins is markedly stimulated by exogenous glucose, more than in any other tissue investigated (Davis & Morris, 1963). The experiments on amino acid incorporation, reviewed by Davis (1969), seemed to indicate that the stimulatory effect of glucose mainly concerned protein synthesis in spermatocytes and spermatids. This was confirmed in a similar study by Means & Hall (1968), who indicated that the glucose effect on the energy-dependent process of protein biosynthesis involved testicular ATP production (Means & Hall, 1969).

In view of the observations that the testis is highly dependent on exogenous glucose, it can be expected that factors which interfere with testicular glucose supply also will affect spermatogenesis. One indirect indication for this might be that streptozotocin-induced experimental diabetes in rats results in quantitative impairment of spermatogenesis (Oksanen, 1975; Anderson & Thliveris, 1986). However, the effects of experimentally induced diabetes on the testis are not very pronounced, and may not involve a direct biochemical effect of diabetes on the testis (Ford & Hamilton, 1984). In this context, the compound 5-thio-D-glucose, an analogue of D-glucose that inhibits active glucose uptake but not passive glucose transport (Whistler & Lake, 1972), exerts a pronounced antispermatogenic activity (Zysk *et al.*, 1975; Homm *et al.*, 1977). The mechanism of action of 5-thio-D-glucose on the testis is not clear, but it seems possible that it affects spermatogenesis through an action on testicular glucose supply.

All the early investigations have been performed with total testis tissue, which is composed of different cell types with widely different functions. Not surprisingly, more recent studies using isolated cells have indicated that energy metabolism by total testis tissue and by isolated testicular cell types are different.

Carbohydrate metabolism by Sertoli cells and spermatids

Experiments using isolated cell types involve a number of disadvantages. It is difficult to obtain conclusive evidence that metabolic pathways which are active in the isolated cells also are active *in situ*. Notwithstanding this, data on cultured cells can be fitted into a model, the validity of which may become apparent from data of *in vivo* experiments.

Sertoli cells, isolated from immature rats and incubated in a glucose-containing medium under aerobic conditions, convert glucose to pyruvate and lactate. There is very little mitochondrial oxidation of the endogenous pyruvate (Robinson & Fritz, 1981; Mita *et al.*, 1982), although this depends on the extracellular pyruvate and lactate concentrations (Grootegoed *et al.*, 1986b). It is not clear if the aerobic glycolysis by Sertoli cells mainly serves the purpose of ATP production. Sertoli cells can produce much ATP via mitochondrial metabolism of fatty acids and amino acids (Jutte *et al.*, 1985a; Grootegoed *et al.*, 1986b). The rate of glucose metabolism *via* the glycolytic pathway by cultured Sertoli cells is controlled by many factors, including the hormones FSH, insulin and IGF-I (Jutte *et al.*, 1982, 1983; Mita *et al.*, 1982; Oonk *et al.*, 1985; Oonk & Grootegoed, 1987, 1988). However, under certain incubation conditions the glycolytic pathway in Sertoli cells will stop, such as during long-term incubation without medium renewal and further hormone additions (Oonk *et al.*, 1985; Oonk, 1987). It seems likely that the rate of glycolysis and lactate production by Sertoli cells *in situ* is carefully controlled, so that a steady state concentration of lactate in the spermatogenic microenvironment is obtained. This concentration is not known. Similarly, in the brain, there is a continuous, but controlled supply, of lactate *via* the glycolytic pathway, and toxic concentrations of lactate are not reached (Cohen, 1985). The high concentration of lactate that is observed in testicular fluids collected postmortem is not a true reflection of the normal *in vivo* situation (Free, 1970). Data on rete testis fluid collected from living rams indicate a very low glucose concentration and a lactate concentration about equal to that of plasma (Setchell *et al.*, 1969).

Spermatids contain all the enzyme activities which constitute the glycolytic pathway (Nakamura *et al.*, 1982; Tvermyr *et al.*, 1984). Furthermore, the isolated cells convert glucose to lactate (Mita & Hall, 1982; Nakamura *et al.*, 1982). However, this glucose metabolism does not result in the maintenance of the cellular ATP content. Rather, exposure of isolated spermatids to glucose results in ATP depletion (Grootegoed *et al.*, 1986a). This is an interesting but not fully understood phenomenon, as will be discussed below. At this point, it is sufficient to note that pyruvate and lactate are excellent energy-yielding substrates for isolated spermatids from different species, including rat, mouse and hamster.

The spermatids convert exogenous lactate to pyruvate and produce ATP via mitochondrial oxidation of pyruvate. A low concentration of glucose in combination with the presence of an extracellular lactate pool, which may be the actual situation in the spermatogenic microenvironment, is compatible with the energy requirements of the spermatogenic cells. The oxygen tension in the lumen of the testicular tubules also is consistent with an oxygen-dependent mode of ATP production (Free *et al.*, 1976). The lactate requirement of the advanced spermatogenic cells could explain the above described effect of glucose on testicular ATP and protein synthesis. It is probable that factors which interfere with testicular glucose supply and lactate production by Sertoli cells at the same time interfere with spermatogenesis.

Of course, the metabolism of glucose by different testicular cell types is more complex than merely the glycolytic conversion of glucose to pyruvate and lactate. First, glucose is also important for the production of NADPH and ribose 5-phosphate via the pentose phosphate pathway. The activity of this pathway in isolated spermatogenic cells and Sertoli cells has been estimated (Robinson & Fritz, 1981; Grootegoed *et al.*, 1984, 1986b). Second, glucose and other monosaccharides and derivatives are required for the biosynthesis of oligosaccharide side chains of glycoproteins (Grootegoed *et al.*, 1982; Gerton & Millette, 1986). Third, glucose can be converted into glycogen and lipids, and these aspects of glucose metabolism have been more extensively studied using cultured Sertoli cells (Robinson & Fritz, 1981; Jutte *et al.*, 1985a).

In this context, it is also of interest that Sertoli cells contain aldose reductase (Ludvigson *et al.*, 1982), an enzyme of the polyol pathway which converts a number of aldoses to the corresponding ketoses (glucose to fructose via sorbitol, catalyzed by aldose reductase and sorbitol dehydrogenase). There is sorbitol dehydrogenase activity in the testis, and this enzyme has been used as a marker for spermatogenesis (Bishop, 1968). However, no significant amounts of sorbitol, fructose or other hexoses have been detected in testicular fluids (see Waites & Gladwell, 1982), and the actual occurrence of the polyol pathway in the testis can be questioned.

Finally, there is information concerning the production of the cyclic alcohol *myo*-inositol from glucose by Sertoli cells (Robinson & Fritz, 1979). The concentration of *myo*-inositol in testis fluid is very high (approximately 2 mM in the rat; Hinton *et al.*, 1980). A possible role for the high *myo*-inositol concentration has not yet been defined, and there is no information on possible effects of *myo*-inositol on spermatids and spermatozoa.

The mitochondria of spermatids

It is probable that ATP production in spermatids is dependent on a faultless performance of their mitochondria. One indication for this is that agents which interfere with mitochondrial ATP production cause degeneration of the spermatogenic cells *in vivo*, at doses which do not exert pronounced effects on other cell types. For example, in rats spermatogenesis is severely impaired by feeding the animals low doses of fluoroacetate or fluoroacetamide (Mazzanti *et al.*, 1964; Novi, 1968; Steinberger & Sud, 1970; Sullivan *et al.*, 1979). The mechanism of action of these compounds on spermatogenesis most likely involves conversion of fluoroacetate and fluoroacetamide to fluorocitrate, which blocks the citric acid cycle through inhibition of the enzyme aconitase, resulting in inhibition of mitochondrial ATP production. This may illustrate that compounds with an overall toxic effect can more or less specifically affect spermatogenic cells, possibly because the advanced spermatogenic cells are remarkably sensitive to toxic agents. Gossypol, the extensively studied antifertility agent from cottonseed, seems to be another example. There are a number of indications that gossypol exerts its major effect(s) through a direct effect on mitochondrial ATP production (Reyes *et al.*, 1984; Den Boer & Grootegoed, 1987). The action of gossypol *in vivo* on spermatogenic cells might also be explained by the very high sensitivity of these cell types to inhibitors of mitochondrial function.

The mitochondria of spermatids have special properties. Firstly, mitochondria in round spermatids are located near the periphery of the cells, but the significance of this is unknown. Secondly, during spermatid elongation, mitochondria migrate to form a helical arrangement around the middle piece of the developing sperm. Lastly, mitochondria of spermatids are in a condensed configuration, the result of a series of morphological changes which take place in spermatocytes. During meiotic prophase the mitochondrial cristae swell until the matrix is flattened towards the outer mitochondrial membrane (De Martino *et al.*, 1979). The condensed type of mitochondria is observed in spermatocytes and round spermatids, and also in oocytes and early embryos (Stern *et al.*, 1971). This is of interest, because oocytes and early embryos require exogenous pyruvate and lactate for ATP production (Biggers *et al.*, 1967; Zeilmaker *et al.*, 1972; Barbehenn *et al.*, 1974). In oocytes and early embryos, the block in the glycolytic pathway has been assigned to the enzyme phosphofructokinase (Barbehenn *et al.*, 1974).

The condensed configurational state of the mitochondria might be somehow related to their functional state. Quite similar changes in the configurational state of the cristal membranes of mitochondria in different cell types is induced by energizing conditions (Harris *et al.*, 1969; Hackenbrock *et al.*, 1971). This energized configuration involves condensation of the inner compartment and dilatation of the cristae.

From this, one might ask whether or not ATP production in the mitochondria of spermatids occurs at a very high and perhaps maximal rate. The answer to this question is not readily available. However, it has been observed that the rate of pyruvate decarboxylation *via* the pyruvate dehydrogenase complex in isolated rat spermatogenic cells was not enhanced by incubating the cells in the presence of uncoupling agents (Grootegoed *et al.*, 1984). The conclusion is that mitochondrial pyruvate oxidation, and hence mitochondrial ATP production, are maximal during incubation of the cells in the presence of pyruvate and in the absence of the uncoupling agents. It should be noted, however, that the addition of exogenous pyruvate to the incubation medium of isolated spermatids results in awkward metabolic behaviour of the cells. The exogenous pyruvate is very rapidly converted to lactate, at the expense of reducing equivalents from cellular NADH (Grootegoed *et al.*, 1984). Due to the urgent need for NADH, the pyruvate dehydrogenase complex and the citric acid cycle are maximally activated.

With respect to lactate metabolism, the situation is somewhat different.

It can be calculated from published data (Grootegoed *et al.*, 1984) that the pyruvate dehydrogenase complex in rat spermatids attains approximately a third of its maximal activity during incubation of the cells in the presence of lactate. Spermatids *in situ* may metabolize exogenous lactate rather than pyruvate, as will be further discussed below, and it would appear that mitochondrial ATP production in round spermatids under physiological conditions is not rate-limited by the pyruvate dehydrogenase complex. However, there may be little reserve capacity of this mitochondrial enzyme complex to meet occasional higher energy demands. There are no indications, although there is also little information available, that the capacities of the citric acid cycle and the electron transport chain could become rate-limiting. At the same time, it cannot be excluded that there is another rate-limiting step, such as the mitochondrial F_oF_1–ATPase complex.

One other mitochondrial protein may play a particular role in the energy metabolism of spermatogenic cells. A testis-specific cytochrome c has been discovered in the mouse, which differs in 13 of its 104 amino acid residues from the ubiquitous cytochrome c (Hennig, 1975). This cytochrome c_t is detected in spermatocytes and the more advanced spermatogenic cells (Goldberg *et al.*, 1977). The specific function of cytochrome c_t is not known, but its presence is one more indication that the mitochondria in spermatids and spermatozoa have not yet disclosed all their properties.

Energy-yielding substrates for spermatids

Many authors agree that lactate is the best energy-yielding substrate for isolated round spermatids from rats (Jutte *et al.*, 1981; Nakamura *et al.*, 1981a, b, 1984b, 1986b; Mita & Hall, 1982; Grootegoed *et al.*, 1984, 1986a). This is also true for spermatogenic cells from mice and hamsters (Den Boer & Grootegoed, unpublished results), and it would be of great interest to obtain information on the metabolism of energy-yielding substrates by the germ cells from other mammalian species including man.

The energy metabolism of male germ cells at the early stages of their development has not been extensively studied. The effects of lactate on ATP synthesis have been demonstrated not only for spermatids, but also for pachytene spermatocytes (Jutte *et al.*, 1981, 1982; Grootegoed *et al.*, 1984; Nakamura *et al.*, 1984a, c). In this respect, however, there is no information on spermatogonia and early spermatocytes. Interesting observations on metabolic characteristics of fetal germ cells of the mouse have been reported (Brinster & Harstad, 1977). These observations indicate that early germ cells show a relatively high rate of pyruvate oxidation and contain lactate dehydrogenase (LDH) composed of predominantly B subunits, whereas other embryonic tissues have predominantly A subunits. Perhaps, some type of metabolic differentiation of male and female germ cells is initiated during prenatal gonadal development, or is an inherent property of cells of the germ line and the early embryo. Thus, the energy metabolism of the advanced spermatogenic cells may involve many highly specialized processes which do not occur in early germ cells, growing oocytes or early embryos.

As indicated above, the metabolism of exogenous pyruvate by isolated rat spermatids involves a high activity of the mitochondrial machinery to produce NADH not only for the electron transport chain but also for the conversion of pyruvate to lactate. Under these conditions, the cellular NAD^+/NADH system in the spermatids is probably highly oxidized, whereas during metabolism of exogenous lactate this redox system may be much more reduced (Grootegoed *et al.*, 1985a; Nakamura *et al.*, 1986b). Pyruvate and lactate have similar effects on the NAD^+/NADH system in spermatozoa (Brooks & Mann, 1971, 1972; Milkowski & Lardy, 1977).

Conflicting results on the effect of exogenous pyruvate on the cellular ATP content of rat spermatids have been reported. The ATP content of the isolated spermatids either was maintained (Grootegoed *et al.*, 1984), or was not maintained (Mita & Hall, 1982; Nakamura *et al.*, 1984b, 1986a), when the cells were incubated in the presence of pyruvate. There is no clear explanation why the spermatids might not be capable of producing ATP from exogenous pyruvate. Most likely, the discrepancy is caused by experimental errors (Grootegoed *et al.*, 1984). Indeed, Nakamura *et al.* (1986d) have observed that the ATP content of rat spermatids was maintained by pyruvate when the experimental conditions included isolation of the cells in the presence of proper energy substrates.

When isolated spermatocytes and spermatids are incubated in the presence of pyruvate and/or lactate, it is possible to study various processes such as biosynthesis of RNA and proteins, fucosylation of glycoproteins, transferrin-mediated iron uptake, and long-term effects of toxic compounds (Grootegoed *et al.*, 1977, 1982; Jutte *et al.*, 1985b; Gerton & Millette, 1986; Den Boer & Grootegoed, 1987; Toebosch *et al.*, 1987).

It has become apparent that lactate rather than pyruvate is the energy-yielding substrate which is most likely used by the advanced spermatogenic cells *in situ* (Grootegoed *et al.*, 1984, 1985a; Nakamura *et al.*, 1984b, 1986b). This is also substantiated by the following. After culture of Sertoli cells at a high cell density for 24 h, the lactate concentration in the spent incubation medium was 2.5–3.0 mM, and the [lactate]/[pyruvate] ratio was approximately 10, which is the equilibrium ratio that is also observed for other cell systems (Grootegoed *et al.*, 1985b). In other words, the pyruvate concentration in the spermatogenic microenvironment is probably very low, the more so because spermatogenic cells convert exogenous pyruvate to lactate until the [lactate]/[pyruvate] ratio has become very high (Grootegoed *et al.*, 1984). Another indication that lactate rather than pyruvate is the main energy-yielding substrate used under physiological conditions, is that rat spermatogenic cells *in situ* reduce α-ketoisocaproate to α-hydroxyisocaproate, which is indicative of lactate metabolism (Grootegoed *et al.*, 1985a).

The choices for the advanced spermatogenic cells to extract energy from other substrates are limited. Isolated spermatids cannot produce a significant net amount of ATP from glucose and amino acids (see above). The substrates succinate, acetate and glycerol also appeared to have no effect on the cellular ATP content, in spite of the relative simplicity of the metabolism of these compounds (unpublished observations). There is some β-oxidation of endogenous fatty acids, and the spermatids also contain a small amount of acetyl- and acyl-carnitine (Grootegoed *et al.*, 1986a). Spermatozoa show ATP-yielding endogenous metabolism, probably including β-oxidation of fatty acids (Setchell *et al.*, 1969; Ford & Harrison, 1981b; Inskeep & Hammerstedt, 1985). Spermatocytes and spermatids contain a marked activity of carnitine acetyltransferase (Vernon *et al.*, 1971), and acetyl-carnitine could have an energy storage function in advanced spermatogenic cells, as described for spermatozoa (Milkowski *et al.*, 1976; Van Dop *et al.*, 1977). In isolated spermatids, however, fatty acid oxidation and the cellular content of acetyl-carnitine are not sufficient to maintain cellular ATP content in the absence of lactate (Grootegoed *et al.*, 1986a).

In this respect, it is relevant to note that pachytene spermatocytes and round spermatids, isolated from mice, have been found to incorporate exogenous acetate and arachidonate into fatty acids (Grogan & Lam, 1982; Grogan & Huth, 1983). Moreover, mouse testicular cells incorporate acetate into cholesterol and the polyisoprenoid alcohol dolichol (Potter *et al.*, 1981). Dolichol functions in the form of dolichyl phosphate as carrier of carbohydrates in the biosynthesis of glycoproteins. These results indicate that complex synthetic pathways may occur in spermatogenic cells for which acetyl CoA is needed, so that acetyl CoA is not exclusively used as a fuel for mitochondrial ATP production.

Effects of glucose on spermatids

Isolated round spermatids from rats convert glucose to lactate (Mita & Hall, 1982; Nakamura *et al.*, 1982). Moreover, the spermatids convert ^{3}H-labelled glucose to ^3H_2O at a considerable rate, which reflects metabolism of glucose via the glycolytic pathway (Grootegoed *et al.*, 1984). In spite of this, glucose is not an effective energy-yielding substrate for spermatids. The metabolism of glucose by isolated rat spermatids, during incubation in the absence of lactate, results in a high fructose 1,6-biphosphate concentration, but a reduction in the cellular ATP content (Grootegoed *et al.*, 1986a; Nakamura *et al.*, 1986b). The effect on ATP is observed at low glucose and mannose concentrations ($10^{-5} - 10^{-4}$ M). Fructose caused ATP depletion at 100 times higher concentrations ($10^{-3} - 10^{-2}$ M), and galactose had no effect at all (Grootegoed *et al.*, 1986a). This agrees with the substrate specificity and the K_m of hexokinase for the different substrates (Grootegoed *et al.*, 1986a). The effects of hexoses on the ATP content of spermatids indicate that the flow of intermediates through the glycolytic pathway in the spermatids is restricted by a low activity of one or more enzymes which follow phosphofructokinase in the pathway.

It is not known if the high fructose 1,6-biphosphate content of the cells is responsible for the low ATP content. This question has been studied with respect to the action of the compound α-chlorohydrin, which exerts a rapid antifertility effect in male rats and other mammals. This action is believed to involve conversion of α-chlorohydrin to 3-chlorolactaldehyde by an $NADP^+$–dependent dehydrogenase enzyme, and selective inhibition of the glycolytic enzyme glyceraldehyde 3-phosphate dehydrogenase (GA3PDH) in spermatozoa (Ford & Harrison, 1983; Stevenson & Jones, 1985). There is a striking similarity between glucose metabolism by α-chlorohydrin-treated spermatozoa (Ford & Harrison, 1986) and by isolated rat spermatids. This similarity concerns the restricted flow of fructose 1,6- biphosphate through the last part of the glycolytic pathway, concomitant with glucose-induced ATP dephosphorylation. The effect of glucose on the ATP content of α-chlorohydrin-treated spermatozoa was circumvented by giving the spermatozoa pyruvate and lactate (Ford & Harrison, 1985). From experiments using α-chlorohydrin-treated boar and rat spermatozoa, it was concluded that futile substrate cycles were active, but could not fully account for the combined effects of glucose and α-chlorohydrin on the ATP content (Ford & Harrison, 1987).

The factors which play a role in the control of ATP production *via* glycolysis in spermatozoa could be quite complex. In this respect, it is of interest to note that bull spermatozoa have been used as a model system, to study the ATP yield of glycolysis at low and high glucose fluxes (Hammerstedt & Lardy, 1983). It was observed that a low flux of glucose through the glycolytic pathway is accompanied by substrate cycling and ATP dephosphorylation (Hammerstedt & Lardy, 1983). The ATP yield of glycolysis can be zero or negative, especially if the overall net ATP yield of glucose metabolism is positive as a result of mitochondrial oxidation of pyruvate originating from glucose.

Returning to the consideration of the peculiarities of the glycolytic pathway in isolated spermatids. As described above, all the glycolytic enzymes are present in rat spermatids and it was suggested that aldolase (Tvermyr *et al.*, 1984) or GA3PDH (Nakamura *et al.*, 1982) are the rate-limiting enzymes. On the other hand, it cannot be excluded that the apparent block in the second phase of glycolysis is not inherent to the cells, but represents an artefact of cell isolation, induced because of unstable activities or an unstable arrangement of glycolytic enzymes in spermatids.

It might be that GA3PDH is the weak link, especially in view of the experiments discussed above on α-chlorohydrin-treated spermatozoa, but there is no conclusive evidence. The concentrations of glycolytic intermediates have been estimated after incubation of the spermatids in the presence of glucose and in the absence of lactate (Nakamura *et al.*, 1982).

From a biochemical point of view these cells are severely damaged, due to the glucose-induced ATP depletion. Here is a paradox. On the one hand, glucose metabolism causes cell death, and on the other hand lactate-supported ATP production shuts down the glycolytic pathway. When the cells attain a high energy charge from the metabolism of lactate, this will result in inhibition of the glycolytic pathway through regulation of the phosphofructokinase reaction (Bosca *et al.*, 1985). Hence, by using intact cells no reliable information on the flow of metabolites through the glycolytic pathway can be obtained.

It has been suggested that control of glycolysis in spermatocytes and spermatids could be effected through inhibition of GA3PDH by the adenine nucleotides AMP, ADP and ATP in competition with NAD, rather than that glycolysis is regulated by the control enzymes hexokinase, phosphofructokinase and pyruvate kinase (Nakamura *et al.*, 1982, 1984a). However, there are no indications that the NAD content of the spermatogenic cells is sufficiently low to permit this mechanism of regulation. Moreover, there is a dramatic effect of glucose on ATP dephosphorylation in isolated spermatids, indicating that the cells are unsuccessful in controlling glucose metabolism.

The enzyme phosphofructokinase is a good candidate to act as the most important control point in glycolysis in spermatids. Phosphofructokinase can be effectively inhibited by ATP and other factors (Bosca *et al.*, 1985; Hoskins & Stephens, 1969), but in spermatids sufficiently high concentrations of ATP may only be reached when the cells oxidize exogenous pyruvate or lactate. Endogenous pyruvate, produced from glucose, is probably converted to lactate, rather than being oxidized in the mitochondria, because spermatids may find difficulties in transporting reducing equivalents from cytosolic NADH across the inner mitochondrial membrane *via* a shuttle system. NADH is readily oxidized to NAD^+, however, at the conversion of pyruvate to lactate in the cytosol. This deprives the mitochondria of the essential energy source pyruvate. Consequently, the cellular ATP content will be too low to effectively inhibit phosphofructokinase. In this situation, glucose is converted to glycolytic intermediates, but the limited flux of intermediates through the second phase of the glycolytic pathway does not enable the spermatids to produce ATP from glucose.

Spermatids can take up glucose *via* a specific carrier (Nakamura *et al.*, 1987b). The fate of glucose inside cells is that it is phosphorylated by ATP to form glucose 6-phosphate, a reaction catalyzed by hexokinase. Most of the glucose 6-phosphate proceeds down the glycolytic pathway, but a minor catabolic pathway taken by glucose 6-phosphate is the pentose phosphate pathway. This pathway leads to two important products: NADPH and ribose 5-phosphate. The hexokinase reaction is not controlled by the energy status of the cell, and the pentose phosphate pathway can therefore also occur in spermatids when these cells produce ATP from exogenous lactate. However, there are two factors which could limit the metabolism of glucose *via* the pentose phosphate pathway in spermatids. First, the activity of glucose 6-phosphate dehydrogenase is comparatively low (see below), although the cells have some pentose phosphate pathway activity (Grootegoed *et al.*, 1984). Second, an insufficient supply of glucose to the spermatogenic cells *in situ*.

The developing germ cells will require ribose 5-phosphate to meet the needs of the biosynthesis of nucleotides for RNA synthesis, but there is no information on the production of ribose 5-phosphate in spermatocytes and spermatids. Possibly, the amount of glucose available is sufficient to support a low but adequate rate of ribose 5-phosphate production. Moreover, the ribose 5-phosphate that is required can be generated also *via* the non-oxidative-branch of the pentose phosphate pathway, so that an eventual low activity of glucose 6-phosphate dehydrogenase may not play a role in this respect. Spermatogenesis does not occur at great speed, and substantial net amounts of RNA may be produced during the many days that it takes for pachytene spermatocytes to develop into elongating spermatids, especially if the rate of synthesis of RNA precursors is low. However, it cannot be excluded that the pentose phosphate pathway in spermatids *in situ* is virtually completely inhibited due to lack of substrate, and that other mechanisms operate to supply RNA precursors to the germ cells. One such mechanism could be the release of nucleosides by Sertoli cells, and the uptake and phosphorylation of these nucleosides by the germ cells. Isolated spermatocytes and spermatids incorporate exogenous uridine into RNA (Grootegoed *et al.*, 1977).

It should be stressed that very little is known about the *de novo* synthesis of nucleotides and the purine salvage reactions in spermatogenic cells (Grootegoed *et al.*, 1986c). This is highly relevant not only with respect to RNA synthesis, but certainly also in relation to the ATP content of the cells. On the one hand, the metabolism of glucose by spermatocytes and spermatids results in ATP dephosphorylation (see above), whereas glucose could be important for ribose 5-phosphate synthesis and the *de novo* synthesis of adenine nucleotides.

Finally, it may be relevant to point out that in view of the more recent data on the effects of glucose on spermatids, it is difficult to explain earlier observations on the effects of glucose and 5-thio-D-glucose on protein synthesis by isolated spermatids from rats (Nakamura & Hall, 1976, 1977; Nakamura *et al.*, 1978). In these experiments, it was observed that the incorporation of radioactively labeled amino acids into proteins was stimulated by glucose and inhibited by 5-thio-D-glucose. It would be important to establish if this incorporation represented the actual rate of protein synthesis, rather than changes in the specific radioactivity of the precursor pools. It cannot be excluded that glucose exerts a stimulatory effect on protein synthesis independent of the effect on ATP, although protein synthesis would be strongly suppressed by lack of ATP. The rate of amino acid incorporation into proteins of spermatocytes and spermatids is markedly increased when the glucose-containing incubation medium is supplemented with lactate (Jutte *et al.*, 1981; Nakamura *et al.*, 1981a).

Glycolytic enzymes in spermatids

There is a considerable amount of information on glycolytic enzymes in spermatids and spermatozoa including the expression of different isoenzymes of the glycolytic pathway. These include the isoenzymes of phosphoglycerate kinase (PGK) and enolase (ENO), and also of lactate dehydrogenase (LDH) which in the present discussion is considered as a reaction involved in glycolysis.

During spermiogenesis several glycolytic enzymes appear to be subjected to testis-specific post-translational modifications, which result in 'sperm-type' enzymes or sperm-specific behaviour. These are hexokinase (Katzen *et al.*, 1968; Harrison, 1971; Sosa *et al.*, 1972), phosphoglucose isomerase (Buehr & McLaren, 1981) and aldolase (Gillis & Tamblyn, 1984). The kinetics of the enzymes hexokinase, phosphofructokinase and pyruvate kinase of rat spermatids have been studied, but no abnormalities of the kinetic behaviour of these enzymes have been observed (Nakamura *et al.*, 1984d, 1986c, 1987a).

A side issue of a discussion on PGK, is that the X chromosome is genetically inactivated early in spermatogenesis (Monesi, 1971). During meiotic prophase, the X and Y chromosomes are condensed and form a heterochromatic body, the sex vesicle. It is not clear what purpose is served by the formation of the sex vesicle, but many chromosomal anomalies which disturb spermatogenesis seem to involve the X chromosome (de Boer & Searle, 1980). As a consequence of X chromosome inactivation, the spermatogenic cells may fail to maintain high activity levels of enzymes encoded by genes on the X chromosome, such as phosphoglycerate kinase A (PGK-A) and glucose 6-phosphate dehydrogenase (G6PDH). There is probably no autosomal enzyme synthesized in spermatogenic cells to substitute for G6PDH, the first enzyme of the pentose phosphate pathway. Consequently, in spermatozoa the pentose phosphate pathway is probably almost completely absent (Setchell *et al.*, 1969; Grootegoed *et al.*, 1986c). With respect to PGK, the spermatogenic cells follow a different strategy. The mammalian genome contains an autosomal gene which encodes the isoenzyme PGK-B. Expression of this autosomal gene is restricted to the spermatogenic cells (VandeBerg *et al.*, 1973). Recent information on the Pgk-2 gene (encoding PGK-B), shows that it is peculiar, in that it completely lacks introns and has the characteristics of a processed gene (McCarrey & Thomas, 1987; Boer *et al.*, 1987). A functional processed gene is rare, and it seems that the spermatogenic cells have evolved a specific and original solution to solve the threatening problem of a missing link in the glycolytic pathway. This by itself indicates the importance of the glycolytic pathway for spermatozoa.

Enolase is a dimer coded for by three genes, producing subunits α-, β- and γ-, and there are developmental changes in the enolase isoenzyme patterns in different tissues (Rider & Taylor, 1974). The testis, and in particular late spermatids and spermatozoa, contain an unusual enolase isoenzyme (Edwards & Grootegoed, 1983). This ENO-S is distinguished from somatic enolases by physical properties. It is unknown if ENO-S contains subunits encoded by another gene, and there is also no information which could explain the specific function of ENO-S.

Although we have learned a lot about the testis-specific lactate dehydrogenase isoenzyme LDH-C_4 since its discovery in 1963 (Blanco & Zinkham, 1963; Goldberg, 1963, 1977), there is still uncertainty on why spermatogenic cells and spermatozoa would need LDH-C_4. The isoenzyme is composed of four C subunits, coded for by a gene different from the A and B subunit genes. In different species, LDH-C_4 shows a broad substrate specificity towards different α-keto acids. This is true especially for rats and mice (Coronel *et al.*, 1983). The LDH-C_4 enzyme from these animals converts branched chain α-keto acids to the corresponding α-hydroxy acids. The branched chain α-keto acids are derived from transamination of the branched chain amino acids leucine, isoleucine and valine. Cultured Sertoli cells from rats convert leucine to α-ketoisocaproate, and spermatogenic cells reduce this compound to α-hydroxyisocaproate (Grootegoed *et al.*, 1983, 1985a). This represents a unique intercellular pathway of leucine catabolism, but it is not known if a function can be assigned to it. Possibly, the catalytic properties of LDH-C_4 are not the main justification for its existence. The possibilities for a function which is directly related to specific physical properties of the isoenzyme seem to be unlimited, in view of the recent observation that lactate dehydrogenase can act as a structural protein in avian and crocodilian lenses (Wistow *et al.*, 1987).

In spermatids, most of the LDH-C_4 is found in the cytosol (Hintz & Goldberg, 1977). This isoenzyme is also located in the mitochondrial matrix of advanced spermatogenic cells and spermatozoa (Montamat & Blanco, 1976). Many authors have indicated the presence of LDH-C_4 inside the mitochondria, in particular in spermatozoa, where it could be involved in a shuttle mechanism for the transport of reducing equivalents into and out of the mitochondria, or could catalyze direct intramitochondrial lactate oxidation (Milkowski & Lardy, 1977; Storey & Kayne, 1977; Gerez de Burgos, 1978). The transport of a protein into the mitochondrial matrix involves a specific sequence of events, which is dependent on the synthesis of a precursor protein with an N-terminal extension (Schatz & Butow, 1983). This has not been shown for LDH-C subunits. However, results from the molecular cloning of the mouse C-subunit gene indicates the presence of multiple LDH-C gene-related sequences (Sakai *et al.*, 1987). Moreover, it cannot be excluded that one gene codes for two proteins on opposite sides of the mitochondrial inner membrane (Fox, 1982). All in all, there is still no conclusive evidence for the presence of LDH-C_4 in the mitochondrial matrix, such as could be provided by an immunocytochemical study at the ultrastructural level. Independent of this, the fact that most if not all of the enzymatic activity is outside the mitochondrial inner compartment indicates that the enzyme could very well catalyze the last reaction in the aerobic glycolysis of spermatozoa.

The energy budget of spermatids

In developing germ cells, the meiotic divisions are the last great expense of energy needed for the generation of four spermatids from one spermatocyte. The energy budget of spermatids does not include expenditures for DNA synthesis and the process of cell division. In round spermatids, there is still RNA and protein synthesis, but the rates must be modest since the cells do not undergo a new round of cell division. In elongating spermatids, RNA synthesis becomes completely arrested when the DNA is compacted in the nucleus, and the last major act of protein biosynthesis is the synthesis of the nuclear proteins involved in this compaction process (Monesi, 1971; see the Chapter by Hecht, this volume).

There are other processes that consume ATP. In most cells, much energy is expended on active transport of ions across the plasma membrane. The transport of Na^+ and K^+ by a specific transport system, the Na^+-K^+ pump, results in a high concentration of K^+ and a low concentration of Na^+ in the cells relative to the external medium. The Na^+-K^+ pump contains ATPase activity, and it has been reported that inhibition of the Na^+-K^+ pump by the toxic glucoside ouabain results in a major reduction in the energy expenditure of tumor cells (Racker *et al.*, 1983). Spermatids are in a unique situation. There is a remarkable difference between the ionic composition of tubular fluid and plasma, most importantly in the K^+ concentration, which at approximately 40 mM in tubular fluid, is 10 times higher than that in plasma (Tuck *et al.*, 1970; Levine & Marsh, 1971; Jenkins *et al.*, 1980; see Waites & Gladwell, 1982). Concomitantly, the Na^+ concentration in the tubular fluid is relatively low. In primary tubular fluid, which may more closely represent the fluid that surrounds the spermatogenic cells, the K^+ concentration is even higher than in free-flow tubular fluid (Tuck *et al.*, 1970). From this, it can be speculated that spermatids are in an ionic microenvironment that approximates to the intracellular ionic composition, and may not have to pay much attention to the pumping of Na^+ and K^+. Possibly, the active transport of Na^+ and K^+ across the plasma membrane is a minor item of the energy budget of spermatids.

Living cells also utilize the chemical energy of ATP for the performance of mechanical work. Early spermatids contain two motile components: the primary flagellum, and rhythmic cytoplasmic movements (Walt & Hedinger, 1983). In late spermatids, the motility of the flagellum and the cytoplasm is reduced (Walt & Hedinger, 1983). The spermatids contain F-actin, myosin and tubulin (Campanella *et al.*, 1979; Hecht *et al.*, 1984; Russell *et al.*,

1986). F-actin and myosin together can use the chemical energy stored as ATP for the generation of force, and tubulin in the axonemal microtubules interacts with the ATPase activity of dynein arms. However, it is not possible to give an estimate of how much ATP is consumed in reactions generating force. Possibly, mechanical work is the most important process of ATP turnover in the developing spermatids.

From spermatid to spermatozoon

The middle piece of spermatozoa contains a mitochondrial system which is capable of producing ATP through oxidative phosphorylation (Ford & Harrison, 1981b). Yet, aerobic glycolysis seems to be important in the energy metabolism of mature spermatozoa (Peterson & Freund, 1970). In spermatozoa, energy metabolism is geared to producing ATP for the contractile work of motility, and it has been suggested that the energy metabolism of spermatozoa is similar to that of muscle cells (Storey & Kayne, 1980). This means that spermatozoa depend on a high rate of glycolysis to produce ATP for motility. It has been shown that rabbit spermatozoa contain a muscle-type pyruvate kinase isoenzyme that is suited for a high rate of glycolysis (Storey & Kayne, 1980). Furthermore, the potential catalytic activities of a number of glycolytic enzymes in spermatozoa of various species, including rat spermatozoa, are considerably in excess of the flux of intermediates through the glycolytic pathway (Harrison, 1971; Ford *et al.*, 1981). While a detailed discussion of energy metabolism in spermatozoa is not practicable within the limits of the present text, particularly in view of marked species differences, it can be suggested that spermatozoa acquire glycolytic capacity compared to spermatids, and learn how to control glycolysis. This is not to say that under all circumstances glucose is the preferred energy-yielding substrate of spermatozoa. The adenine nucleotide concentrations in and motility of rat spermatozoa are supported by pyruvate and lactate better than by glucose (Ford & Harrison, 1981a), and high concentrations of glycolytic intermediates accumulate in rat spermatozoa incubated in the presence of glucose. Furthermore, it has been suggested that lactate could be the preferred energy source of rabbit spermatozoa (Storey & Kayne, 1977). Yet, the behaviour of the glycolytic enzymes may be changed during late spermiogenesis, so that a more efficient glycolytic apparatus is obtained.

A number of glycolytic enzymes remain bound to the sperm middle piece, during treatment of spermatozoa with hypotonic media (Harrison & White, 1972; Storey & Kayne, 1975; Gillis & Tamblyn, 1984). This could indicate that specific mechanisms operate that influence the locations and interactions of the enzymes. Little is known about the intracellular transport and structural interactions of glycolytic enzymes in most cells. However, it is not a hard and fast rule that the soluble enzymes of glycolysis are freely diffusing in the cytosol (Fulton, 1982; Srere, 1987). For example, enolase is present in a complex with other proteins, so that it can be transported within nerve cells (Brady & Lasek, 1981). Noteworthy also, in view of the comparison between spermatozoa and flagellated protozoa (see the Chapter by Docampo, this volume), is that in *Trypanosoma brucei* the glycolytic enzymes are located in an organelle, the glycosome (Opperdoes & Borst, 1977). This could enable this organism to obtain all its energy through glycolysis. There is little information on the exact location of the glycolytic enzymes in spermatozoa. Possibly, the glycolytic machinery is mainly associated with the mitochondria in the midpiece, and this could be effected also through the formation of enzyme complexes.

In sea urchin sperm, as in muscle cells, a phosphorylcreatine shuttle mediates the transport of high energy phosphate from the mitochondrion down the sperm tail (Tombes & Shapiro, 1985). It is not known whether a similar mechanism is present in mammalian sperm, or whether diffusion of ATP can account for the energy needed for the dynein-induced sliding of microtubules. However, it seems likely that special adaptations must take place during the last phase of spermiogenesis to ensure an optimal flow of energy from energy-yielding reactions to the motile proteins of the sperm axoneme.

During late spermiogenesis, a series of events may occur which end the random arrangements of glycolytic enzymes. First, a number of genetic and post-translational isoenzymes partially or completely take the place of the somatic glycolytic enzymes. Second, further post-translational modification of enzymes might occur via phosphorylation of serine and threonine residues, and also tyrosine residues (see the Chapter by Meijer *et al.*, this volume). Third, during the final stages of spermatid elongation, much of the spermatid cytoplasm flows from the flagellum to the spermatid head where it is removed *via* so-called tubulobulbar complexes (Russell, 1979). This could be accompanied by an increase in the concentration of cytosolic enzymes and an altered intracellular location. The concentration of enzymes in the cell is an important factor in the control of metabolite transfer between two sequential enzyme sites (Srivastava & Bernhard, 1986a, b). It can be speculated that during late spermiogenesis, even very shortly before spermiation, the enzyme–enzyme interactions among glycolytic enzymes are altered. This could result in proper control of glycolysis in spermatozoa.

Synopsis

The following working hypothesis is proposed: Glucose metabolism by isolated spermatids initiates a sequence of events that involves accumulation of glycolytic intermediates and ATP dephosphorylation. For unknown reasons, glucose metabolism goes out of control. In the spermatogenic epithelium, a similar response of spermatids to glucose does not occur, because the local glucose concentration is probably very low and the cells do not depend on glycolysis for ATP production. Rather, developing spermatids rely on the mitochondrial machinery to produce ATP. The advanced spermatogenic cells have specialized to utilize exogenous lactate for ATP production, and these cells have largely or completely given up the capacity to metabolize other energy-yielding substrates.

Inhibitors of mitochondrial ATP production have lethal effects on spermatids. On the other hand, compounds such as α-chlorohydrin, or other factors that interfere with the second energy-conserving phase of glycolysis, will affect sperm cells but not the developing germ cells in the testis. Spermatozoa can produce ATP *via* a high rate of lactate production by glycolysis, but also *via* oxidation of exogenous lactate and endogenous substrates.

During late spermiogenesis, there are tumultuous changes in the interactions between cytosolic enzymes. These enzymes include genetic and post-translational isoenzymes. Non-random arrangement of glycolytic enzymes may allow for stable control of glycolysis.

The regulation of energy metabolism in spermatids and spermatozoa could be much more complex than we can now anticipate. From the views presented, it may be inferred that impairment of sperm glycolysis, induced during spermiogenesis, or later, is an approach to interfere with sperm function without an effect on spermatogenesis. However, there are as yet no clear indications that defined compounds can act upon glycolytic enzymes in spermatozoa in a highly cell-specific manner. Nevertheless, research in this area should continue, especially to compile more data needed to attack the problem of cell-specific inhibition of sperm function.

Acknowledgements

This work received financial support from the Special Programme of Research, Development, and Research Training in Human Reproduction, World Health Organization.

References

Anderson, J.E. & Thliveris, T. (1986). Testicular histology in streptozotocin-induced diabetes. **Anatomical Record** 214:378–382.

Barbehenn, E.K., Wales, R.G. & Lowry, O.H. (1974). The explanation for the blockade of glycolysis in early mouse embryos. **Proceedings of the National Academy of Sciences U.S.A.** 71:1056–1060.

Biggers, J.D., Whittingham, D.G. & Donahue, R.P. (1967). The pattern of energy metabolism in the mouse oocyte and zygote. **Proceedings of the National Academy of Sciences, U.S.A.** 58:560–567.

Bishop, D.W. (1968). Sorbitol dehydrogenase in relation to spermatogenesis and fertility. **Journal of Reproduction and Fertility** 17:410–411.

Blanco, A. & Zinkham, W.H. (1963). Lactate dehydrogenases in human testes. **Science** 139:601–602.

Boer, P.H., Adra, C.N., Lau, Y-F. & McBurney, M.W. (1987). The testis-specific phosphoglycerate kinase gene pgk-2 is a recruited retroposon. **Molecular and Cellular Biology** 7:3107–3112.

Boer, P. de & Searle, A.G. (1980). Workshop on chromosomal aspects of male sterility in mammals: Summary and synthesis. **Journal of Reproduction and Fertility** 60:257–265.

Bosca, L., Aragon, J.J. & Sols, A. (1985). Modulation of muscle phosphofructokinase at physiological concentration of enzyme. **The Journal of Biological Chemistry** 260:2100–2107.

Brady, S.T. & Lasek, R.J. (1981). Nerve-specific enolase and creatine phosphokinase in axonal transport: Soluble proteins and the axoplasmic matrix. **Cell** 23:515–523.

Brinster, R.L. & Harstad, H. (1977). Energy metabolism in primordial germ cells of the mouse. **Experimental Cell Research** 109:11–117.

Brooks, D.E. & Mann, T. (1971). NAD in the metabolism of motile spermatozoa. **Nature (London)** 234:301–302.

Brooks, D.E. & Mann, T. (1972). Relation between the oxidation state of nicotinamide-adenine dinucleotide and the metabolism of spermatozoa. **The Biochemical Journal** 129:1023–1034.

Buehr, M. &, McLaren, A. (1981). An electrophoretically detectable modification of glucosephosphate isomerase in mouse spermatozoa. **Journal of Reproduction and Fertility** 63:169–173.

Campanella, C., Gabbiani, G., Baccetti, B., Burrini, A.G. & Pallini, V. (1979). Actin and myosin in the vertebrate acrosomal region. **Journal of Submicroscopical Cytology** 11:53–71.

Cohen, S.R. (1985). Why does brain make lactate? **Journal of Theoretical Biology** 112:429–432.

Coronel, C.E., Burgos, C., Gerez de Burgos, N.M., Rovai, L.E. & Blanco, A. (1983). Catalytic properties of the sperm-specific lactate dehydrogenase (LDH X or C_4) from different species. **The Journal of Experimental Zoology** 225:379–385.

Davis, J.R. & Morris, R.N.(1963). Effect of glucose on incorporation of L-lysine-U-C^{14} into testicular proteins. **American Journal of Physiology** 205:833–836.

Davis, J.R. (1969). Metabolic aspects of spermatogenesis. **Biology of Reproduction** 1:93–118.

De Martino, C., Floridi, A., Marcante, M.L., Malorni, W., Scorza Barcellona, P., Bellocci, M. & Silvestrini, B. (1979). Morphological, histochemical and biochemical studies on germ cell mitochondria of normal rats. **Cell and Tissue Research** 196:1–22.

Den Boer, P.J. & Grootegoed, J.A. (1987). Effect of (−) and (+) enantiomers of gossypol on isolated round spermatids. **Annals of the New York Academy of Sciences** 513:535–537.

Edwards, Y.H. & Grootegoed, J.A. (1983). A sperm-specific enolase. **Journal of Reproduction and Fertility** 68:305–310.

Ford, W.C.L. & Hamilton, D.W. (1984). The effect of experimentally induced diabetes on the metabolism of glucose by seminiferous tubules and epididymal spermatozoa from the rat. **Endocrinology** 115:716–721.

Ford, W.C.L. & Harrison, A. (1981a). The effect of 6-chloro-6- deoxysugars on adenine nucleotide concentrations in and motility of rat spermatozoa. **Journal of Reproduction and Fertility** 63:75–79.

Ford, W.C.L. & Harrison, A. (1981b). The role of oxidative phosphorylation in the generation of ATP in human spermatozoa. **Journal of Reproduction and Fertility** 63:271–278.

Ford, W.C.L. & Harrison, A. (1983). The activity of glyceraldehyde 3-phosphate dehydrogenase in spermatozoa from different regions of the epididymis in laboratory rodents treated with α-chlorohydrin or 6-chloro-deoxyglucose. **Journal of Reproduction and Fertility** 69:147–156.

Ford, W.C.L. & Harrison, A. (1985). The presence of glucose increases the lethal effect of α-chlorohydrin on ram and boar spermatozoa *in vitro*. **Journal of Reproduction and Fertility** 73:197–206.

Ford, W.C.L. & Harrison, A. (1986). The concerted effect of α-chlorohydrin and glucose on the ATP concentration in spermatozoa is associated with the accumulation of glycolytic intermediates. **Journal of Reproduction and Fertility** 77:537–545.

Ford, W.C.L. & Harrison, A. (1987). Futile substrate cycles in the glycolytic pathway of boar and rat spermatozoa and the effect of α-chlorohydrin. **Journal of Reproduction and Fertility** 79:21–32.

Ford, W.C.L., Harrison, A. & Waites, G.M.H. (1981). Effects of 6-chloro-6-deoxysugars on glucose oxidation in rat spermatozoa. **Journal of Reproduction and Fertility** 63:67–73

Fox, T. (1982). Do mitochondria share enzymes with the rest of the cell? **Nature** 298:323.

Free, M.J. (1970). Carbohydrate metabolism in the testis. In: Johnson, A.D., Gomes, W.R. & Van Demark, D., eds. **The Testis**, Volume II. New York and London, Academic Press, pp.125–192.

Free, M.J., Schluntz, G.A. & Jaffe, R.A. (1976). Respiratory gas tensions in tissues and fluids of the male rat reproductive tract. **Biology of Reproduction** 14:481–488.

Fulton, A.B. (1982). How crowded is the cytoplasm? **Cell** 30:345–347.

Gerez de Burgos, N.M., Burgos, C., Montamat, E.E., Moreno, J. & Blanco, A. (1978). A shuttle system for the transfer of reducing equivalents in mouse sperm mitochondria. **Biochemical and Biophysical Research Communications** 81:644–649.

Gerton, G.L. & Millette, C.F. (1986). Stage-specific synthesis and fucosylation of plasma membrane proteins by mouse pachytene spermatocytes and round spermatids in culture. **Biology of Reproduction** 35:1025–1035.

Gillis, B.A.& Tamblyn, T.M. (1984). Association of bovine sperm aldolase with sperm subcellular components. **Biology of Reproduction** 31:25–35.

Goldberg, E. (1963). Lactic and malic dehydrogenases in human spermatozoa. **Science** 139:602–603.

Goldberg, E. (1977). Isozymes in testes and spermatozoa. In: Rattazzi, M.C., Scandalios, J.G. & Whitt, G.S., eds. **Isozymes: Current Topics in Biological and Medical Research**, Volume 1. New York, Alan R. Liss, Inc., pp.79–124.

Goldberg, E., Sberna, D., Wheat, T.E., Urbanski, G.J. & Margoliash, E. (1977). Cytochrome c: Immunofluorescent localization of the testis-specific form. **Science** 196:1010–1012.

Grogan, W.M. & Lam, J.W. (1982). Fatty acid synthesis in isolated spermatocytes and spermatids of mouse testis. **Lipids** 17:604–611.

Grogan, W.M. & Huth, E.G. (1983). Biosynthesis of long-chain polyenoic acids from arachidonic acid in cultures of enriched spermatocytes and spermatids from mouse testis. **Lipids** 18:275–284.

Grootegoed, J.A., Grolle-Hey, A.H., Rommerts, F.F.G.& van der Molen, H.J. (1977). Ribonucleic acid synthesis *in vitro* in primary spermatocytes isolated from rat testis. **The Biochemical Journal** 168:23–31.

Grootegoed, J.A., Kruger-Sewnarain, B.C., Jutte, N.H.P.M., Rommerts, F.F.G. & van der Molen, H.J. (1982). Fucosylation of glycoproteins in rat spermatocytes and spermatids. **Gamete Research** 5:303–315.

Grootegoed, J.A., Jutte, N.H.P.M., Jansen, R., van der Molen, H.J. (1983). Hormonal activation of the supporting role of Sertoli cells in spermatogenesis. In: Dumont, J.E., Nunez, J. & Denton, R.M., eds. **Hormones and Cell Regulation**, Volume 7. Amsterdam, Elsevier Biomedical Press, pp.299–316.

Grootegoed, J.A., Jansen, R. & van der Molen, H.J. (1984). The role of glucose, pyruvate and lactate in ATP production by rat spermatocytes and spermatids.

Biochimica et Biophysica Acta 767:248–256.

Grootegoed, J.A., Jansen, R. & van der Molen, H.J. (1985a). Intercellular pathway of leucine catabolism in rat spermatogenic epithelium. **The Biochemical Journal** 226:889–892.

Grootegoed, J.A., Oonk, R.B., Jansen, R. & van der Molen, H.J. (1985b). Spermatogenic cells utilize metabolic intermediates from Sertoli cells. In: Rolland, R., ed. **Gamete Quality and Fertility Regulation**, Amsterdam, Excerpta Medica, pp. 225–237.

Grootegoed, J.A., Jansen, R. & van der Molen, H.J. (1986a). Effect of glucose on ATP dephosphorylation in rat spermatids. **Journal of Reproduction and Fertility** 77:99–107.

Grootegoed, J.A., Oonk, R.B., Jansen, R. & van der Molen, H.J. (1986b). Metabolism of radiolabelled energy-yielding substrates by rat Sertoli cells. **Journal of Reproduction and Fertility** 77:109–118.

Grootegoed, J.A., Oonk, R.B., Toebosch, A.M.W. & Jansen, R. (1986c). Extracellular factors that contribute to the development of spermatogenic cells. In: Stefanini, M., Conti, M., Geremia, R. & Ziparo, E., eds. **Molecular and Cellular Endocrinology of the Testis**. Amsterdam, Excerpta Medica, pp. 215–225.

Hackenbrock, C.R., Rehn, T.G., Weinbach, E.C. & Lemasters, J.J. (1971). Oxidative phosphorylation and ultrastructural transformation in mitochondria in the intact ascites tumor cell. **The Journal of Cell Biology** 51:123–137.

Hammerstedt, R.H. & Lardy, H.A. (1983). The effect of substrate cycling on the ATP yield of sperm glycolysis. **The Journal of Biological Chemistry** 258:8759–8768.

Harris, R.A., Williams, C.H., Caldwell, M., Green, D.E. & Valdivia, E. (1969). Energized configurations of heart mitochondria *in situ*. **Science** 165:700–703.

Harrison, R.A.P. (1971). Glycolytic enzymes in mammalian spermatozoa. Activities and stabilities of hexokinase and phosphofructokinase in various fractions from sperm homogenates. **The Biochemical Journal** 124:741–750.

Harrison, R.A.P. & White, I.G. (1972). Glycolytic enzymes in the spermatozoa and cytoplasmic droplets of bull, boar and ram, and their leakage after shock. **Journal of Reproduction and Fertility** 30:105–115.

Hecht, N.B., Kleene, K.C., Distel, R.J. & Silver, L.M. (1984). The differential expression of the actins and tubulins during spermatogenesis in the mouse. **Experimental Cell Research** 153:275–280.

Hennig, B. (1975). Change of cytochrome c structure during development of the mouse. **European Journal of Biochemistry** 55:167–183.

Hinton, B.T., White, R.W. & Setchell, B.P. (1980). Concentrations of myo-inositol in the luminal fluid of the mammalian testis and epididymis. **Journal of Reproduction and Fertility** 58:395–399.

Hintz, M. & Goldberg, E. (1977). Immunohistochemical localization of LDH-X during spermatogenesis in mouse testes. **Developmental Biology** 57:375–384.

Homm, R.E., Rusticus, C. & Hahn, D.W. (1977). The antispermatogenic effects of 5-thio-D-glucose in male rats. **Biology of Reproduction** 17:698–700.

Hoskins, D.D. & Stephens, D.T. (1969). Regulatory properties of primate sperm phosphofructokinase. **Biochimica et Biophysica Acta** 191:292–302.

Inskeep, P.B. & Hammerstedt, R.H. (1985). Endogenous metabolism by sperm in response to altered cellular ATP requirements. **Journal of Cellular Physiology** 123:180–190.

Jenkins, A.D., Lechene, C.P. & Howards, S.S. (1980). Concentrations of seven elements in the intraluminal fluids of the rat seminiferous tubules, rete testis, and epididymis. **Biology of Reproduction** 23:981–987.

Jutte, N.H.P.M., Grootegoed, J.A., Rommerts, F.F.G. & van der Molen, H.J. (1981). Exogenous lactate is essential for metabolic activities in isolated rat spermatocytes and spermatids. **Journal of Reproduction and Fertility** 62:399–405.

Jutte, N.H.P.M., Jansen, R., Grootegoed, J.A., Rommerts, F.F.G., Clausen, O.P.F. & van der Molen, H.J. (1982). Regulation of survival of rat pachytene spermatocytes by lactate supply from Sertoli cells. **Journal of Reproduction and Fertility** 65:431–438.

Jutte, N.H.P.M., Jansen, R., Grootegoed, J.A., Rommerts, F.F.G. & van der Molen, H.J. (1983). FSH stimulation of the production of pyruvate and lactate by rat Sertoli cells may be involved in hormonal regulation of spermatogenesis. **Journal of Reproduction and Fertility** 68:219–226.

Jutte, N.H.P.M., Eikvar, L., Levy, F.O. & Hansson, V. (1985a). Metabolism of palmitate in cultured rat Sertoli cells. **Journal of Reproduction and Fertility** 73:497–503.

Jutte, N.H.P.M., Jansen, R., Grootegoed, J.A., Rommerts, F.F.G. & van der Molen, H.J. (1985b). Protein synthesis by isolated pachytene spermatocytes in the absence of Sertoli cells. **The Journal of Experimental Zoology** 233:285–290.

Katzen, H.M., Soderman, D.D. & Cirillo, V.J. (1968). Tissue distribution and physiological significance of multiple forms of hexokinase. **Annals of the New York Academy of Sciences** 151:351–358.

Levine, N. & Marsh, D.J. (1971). Micropuncture studies of the electrochemical aspects of fluid and electrolyte transport in individual seminiferous tubules, the epididymis and the vas deferens in rats. **The Journal of Physiology (London)** 213:557–570.

Ludvigson, M.A., Waites, G.M.H. & Hamilton, D.W. (1982). Immunocytochemical evidence for the specific localization of aldose reductase in rat Sertoli cells. **Biology of Reproduction** 26:311–317.

Mazzanti, L., Lopez, M. & Berti, M.G. (1964). Selective destruction in testes induced by fluoroacetamide. **Experientia (Basel)** 21:492–493.

McCarrey, J.R. & Thomas, K. (1987). Human testis-specific PGK gene lacks introns and possesses characteristics of a processed gene. **Nature** 326:501–505.

Means, A.R. & Hall, P.F. (1968). Protein biosynthesis in the testis: I.Comparison between stimulation by FSH and glucose. **Endocrinology** 82:597–602.

Means, A.R. & Hall, P.F. (1969). Protein biosynthesis in the testis: II.Role of adenosine triphosphate (ATP) in stimulation by glucose. **Endocrinology** 83:86–96.

Meijer, D., von Lindern, M., van Agthoven, T., Grosveld, G., Talarico, D., Della Valle, G., Mackenbach, P. & Grootegoed, J.A. (1989). Expression of the c-*abl* oncogene in mouse spermatids. This volume.

Milkowski, A.L., Babcock, D.F. & Lardy, H.A. (1976). Activation of bovine epididymal sperm respiration by caffeine: its transient nature and relationship to the utilization of acetylcarnitine. **Archives of Biochemistry and Biophysics** 176:250–256.

Milkowski, A.L. & Lardy, H.A. (1977). Factors affecting the redox state of bovine epididymal spermatozoa. **Archives of Biochemistry and Biophysics** 181:270–277.

Mita, M. & Hall, P.F. (1982). Metabolism of round spermatids from rats: Lactate as the preferred substrate. **Biology of Reproduction** 26:445–455.

Mita, M., Price, J.M. & Hall, P.F. (1982). Stimulation by follicle-stimulating hormone of synthesis of lactate by Sertoli cells from rat testis. **Endocrinology** 110:1535–1541.

Monesi, V. (1971). Chromosome activities during meiosis and spermiogenesis. **Journal of Reproduction and Fertility** Suppl. 13:1–14.

Montamat, E.E. & Blanco, A. (1976). Subcellular distribution of the lactate dehydrogenase isozyme specific for testis and sperm. **Experimental Cell Research** 103:241–245.

Nakamura, M. & Hall, P.F. (1976). Inhibition by 5-thio-D-glucopyranose of protein biosynthesis in vitro in spermatids from rat testis. **Biochimica et Biophysica Acta** 447:474–483.

Nakamura, M. & Hall, P.F. (1977). Effect of 5-thio-D-glucose on protein synthesis *in vitro* by various types of cells from rat testes. **Journal of Reproduction and Fertility** 49:395–397.

Nakamura, M., Romrell, L.J. & Hall, P.F. (1978). The effects of temperature and glucose on protein biosynthesis by immature (round) spermatids from rat testes. **The Journal of Cell Biology** 79:1–9.

Nakamura, M., Hino, A., Yasumasu, I. & Kato, J. (1981a). Stimulation of protein synthesis in round spermatids from rat testes by lactate. **Journal of Biochemistry** 89:1309–1315.

Nakamura, M., Hino, A. & Kato, J. (1981b). Stimulation of protein synthesis in round spermatids from rat testes by lactate. II. Role of adenosine triphosphate (ATP). **Journal of Biochemistry** 90:933–940.

Nakamura, M., Fujiwara, A., Yasumasu, I., Okinaga, S. & Arai, K. (1982). Regulation of glucose metabolism by adenine nucleotides in round spermatids from rat testes. **The Journal of Biological Chemistry** 257:13945–13950.

Nakamura, M., Okinaga, S. & Arai, K. (1984a). Inhibition of glyceraldehyde 3-phosphate dehydrogenase by adenine nucleotides in pachytene primary spermatocytes from rat testes. **Andrologia** 16:446–450.

Nakamura, M., Okinaga, S. & Arai, K. (1984b). Metabolism of round spermatids: Evidence that lactate is preferred substrate. **American Journal of Physiology** 247:E234–E242.

Nakamura, M., Okinaga, S. & Arai, K. (1984c). Metabolism of pachytene primary spermatocytes from rat testes: Pyruvate maintenance of adenosine triphosphate level. **Biology of Reproduction** 30:1187–1197.

Nakamura, M., Suzuki, A., Okinaga, S. & Arai, K. (1984d). Stimulation of spermatid phosphofructokinase by fructose 2,6-biphosphate from rat testes. **Development, Growth and Differentiation** 26:11–16.

Nakamura, M., Kamachi, T., Okinaga, S. & Arai, K. (1986a). Metabolism of round spermatids: Pyruvate cannot maintain the ATP level. **Development, Growth and Differentiation** 28:489–498.

Nakamura, M., Okinaga, S. & Arai, K. (1986b). Studies of metabolism of round spermatids: Glucose as unfavorable substrate. **Biology of Reproduction** 35:927–935.

Nakamura, M., Okinaga, S. & Arai, K. (1986c). Kinetic properties of hexokinase of germ cells in rat testis. **Andrologia** 18:635–638.

Nakamura, M., Yamaguchi, K., Suzuki, A., Okinaga, S. & Arai, K. (1986d). Metabolism of round spermatids: A possible regulation of hexose transport. Development, **Growth and Differentiation** 28:499–504.

Nakamura, M., Okinaga, S. & Arai, K. (1987a). Metabolism of round spermatids: Kinetic properties of pyruvate kinase. **Andrologia** 19:91–96.

Nakamura, M., Okinaga, S. & Arai, K. (1987b). Studies of metabolism of round spermatids: Cytochalasin B binding to cell membrane in relation to glucose transport. **Andrologia** 19:178–182.

Novi, A.M. (1968). Injury of primary spermatocytes during the meiotic prophase and temporary arrest of spermatogenesis induced by fluoroacetamide in the rat. **Virchows Archiv Abt. B, Zellpath**. 1:346–364.

Oksanen, A. (1975). Testicular lesions of streptozotocin diabetic rats. **Hormone Research** 6:138–144.

Oonk, R.B. (1987). Effects of FSH, insulin and IGF-I on rat Sertoli cells. Ph. D. Thesis, Erasmus University Rotterdam.

Oonk, R.B., Grootegoed, J.A. & van der Molen, H.J. (1985). Comparison of the effects of insulin and follitropin on glucose metabolism by Sertoli cells from immature rats. **Molecular and Cellular Endocrinology** 42:39–48.

Oonk, R.B. & Grootegoed, J.A. (1987). Identification of insulin receptors on rat Sertoli cells. **Molecular and Cellular Endocrinology** 49:51–62.
Oonk, R.B. & Grootegoed, J.A. (1988). Insulin-like growth factor-I (IGF-I) receptors on Sertoli cells from immature rats and age-dependent testicular binding of IGF-I and insulin. **Molecular and Cellular Endocrinology** 55:33–43.
Opperdoes, F.R. & Borst, P. (1977). Localization of nine glycolytic enzymes in a microbody-like organelle in *Trypanosoma brucei*: The glycosome. **FEBS Letters** 80:360–370.
Peterson, R.N. & Freund, M. (1970). ATP synthesis and oxidative metabolism in human spermatozoa. **Biology of Reproduction** 3:47–54.
Potter, J.E.R., Millette, C.F., James, M.J.& Kandutsch, A.A. (1981). Elevated cholesterol and dolichol synthesis in mouse pachytene spermatocytes. **The Journal of Biological Chemistry** 256:7150–7154.
Racker, E., Johnson, J.H. & Blackwell, M.T. (1983). The role of ATPase in glycolysis of Ehrlich ascites tumor cells. **The Journal of Biological Chemistry** 258:3702–3705.
Reyes, J., Allen, J., Tanphaichitr, N., Bellvé, A.R. & Benos, D.J. (1984). Molecular mechanisms of gossypol action on lipid membranes. **The Journal of Biological Chemistry** 259:9607–9615.
Rider, C.C. & Taylor, C.B. (1974). Enolase isoenzymes in rat tissues. Electrophoretic, chromatographic, immunological and kinetic properties. **Biochimica et Biophysica Acta** 365:285–300.
Robinson, R. & Fritz, I.B. (1979). Myoinositol biosynthesis by Sertoli cells, and levels of myoinositol biosynthetic enzymes in testis and epididymis. **Canadian Journal of Biochemistry** 57:962–967.
Robinson, R. & Fritz, I.B. (1981). Metabolism of glucose by Sertoli cells in culture. **Biology of Reproduction** 24:1032–1041.
Russell, L.D. (1979). Further observations on tubulobulbar complexes formed by late spermatids and Sertoli cells in the rat testis. **The Anatomical Record** 194:213–232.
Russell, L.D., Weber, J.E. & Vogl, A.W. (1986). Characterization of filaments within the subacrosomal space of rat spermatids during spermiogenesis. **Tissue and Cell** 18:887–898.
Sakai, I., Sharief, F.S. & Li, SS-L. (1987). Molecular cloning and nucleotide sequence of the cDNA for sperm-specific lactate dehydrogenase-C from mouse. **The Biochemical Journal** 242:619–622.
Schatz, G. & Butow, R.A. (1983). How are proteins imported into mitochondria? **Cell** 32:316–318.
Setchell, B.P., Scott, T.W., Voglmayr, J.K. & Waites, G.M.H. (1969). Characteristics of testicular spermatozoa and the fluid which transports them into the epididymis. **Biology of Reproduction** 1:40–66.
Sosa, A., Altamirano, E., Hernandez, P. & Rosado, A. (1972). Developmental pattern of rat testis hexokinase. **Life Sciences** 11:499–510.
Srere, P.A. (1987). Complexes of sequential metabolic enzymes. **Annual Review of Biochemistry** 56:89–124.
Srivastava, D.K. & Bernhard, S.A. (1986a). Metabolite transfer *via* enzyme-enzyme complexes. **Science** 234:1081–1086.
Srivastava, D.K. & Bernhard, S.A. (1986b). Enzyme-enzyme interactions and the regulation of metabolic reaction pathways. **Current Topics in Cellular Regulation** 28:1–68.
Steinberger, E. & Sud, B.N. (1970). Specific effect of fluoroacetamide on spermiogenesis. **Biology of Reproduction** 2:369–375.
Stern, S., Biggers, J.D. & Anderson, E. (1971). Mitochondria and early development of the mouse. **Journal of Experimental Zoology** 176:179.

Stevenson, D. & Jones, A.R. (1985). Production of (S)-3-chlorolactaldehyde from (S)- α-chlorohydrin by boar spermatozoa and the inhibition of glyceraldehyde 3-phosphate dehydrogenase *in vitro*. **Journal of Reproduction and Fertility** 74:157–165.

Storey, B.T. & Kayne, F.J. (1975). Energy metabolism of spermatozoa. V. The Embden-Meyerhof pathway of glycolysis, activities of pathway enzymes in hypotonically treated rabbit epididymal spermatozoa. **Fertility and Sterility** 26:1257–1265.

Storey, B.T. & Kayne, F.J. (1977). Energy metabolism of spermatozoa. VI. Direct intramitochondrial lactate oxidation by rabbit sperm mitochondria. **Biology of Reproduction** 16:549–556.

Storey, B.T. & Kayne, F.J. (1980). Properties of pyruvate kinase and flagellar ATPase in rabbit spermatozoa: Relation to metabolic strategy of the sperm cell. **The Journal of Experimental Zoology** 211:361–367.

Sullivan, J.L., Smith, F.A. & Garman, R.H. (1979). Effects of fluoroacetate on the testis of the rat. **Journal of Reproduction and Fertility** 56:201–207.

Toebosch, A.M.W., Kroos, M.J. & Grootegoed, J.A. (1987). Transport of transferrin-bound iron into rat Sertoli cells and spermatids. **International Journal of Andrology** 10:753–764.

Tombes, R.M. & Shapiro, B.M. (1985). Metabolite channeling: A phosphorylcreatine shuttle to mediate high energy phosphate transport between sperm mitochondrion and tail. **Cell** 41:325–334.

Tuck, R.R., Setchell, B.P., Waites, G.M.H. & Young, J.A. (1970). The composition of fluid collected by micropuncture and catheterization from the seminiferous tubules and rete testis of rats. **European Journal of Physiology** 318:225–243.

Tvermyr, S.M., Froysa, A., Jutte, N.H.P.M. & Hansson, V. (1984). Glucose metabolism in rat germ cells : Mechanism of action of gossypol. **Annals of the New York Academy of Sciences** 438:543–545.

Van de Berg, J.L., Cooper, D.W. & Close, P.J. (1973). Mammalian testis phosphoglycerate kinase. **Nature (New Biology)** 243:48–50.

Van Dop, C., Hutson, S.M. & Lardy, H.A. (1977). Pyruvate metabolism in bovine epididymal spermatozoa. **The Journal of Biological Chemistry** 252:1303–1308.

Vernon, R.G., Go, V.L.W. & Fritz, I.B. (1971). Studies on spermatogenesis in rats. II. Evidence that carnitine acetyltransferase is a marker enzyme for the investigation of germ cell differentiation. **Canadian Journal of Biochemistry** 49:761–767.

Waites, G.M.H. & Gladwell, R.T. (1982). Physiological significance of fluid secretion in the testis and blood-testis barrier. **Physiological Reviews** 62:624–671.

Walt, H. & Hedinger, C. (1983). Motile components in spermatids as related to transport of spermatids and spermatozoa. **Andrologia** 15:34–39.

Whistler, R.L. & Lake, W.C. (1972). Inhibition of cellular transport processes by 5-thio-D-glucopyranose. **The Biochemical Journal 130:919–925.**

Wistow, G.J., Mulders, J.W.M. & de Jong, W.W. (1987). The enzyme lactate dehydrogenase as a structural protein in avian and crocodilian lenses. **Nature** 326:622–624.

Zeilmaker, G.H., Hulsmann, W.C., Wensinck, F. & Verhamme, C. (1972). Oxygen-triggered mouse oocyte maturation *in vitro* and lactate utilization by mouse oocytes and zygotes. **Journal of Reproduction and Fertility** 29:151–152.

Zysk, J.R., Bushway, A.A., Whistler, R.L. & Carlton, W.W. (1975). Temporary sterility produced in male mice by 5-thio-D-glucose. **Journal of Reproduction and Fertility** 45:69–72.

Mammalian spermiogenesis: assembly and fate of perinuclear theca constitutents

ANTHONY R. BELLVÉ[1], KELWYN H. THOMAS[3], R. CHANDRIKA[2], THOMAS M. WILKIE[3] AND MELVIN I. SIMON[3]

Departments of Anatomy and Cell Biology[1], Urology[1], and Obstetrics and Gynecology[2], and Center for Reproductive Sciences [1], [2], College of Physicians and Surgeons, Columbia University, West 168th Street, New York, NY 10032,

and

Division of Biology[3], 147–75, California Institute of Technology, 391 South Holliston Avenue, Pasadena, CA 91125, USA.

Abstract

During mammalian spermiogenesis gene expression occurs in a precise, developmentally-regulated manner. In this chapter we discuss regulation of expression of a number of genes by looking at the abundance of their mRNAs during spermatogenesis with Northern analysis using [^{32}P]-labelled cDNA probes. We identify and discuss four major periods of stage-specific gene expression during mouse spermatogenesis. Using monoclonal antibodies against a group of immunologically-related proteins, we demonstrate their ontogeny and polar distribution in the perinuclear theca of mouse spermatids and discuss their biochemistry.

Keywords

Spermiogenesis, spermatogenesis, mammalian, perinuclear theca, spematid, spermatocyte, spermatogonium, antigenic determinants, meiosis, monoclonal antibodies, Northern analysis, cDNA, mRNA.

Introduction

Mammalian spermatogenesis involves the proliferation and renewal of spermatogonia, the cytodifferentiation and divisions of spermatocytes, and the transformation of haploid spermatids to form spermatozoa (Fig. 1). After the period of spermatogonial proliferation, the primary spermatocytes pass through a brief quiescent phase and then each grows 10-fold in volume during the next 12 to 20 days, the period varying with the species. In this phase the cell's genome transcribes unique gene sequences, chromosomes condense as homolog pairing is facilitated by the synaptonemal complexes (SCs), and the sex chromosomes form the XY heterochromatin (Bellvé, 1979). Following a transient diplotene stage the spermatocytes progress

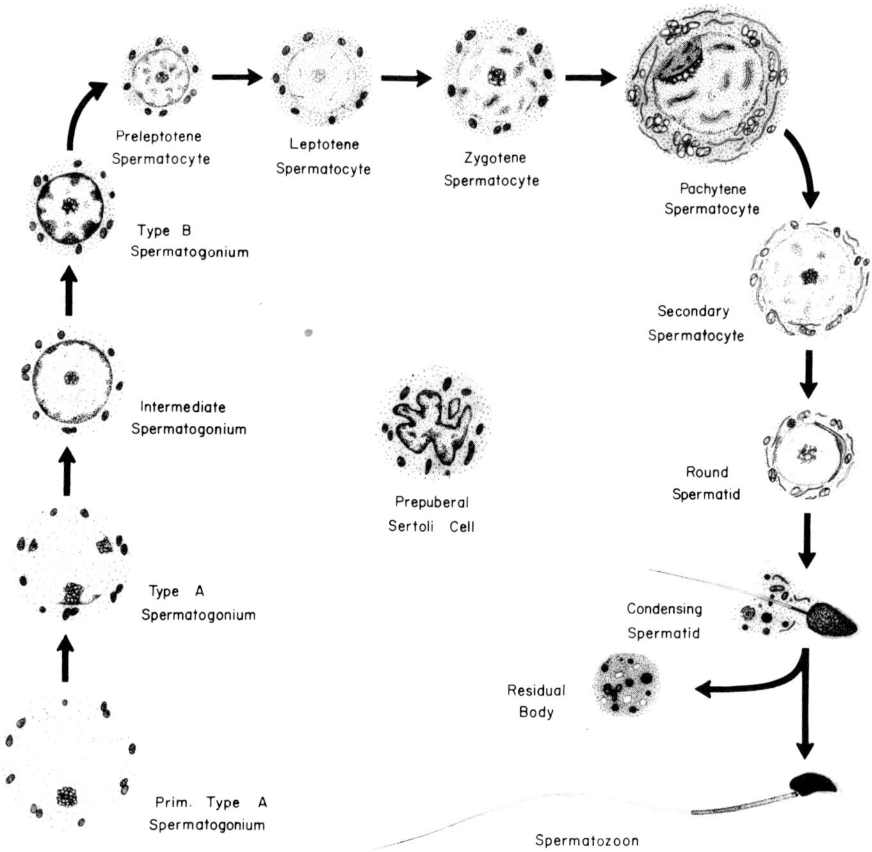

Fig. 1 Schematic diagram depicting germ cell differentiation during the onset of spermatogenesis in the developing mouse testes. Initially, primitive type A spermatogonia renew proliferation at ~days 4 to 6 of postnatal development to produce the series of types A_{1-4}, intermediate, and type B spermatogonia (ascending axis). The latter undergo a mitotic division to generate the first preleptotene spermatocytes on days 9 to 10. These cells pass through a phase of semi-conservative, bulk DNA synthesis, and progress further into meiotic prophase, forming in sequence leptotene, zygotene, pachytene and diplotene spermatocytes (horizontal axis). During this period, the cells undergo extensive gene transcription, synthesize novel proteins, pair homologous chromosomes, and chromosomes exchange gene sequences. Thereafter, the primary spermatocytes pass rapidly through two meiotic divisions, starting on ~day 18, each to yield four haploid spermatids. During spermiogenesis the spermatids differentiate, morphologically and biochemically, to form the spermatozoa (descending axis). Concurrently the prepuberal Sertoli cell

rapidly through two reduction divisions to form spermatids. Then, during spermiogenesis, coincident with the gradual repression of gene transcription, the germ cell undergoes a sequence of morphological changes. These events include assembly of the acrosome over the apical pole of the nucleus, polymerization of the tail from the centriolar complex at the dorsal pole, condensation of chromatin, shaping of the nucleus, and morphogenesis of the perinuclear theca (Bellvé & O'Brien, 1983).

The genome facilitates these differentiative processes through an ordered sequence of transcriptional and translational events. Those proteins destined to be assembled into the mature spermatozoon are synthesized primarily during meiotic prophase and at declining levels throughout spermiogenesis. This pattern of protein synthesis generally reflects the temporal profile of gene transcription, although certain transcripts are stored for prolonged periods prior to their translation. It is in response to this regulatory cascade that the haploid spermatid undergoes a complex sequence of morphological transformations to assume the highly polarized configuration of the mature spermatozoon.

Patterns of protein synthesis

The profile of protein synthesis that occurs during meiosis and spermiogenesis has been defined from the incorporation of isotopic precursors at successive stages of differentiation. The two principal synthetic periods, based on autoradiographic and biochemical studies, occur during the zygotene and pachytene stages of meiotic prophase and during the initial steps of spermiogenesis (Monesi, 1965, 1971; Bellvé 1979). Both periods involve the *de novo* synthesis of a diverse population of cellular proteins, including constituents of the plasmalemma, cytoplasm, mitochondria and nucleus (O'Brien & Bellvé, 1980a, b). During spermiogenesis, however, many proteins are degraded or sequestered within the residual body just prior to spermiation (Bellvé, 1979; Millette & Bellvé, 1980). Presumably these particular polypeptides serve transient functions in assembling the spermatozoon.

Other constituents of the differentiating germ cells, based on the expression of enzyme activities and antigenic determinants, are integrated into the mature spermatozoon (Bellvé & O'Brien, 1983). Direct evidence for the conservation of proteins from the synthetic activities of spermatogenesis has been obtained by quantifying incorporated isotopic amino acids, later

Caption for fig. 1 (*cont.*)

undergoes substantial growth and differentiation, forming intercellular tight junctions and acquiring hormonal responsiveness by days 16 to 18. [Figure taken from Bellvé *et al.*, (1977a), reproduced by permission of the Editor, **Journal of Cell Biology**.]

found to be present in spermatozoa recovered from the caudae epididymides and vasa deferentia. Thus, [^{3}H]leucine incorporated by germ cells during meiotic prophase and spermiogenesis is conserved within a variety of proteins present in spermatozoa (O'Brien & Bellvé, 1980b). On this basis, the complex, SDS-soluble constituents of the sperm cell are synthesized primarily during meiotic prophase (Fig. 2). This form of protein synthesis continues in the haploid spermatids, but occurs at declining levels as spermiogenesis progresses. By contrast, the SDS-insoluble proteins of the spermatozoon are synthesized almost exclusively during spermiogenesis, many coincident with the final assembly of the nucleus, the perinuclear theca, and the flagellum (Fig. 2) (O'Brien & Bellvé, 1980b). On this basis, most of the

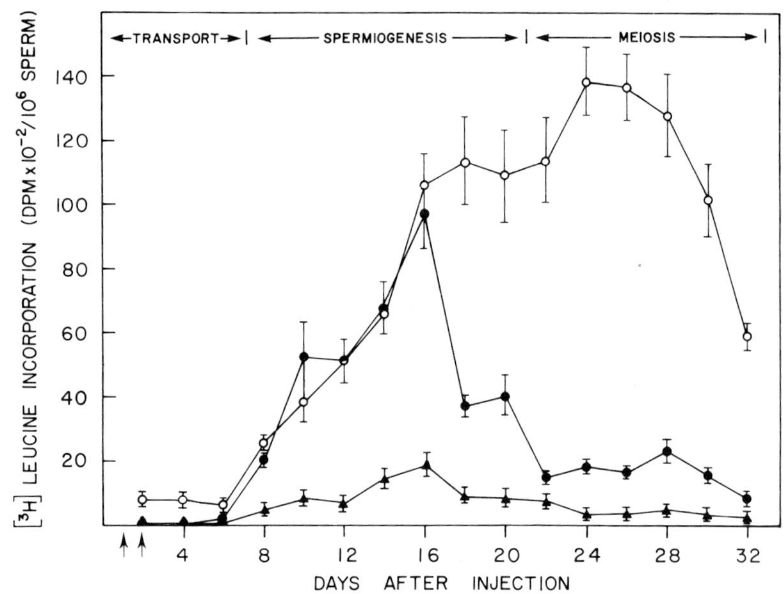

Fig. 2 Temporal synthesis of proteins during mouse spermatogenesis that are destined for assembly into the spermatozoon. Mice were given two intratesticular injections of [^{3}H]leucine, 24 h apart, in both testes (arrows). Three animals were sacrificed on every alternate day during the next 32-day period. Spermatozoa were recovered separately from each pair of caudae epididymides and then processed to quantify the [^{3}H]leucine incorporation and retention by three different intracellular compartments. The data demonstrated the following: SDS-soluble proteins of the sperm cell, principally constituents of the plasmalemma, acrosome, axoneme, and mitochondrial matrices and cristae were synthesized primarily during meiotic prophase and at declining levels throughout spermiogenesis. SDS-insoluble tail proteins, including structural elements of the connecting piece, outer dense fibers, fibrous sheath, and mitochondrial membranes were synthesized in low amounts during meiotic prophase, and then at

covalently-linked, structural proteins of the spermatozoon are produced during spermiogenesis.

Some of the proteins synthesized during spermiogenesis have been identified. Protamines, based on the incorporation of [^{3}H]arginine, are known to be synthesized during the final steps of spermiogenesis, concomitant with the terminal stages of chromatin condensation (Bellvé *et al.*, 1975, 1988b; Mayer & Zirkin, 1979; Mayer *et al.*, 1981). Other nuclear and flagellar proteins undergo *de novo* synthesis coincident with, and even after that of the protamines (O'Brien & Bellvé, 1980b). The identities of these proteins have yet to be resolved. It is not known whether any cytosolic and/or structural proteins, nuclear, acrosomal or flagellar, are synthesized after spermiation or during the period of sperm maturation in the excurrent ducts. Other evidence suggests that a few mitochondrial proteins may be synthesized by the mature spermatozoon (Bragg & Handel, 1979).

During meiosis and spermiogenesis many genes are transcribed for the purpose of producing proteins needed for the differentiation of the spermatozoon. The transcriptional events of meiotic prophase generate mRNAs encoding the germ cell-specific lactate dehydrogenase-C (LDH-C) (Wieben, 1981), phosphoglycerate kinase-2 (PGK-2) (Gold *et al.*, 1983), heat shock protein, hsp 70.2 (Zakeri *et al.*, 1988), and those containing the homeobox, Hox 1.4 (Wolgemuth *et al.*, 1986, 1987; Rubin *et al.*, 1986). The respective periods of mRNA translation are known in some cases. The polypeptides LDH-C (Meistrich *et al.*, 1977) and PGK-2 (Erickson *et al.*, 1980) are synthesized concurrent with the respective transcriptional processes during late meiotic prophase and spermiogenesis. However, for the Hox 1.4 and hsp 70.2 products, it is not known when translation occurs, nor when each protein serves its function during spermatogenesis. Presumably, both polypeptides serve important and perhaps unique regulatory roles in germ cell differentiation.

Haploid gene expression has been shown to occur at different steps of spermiogenesis. Unique transcripts exist for the germ cell-specific proteins, α and β tubulin (Distel *et al.*, 1984; Hecht *et al.*, 1984; Villisante *et al.*, 1986, see chapter by Hecht in this volume), actin (Waters *et al.*, 1985), t-haplotype protein, TCP-1 (Willison *et al.*, 1986), hsp 70.1 (Zakeri & Wolgemuth, 1987), c-*mos* (Goldman *et al.*, 1987), and the protamines, P1 and P2 (Kleene *et al.*, 1985; Yelick *et al.*, 1987). The encoded proteins are

Caption for Fig. 2 (cont.)

relatively high levels during spermiogenesis. SDS-insoluble nuclear proteins, inclusive of the nucleus and the perinuclear theca, were synthesized exclusively during spermiogenesis. (o) SDS-soluble sperm proteins, (●) SDS-insoluble tail proteins, (▲) SDS-insoluble sperm head proteins. [Data modified after O'Brien & Bellvé (1980b), reproduced by permission of the Editor, **Developmental Biology**.]

synthesized primarily, if not exclusively, during mid to late spermiogenesis, when the haploid germ cells are transforming in shape. The two structural proteins, actin and tubulin, presumably play distinct roles in the transformation of cell shape and/or in the assembly of the flagellum. In addition to these products, unusual sequences of the proto-oncogene, c-*abl* (Ponzetto & Wolgemuth, 1985; Oppi *et al.*, 1987), are transcribed during mid spermiogenesis while c-*mos* (Goldman *et al.*, 1987) and *int*-1 (Shackleford & Varmus, 1987) are transcribed during late spermiogenesis. The germ cell-specific proteins encoded by the c-abl, c-mos and int-1 transcripts, each need to be defined in terms of their identity, temporal expression, intracellular distribution and function.

Significantly, the *int*-1 gene is expressed in the neural tube during mid gestation (Shackleford & Varmus, 1987), and so the encoded protein presumably serves an analogous, and perhaps an identical function in the two disparate developmental processes, neurogenesis and spermiogenesis. Conceptually, this developmentally- and spatially-regulated expression of the *int*-1 gene has important implications. It is conceivable, for instance, that many supposedly 'germ cell-specific' proteins may be expressed by somatic cells, each at discrete sites and intervals during early development. In the adult animal, most tissues have reached a terminal state of differentiation, whereas spermatogenesis continues to be an ongoing process of cellular development, each cycle progressing from stem cell through to a specialized spermatozoon. In this context then, spermatogenesis may require the transient expression of 'developmentally-regulated' genes at various stages of the differentiative process.

Expression of antigenic determinants

Transformation of the apolar spermatid to form the highly specialized spermatozoon involves the assembly of many unique macromolecules. These novel constitutents, because of their normal sequestration from the immune system, are not recognized as 'self' and therefore are able to elicit strong syngeneic antibody responses. In addition, the antigenicity of many spermatogenic proteins also ensures avid responses to xenogeneic immunization. On this basis, a number of studies have utilized the specificity of antibodies to study the expression, topographical distribution, biochemical properties and physiological functions of novel sperm constituents. In these efforts, emphasis has been placed primarily on elucidating the ontogeny and polar localization of selected components of the plasmalemma, acrosome and flagellum. Few studies have focused on the question of whether constituents of the spermatozoon nucleus are distributed in a polarized pattern.

Expression of germ cell-specific, nuclear proteins during mouse spermatogenesis has been examined by preparing two panels of monoclonal antibodies (mAbs), both directed primarily against structural elements of the nucleus. For this purpose, nuclear matrices were isolated from pachytene spermatocytes and round spermatids recovered from mice at day 25 of development, and perinuclear matrices from spermatozoa collected from the caudae epididymides and vasa deferentia of adult animals. These structures were prepared by removing DNA (~98%), RNA (~97%), and protein (~85%) from the nucleus, by a sequence of digestions with DNase and RNase, and extractions with monovalent and/or divalent cations (Ierardi *et al.*, 1983, 1989; Bellvé *et al.*, 1989a). The nuclear matrices, comprised of ~15–20% of total nuclear protein, were used to immunize mature female Sprague Dawley rats (Bellvé & Moss, 1983), initially with a single intraperitoneal injection of ~25 μg of protein in complete Freund's adjuvant, followed by a series of injections of antigen with incomplete Freund's adjuvant (Bellvé *et al.*, 1989a).

Activated spleen cells were fused with the nonsecretor, mouse myeloma cell line P3-X63-Ag8.63. The surviving, antibody-producing hybridomas were identified by applying an IgG-oriented, enzyme-assisted, solid-phase immunoassay (ELISA) (Myles *et al.*, 1981), with total spermatogenic nuclear proteins being used as target antigens (Bellvé *et al.*, 1989a). This screening process has identified ~150 IgG-secreting hybridomas producing Ab against specific epitopes on a variety of nuclear proteins. These interspecies cell hybrids were cloned from soft agar and propagated in the presence of 20% serum in HL-1 medium, and then later were weaned successively into 10, 5, 1, 0.5 and 0% serum in HL-1 medium. The latter was used for the *en masse* production of the respective mAbs in serum-free conditions. Individual mAbs were isotyped with specific antibodies by Ouchterlony analysis.

Selected mAbs were purified by affinity chromatography prior to their application to characterize the respective antigenic determinants. Purification was undertaken by concentrating the protein to ≤ 0.5 mg/ml by ultrafiltration, and then by applying the material to a column of silica-conjugated, anti-rat-IgG, mouse IgG (Bellvé *et al.*, 1989a). [The mouse IgG was a secretory product of the high-producing hybridoma MAR 18.5.3, and was purified by affinity chromatography on Protein A-sepharose (Bellvé *et al.*, 1989a)]. Bound rat mAbs were eluted from the anti-rat−IgG, mouse IgG, affinity columns with 0.3 M isothiocyanate in phosphate-buffered saline (pH 7.4).

The purified mAbs were selected more rigorously by applying a combination of immunofluorescence and immunoblot assays. Priority was placed on identifying those mAbs recognizing proteins that are: (a) specific for germ cells, (b) exhibit a stage-dependent pattern of expression, (c) show a polarized distribution, and (d) bind to SDS-denatured determinants on immunoblots. On these bases, two mAbs, denoted as PNT-1 and PNT-2, were selected. Both mAbs did not bind to adult brain, thymus, liver, adrenal, kidney, prostate, or prenatal and adult ovaries, either by indirect immunofluorescence on frozen sections, or by immunoblot analyses of denatured proteins. Only skeletal muscle gave a positive reaction to a single, 43kD protein on immunoblotting with mAb PNT-1 (Bellvé *et al.*, 1989b). Further studies were undertaken to determine the ontogeny, polar distribution and function of the antigens recognized by mAbs PNT-1 and PNT-2.

Reactivity of mAb PNT-1, isotype IgG_{2b}, was observed to occur first, at low levels, within the nucleus of type B spermatogonia at day 10 of postnatal development, as assessed by indirect immunofluorescence analysis of paraformaldehyde-fixed, frozen testis sections (Bellvé *et al.*, 1989b). Thereafter, the intensity of binding within the germ cell nucleus gradually

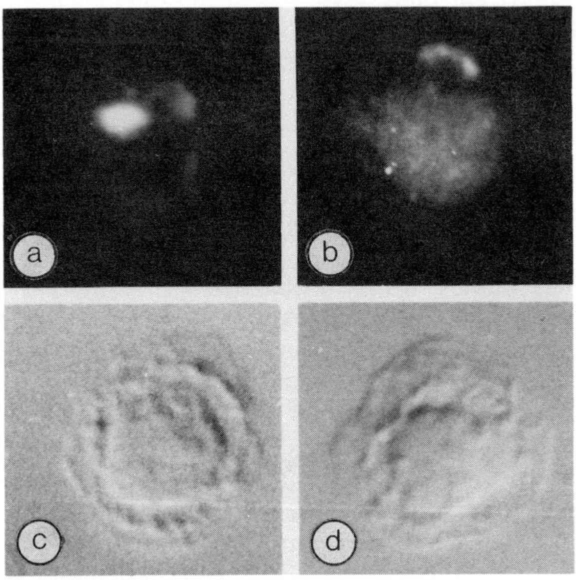

Fig. 3 Immunofluorescence and differential interference micrographs of mAbs PNT-1 and PNT-2 binding to prepuberal mouse pachytene spermatocytes. Testes at day 17 of postnatal development were dissociated by sequential digestion with trypsin and collagenase. Germ cells in the resulting suspension were separated by sedimention velocity at unit gravity (Bellvé *et al.*, 1977a). All purified cell populations were fixed in 4% paraformaldehyde in PBS, exposed to 10% rabbit serum in PBS to block nonspecific binding sites, reacted with the appropriate mAb, and then tagged with a fluorescein-conjugated, anti-rat-IgG, rabbit IgG. Binding of mAb PNT-2 occurred primarily to the XY heterochromatic body, located adjacent to the nuclear envelope. By contrast, mAb PNT-1 bound in a diffuse pattern throughout the nucleus, and more intensely to a discrete body in the cytoplasm. (a) mAb PNT-2 immunofluorescence (b) mAb PNT-1 immunofluorescence; (c, d) corresponding differential interference micrographs.

increased throughout meiotic prophase, reaching a maximum in pachytene spermatocytes. Strong reactivity was also evident to a single, ~2 to 3μm, irregularly-shaped body located within the cytoplasm of advanced pachytene spermatocytes (Fig. 3), a structure tentatively identified to be the Golgi complex.

After the two meiotic reduction divisions, mAbs PNT-1 binding was not detectable throughout the first few steps of spermiogenesis. At approximately step 4 of differentiation, however, PNT-1 was observed to bind intensely to a clearly delineated bouton on the apical pole of the spermatid nucleus. This punctate zone gradually expanded peripherally to form a cap enveloping the anterior pole of the spermatid nucleus (Fig. 4). In transverse optical sections the binding pattern was seen as a sharply circumscribed ring, thereby substantiating the cap-like form of the structure (Bellvé *et al.*, 1989b). As spermiogenesis progressed this cap gradually extended caud-

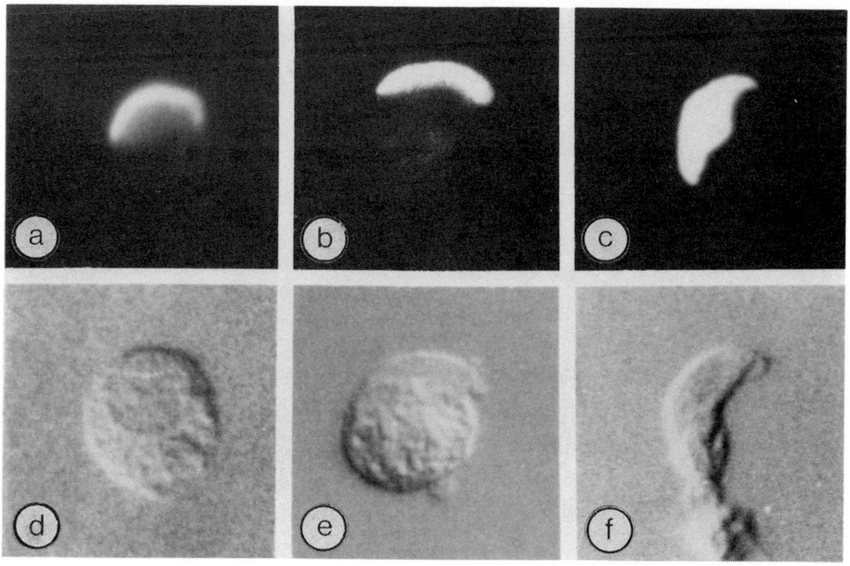

Fig. 4 Immunofluorescence and differential interference micrographs of mAbs PNT-1 and PNT-2 binding to spermatids at different steps of spermiogenesis. The procedures applied were those described previously (Fig. 3, legend). Both mAbs failed to recognize structures within spermatids at steps 1 to 3 (not shown). Thereafter, mAbs PNT-1 and PNT-2 both bound strongly and exclusively to the anterior surface of the spermatid nucleus, initially appearing as a cap-like structure and then gradually extending caudally to envelope most of the condensed and elongated sperm nucleus. Data presented are based on mAb PNT-1. Immunofluorescence (a, b, c); Differential interference (d, e, f). Spermatids at steps 4 (a, d), 12 (b, e), and 16 (c, f).

ally to envelope the anterior pole of the nucleus, approximately down to the limits of the tangentially-orientated, equatorial region. Also, during the condensation phase the cap remained 'firmly adherent' to the nucleus, and therefore gradually acquired a falciform configuration over the anterior pole (Fig. 4) (Bellvé *et al.*, 1989b). Limited binding of mAb PNT-1 also occurred in a narrow zone immediately adjacent to the implantation fossa, extending from the site of tail abutment caudad to the posterior limits of the nucleus.

The mAb PNT-2, isotype IgG_{2a}, showed a different pattern of expression during meiotic prophase. Initially, antibody binding was evident within the pachytene spermatocyte nucleus, as diffuse reactivity throughout the chromatin and, at a more intense level, to a single, round body residing adjacent to the nuclear envelope (Fig. 3) (Bellvé *et al.*, 1989b). The latter was identified, by co-labelling studies with the DNA-intercalating dye Hoechst 33258, to be the condensed XY heterochromatin.

Analogous to mAb PNT-1, no binding of mAb PNT-2 was detected during the early steps of spermiogenesis. However, in step 4 spermatids, localized labelling occurred to the same cap-like structure on the apical pole of the nucleus. This reactive zone expanded peripherally until it enveloped all regions of the nucleus anterior to the equatorial region. In this respect, mAbs PNT-1 and PNT-2 apparently recognized similar if not identical structures at the anterior pole of haploid germ cells (Fig. 4). It is noteworthy, however, that those antigens reacting with the two mAbs differed in their patterns of expression and distribution during meiotic prophase, suggesting that the two mAbs may recognize distinct epitopes (Fig. 3).

Biochemical characterization of the differentiation antigens

Polypeptides recognized by the two mAbs during gonadal development and spermatogenesis were identified and characterized by SDS PAGE-immunoblot analyses, Seminiferous cords or tubules, when isolated at different ages of postnatal development and subjected to immunoblot analyses, were shown to contain a family of reactive proteins with M_rs of 83, 79, 76 and 72kD. These were expressed coordinately at day 24 and then at all later ages of development (Bellvé *et al.*, 1989b). This developmental pattern corresponded with an onset of antigen expression in early spermatids, rather than in primary spermatocytes (cf. Bellvé *et al.*, 1977a, b). Furthermore, analogous studies with isolated populations of germ cells, ranging from primitive type A spermatogonia to mature spermatozoa, revealed that no reactive species were present during spermatogenesis until the same family of proteins was detected in both round (steps 1 to 8) and condensing (steps 12 to 16) spermatids (Fig 5). The onset of expression in round spermatids was consistent with the appearance of these proteins in the seminiferous epithelium at day 24 of development.

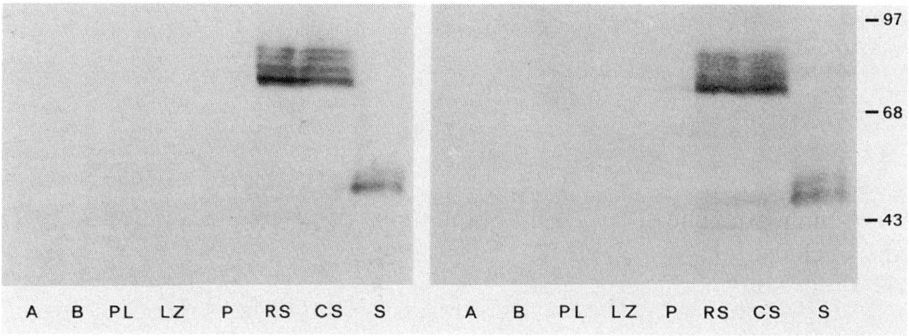

Fig. 5 SDS PAGE-immunoblot analysis of total protein from germ cells representing all stages of mouse spermatogenesis. Testes at days 6, 8, 17 and 84 postnatal were dissociated and the respective germ cells were separated by applying sedimentation velocity at unit gravity (Romrell *et al.*, 1976; Bellvé *et al.*, 1977a). The family of 83–72kD proteins were detected only in the round and condensing spermatids. By contrast, spermatozoa had two polypeptides of 50 and 48kD. Cell types included the following: Type A (A) and type B (B) spermatogonia; preleptotene (PL), leptotene/zygotene (LZ) and pachytene (P) spermatocytes; round spermatids, steps 1 to 8 (RS), condensing spermatids, steps 12 to 16 (CS); and spermatozoa (S) from the caudae epididymides and vasa deferentia.

Significantly, the 83 to 72kD proteins were not detectable in spermatozoa recovered from the caudae epididymides and vasa deferentia. Instead these cells contained two immunoreactive polypeptides with M_rs of 50 and 48kD (Fig 5). Based on two-dimensional, isoelectric-focussing PAGE (IEF), all members of the 83 to 72kD family have pI's of 7.4, whereas the prominent 48kD protein was 7.6. Furthermore, following biochemical fractionation of spermatogenic cells (Ierardi *et al.*, 1988; Bellvé *et al.*, 1989b), the respective proteins were recovered in the nuclear matrix fraction, as opposed to the cytosolic or nucleoplasmic compartments (Fig. 6). All proteins of the 83 to 72kD spermiogenic family and the 50 and 48kD sperm constitutents were removed selectively from the respective nuclei and nuclear matrices by an incubation in DTT and 1% cetyltrimethyl ammonium bromide (CTAB) (Fig 6). The latter observation is consistent with these proteins being constituents of the perinuclear theca that envelops the anterior pole of spermatids and spermatozoa (cf. Bellvé *et al.*, 1989b). Finally, mAbs PNT-1 and PNT-2 gave comparable results on the developmental, spermatogenic, cell-fractionation, and IEF immunoblots, implying that the two antibodies were recognizing a closely related if not an identical family of proteins.

The ontogeny and distribution of these proteins in haploid germ cells suggests an important function in the assembly and maturation of the

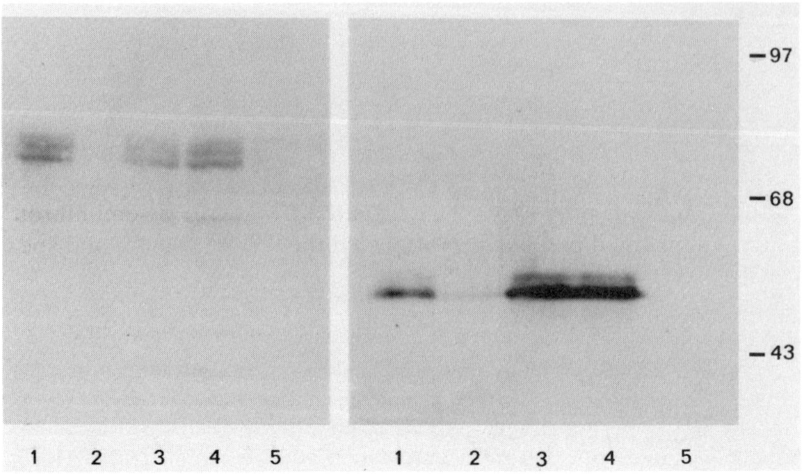

Fig. 6 SDS PAGE-immunoblot analyses of fractionated germ-cell nuclear proteins. Nuclei, isolated from pachytene spermatocytes and spermatids (steps 1 to 16) and from mature spermatozoa were digested with 50 μg DNase and 50 μg RNase/ml and, simultaneously, were extracted stepwise with 125, 150, 175, 200 and 250 mM $CaCl_2$ $MgCl_2$ (3:2; mol:mol) (Bellvé *et al.*, 1988c). The resulting nuclear matrices were extracted with 1% cetyltrimetyl ammonium bromide (CTAB), a reagent that selectively solubilizes the perinuclear theca. Panel A: spermatocyte and spermatid nuclei; Panel B: sperm nuclei. Individual lanes in both panels included the following: (1) total nuclear proteins; (2) soluble nuclear proteins released by $CaCl_2$, $MgCl_2$, DNase, RNase treatments; (3) insoluble nuclear proteins after $CaCl_2$, $MgCl_2$, DNase, RNase; (4) CTAB-soluble proteins; (5) CTAB-insoluble proteins.

perinuclear theca. In this regard, the apparent shifts in M_r from the 83–72kD quadruplet to the 50–48kD doublet during sperm maturation could arise from one of two mechanisms. First, the 83–72kD proteins may be eliminated from the nucleus late in spermiogenesis. These would be replaced by the 'unrelated' 50–48kD, sperm perinuclear theca constituents that share the same epitope. Deposition of the latter two proteins within the nucleus during sperm maturation, a supposed period of translational quiescence, would pose intriguing questions concerning their intracellular origin. Second, it is possible that the 83–72kD protein family may undergo specific proteolysis to yield the 50–48kD products during sperm maturation. This possibility would raise important issues regarding the mechanisms and the presumed enzymes involved in the processing event, and the role such proteins would have in the evolution of the nucleus and the perinuclear theca.

Generation of germ cell cDNA libraries

Mechanisms regulating the temporal expression, distribution and processing of the 83–72kD proteins may be resolved more readily by applying a combination of cellular and molecular approaches. For this and other reasons, both λgt10 and λgt11 cDNA libraries were prepared by using oligo dT affinity-purified, poly[A^+] RNA, isolated from day 15 seminiferous cords and from purified populations of pachytene spermatocytes and spermatids (steps 1–8) (Thomas *et al.*, 1988). These constructs were screened and selected by differential hybridization against cDNAs prepared from liver, pachytene spermatocyte and spermatid poly[A^+] RNA and [^{32}P]-labelled DNA by nick-translation. This protocol was successful in identifying 200 cDNA clones of which 60 were found to hybridize weakly with liver cDNA, 92 hybridized strongly with pachytene spermatocyte cDNA, and 48 were complementary to spermatid cDNA. Those additional cDNA clones that bound strongly to the [^{32}P]-labelled, liver cDNA probes were not studied further.

Characterization of the complementary mRNAs and their patterns of expression during spermatogenesis, was accomplished by undertaking a series of Northern analyses. The first Northern blots were prepared from poly[A^+] RNA isolated from seminiferous epithelia at different ages of development, ranging from days 6 to 60 after birth. Other blots were constructed from total RNA prepared from purified populations of spermatogenic cells (Thomas *et al.*, 1989). Hybridization of these blots with nick-translated, [^{32}P]-labelled, restriction fragments from the selected cDNA clones showed marked differences in the patterns of mRNA expression during postnatal development of the gonads and during spermatogenesis.

The integrity and validity of the Northern blots were checked by hybridizing with cDNA probes against gene products known to be expressed by cells of the seminiferous epithelium. Hybridization of the developmental blots with cDNA probe for the sulphoglycoprotein-2 (SGP-2) mRNA demonstrated that abundant levels of this transcript were present in the seminiferous epithelium throughout development (Fig. 7). Notably, however, SGP-2 mRNA was not detectable in the RNA samples prepared from the purified germ cells (Fig. 8) (Thomas *et al.*, 1989). The absence of SGP-2 transcripts verified the lack of somatic cell contamination among the germ cell populations, including those of type B spermatogonia and leptotene/zygotene spermatocytes (cf. Bellvé *et al.*, 1977a), In addition, three known germ cell-specific cDNA probes, complementary to LDH-C, PGK-2 and P1 mRNAs, each hybridized with the developmental and germ-cell Northern blots with the expected differentiative patterns. The cDNA probe for LDH-C hybridized to gene transcripts expressed initially in leptotene/zygotene spermatocytes, and more abundantly in pachytene spermatocytes

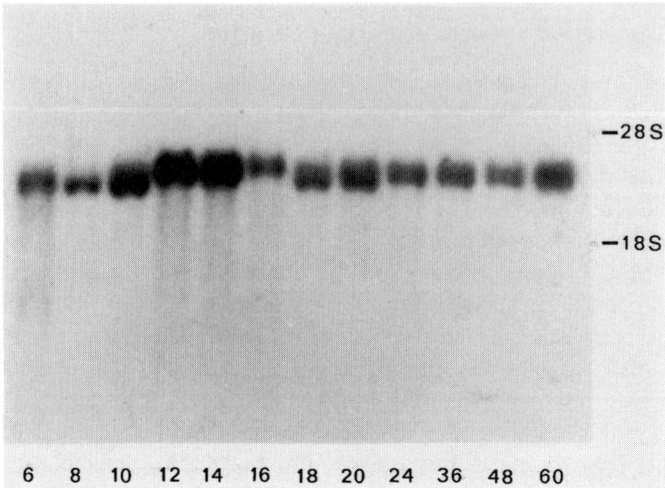

Fig. 7 Northern analysis of sulphoglycoprotein-2 (SGP-2; mRNA accumulation during development of the mouse seminiferous epithelium. Poly[A$^+$] RNA was prepared by oligo dT-affinity chromatography, subjected to denaturing, agarose gel electrophoresis, transferred to Genescreen, and then hybridized with a nick-translated, [^{32}P]-labelled, restriction fragment of cDNA complementary to the SGP-2 mRNA. Within the seminiferous epithelium, SGP-2 is expressed at all stages of development. Postnatal ages included the following: days 6, 8, 10, 12, 14, 16, 18, 20, 24, 36, 48 and 60. [Data are from Thomes *et al.* (1989).]

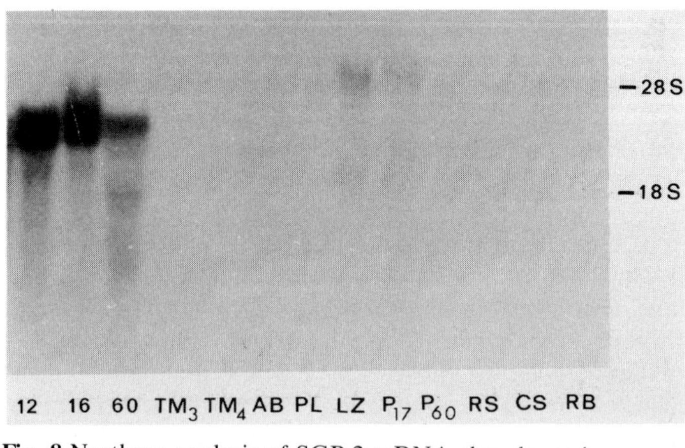

Fig. 8 Northern analysis of SGP-2 mRNA abundance in germ cells at different stages of differentiation. Spermatogenec cells were purified from the seminiferous epithelium at the appropriate developmental ages (Bellvé *et al.*, 1977a). Total RNA was prepared and subjected to denaturing, agarose gel electrophoresis, and then hybridized to the same [^{32}P]-labelled,

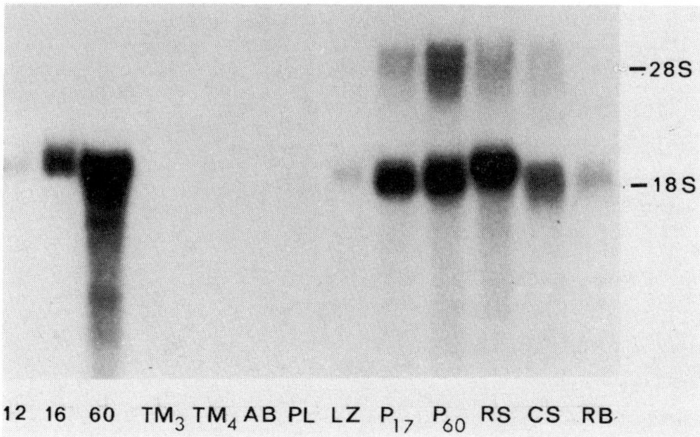

Fig. 9 Northern analysis of lactate dehydrogenase-C (LDH-C) mRNA accumulation in germ cells at different stages of differentiation. The Northern blot of total germ cell RNA was prepared as described previously (Fig. 8, legend) and then probed with a [^{32}P]-labelled, restriction fragment complementary to LDH-C mRNA. Onset of LDH-C expression is evident in leptotene/zygotene and pachytene spermatocytes (P_{17} and P_{60}) and is present at abundant levels in round and elongating spermatids. Samples included the following: seminiferous epithelium at postnatal days 12 (12), 16 (16) and 60 (60); TM_3 Leydig cells (TM_3) and TM_4 Sertoli cells (TM_4); and the germ cells: Types A and B spermatogonia (AB), preleptotene spermatocytes (PL), leptotene/zygotene spermatocytes (LZ), prepuberal pachytene spermatocytes (P_{17}), adult pachytene spermatocytes (P_{60}); round spermatids, steps 1 to 8 (RS) condensing spermatids (CS), and residual bodies (RB). [Data are from Thomas *et al.*, (1988).]

and round spermatids (Fig. 9). The PGK-2 cDNA probe hybridized to gene transcripts first appearing in pachytene spermatocytes and present at more abundant levels in haploid spermatids (Fig. 10). The presence of LDH-C and PGK-2 transcripts in germ cells during meiotic prophase can be attri-

Caption for Fig. 8 (*cont.*)

restriction fragment complementary to SGP-2 mRNA. SGP-2 sequences were abundant only in the developing seminiferous epithelium. No transcripts were detected in TM_3 or TM_4 cells or in the various germ cell populations. Samples included the following: seminiferous epithelium at postnatal days 12 (12), 16 (16) and 60 (60); TM_3 Leydig cells (TM_3) and TM_4 Sertoli cells (TM_4); and the germ cells: Types A and B spermatogonia (AB); preleptotene spermatocytes (PL), leptotene/zygotene spermatocytes (LZ), prepuberal pachytene spermatocytes (P-$_{17}$), adult pachytene spermatocytes (P_{60}); round spermatids, steps 1 to 8 (RS), condensing spermatids, steps 12 to 16 (CS), and residual bodies (RB). [Data are from Thomas *et al.* (1988).]

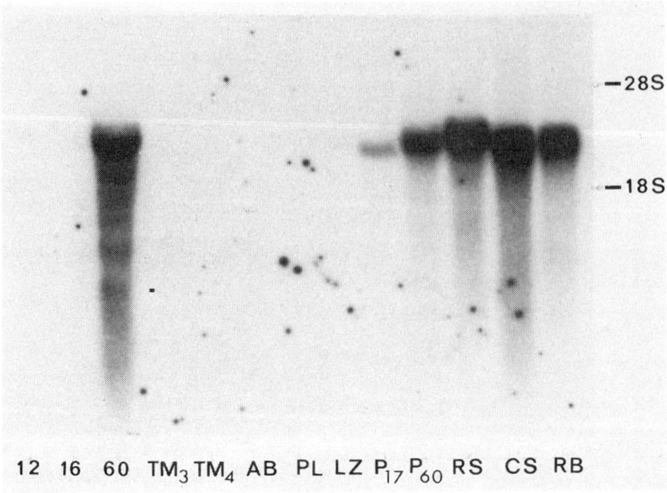

Fig. 10 Northern analysis of PGK-2 transcript abundance in mouse germ cells at different stages of differentiation. The blot of total germ cell RNA was hybridized with a [^{32}P]-labelled, cDNA probe complementary to PGK-2 mRNA. This gene is expressed initially in pachytene spermatocytes (P_{17} and P_{60}) and then at greater levels in spermatids. The samples included the following: seminiferous epithelium at postnatal days 12 (12), 16 (16) and 60 (60); TM_3 Leydig cells (TM_3) and TM_4 Sertoli cells (TM_4); and the germ cells: Types A and B spermatocytes (AB), preleptotene spermatocytes (PL), leptotene/zygotene spermatocytes (LZ), prepuberal pachytene spermatocytes (P_{17}), adult pachytene spermatocytes (P_{60}); round spermatids, steps 1 to 8 (RS) condensing spermatids (CS), and residual bodies (RB). [Data are from Thomas *et al.* (1989).]

buted solely to gene expression in spermatocytes. Possible contributions from haploid gene transcription are negated by the complete absence of spermatid contamination in the prepuberal spermatocyte populations (cf. Figs. 9 and 10; Lane P_{17} *versus* P_{60}). Finally, the P1 cDNA probe only detected expression of the protamine gene during spermiogenesis (Fig 11). The absence of P1 transcripts from the prepuberal pachytene spermatocytes (P_{17}) is consistent with this conclusion. By these criteria, the germ cell Northern blots were shown to contain intact RNA, to be free of contamination by somatic cell transcripts, and primarily to reflect the temporal distribution typical of the designated stages of germ cell differentiation.

Subsequent examination of other cDNA probes by Northern analyses revealed four phases of gene transcription occurring during mouse spermatogenesis (Thomas *et al.*, 1989). First, two cDNA clones hybridized to transcripts present at abundant levels in spermatogonia and at either greatly reduced or undetectable levels in primary spermatocytes and/or spermatids (Fig 12). These gene products were detected in some somatic cells, but at

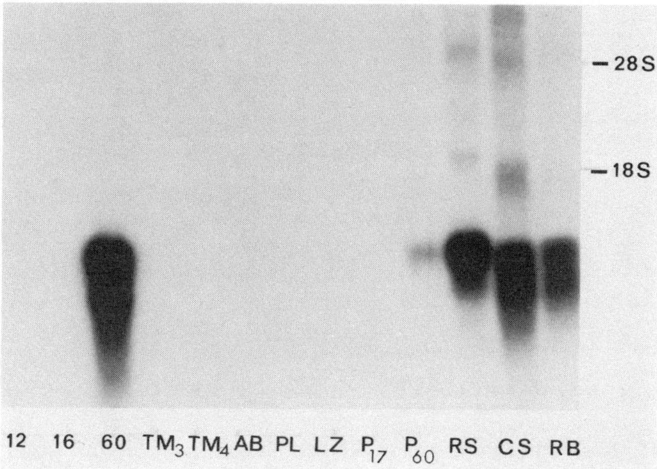

Fig. 11 Nothern analysis of P2 mRNA abundance in mouse germ cells at different stages of differentiation. The Northern blot of total germ cell RNA, was hybridized with a [^{32}P]-labelled, cDNA probe complementary to P2 sequences. P2 mRNA was abundant only in round and elongating spermatids. Hybridization to the P_{60} sample was attributed to spermatid contamination (cf. lane P_{17}). Samples included the following: seminiferous epithelium at postnatal days 12 (12), 16 (16) and 60 (60); TM_3 Leydig cells (TM_3) and TM_4 Sertoli cells (TM_4); and the germ cells: Types A and B spermatogonia (AB), preleptotene spermatocytes (PL), leptotene/ zygotene spermatocytes (LZ), prepuberal pachytene spermatocytes (P_{17}), adult pachytene spermatocytes (P_{60}); round spermatids, steps 1 to 8 (RS) condensing spermatids, steps 12 to 16 (CS), and residual bodies (RB). [Data are from Thomas *et al.* (1989).]

much lower abundancies. Some other cDNA clones that hybridized to spermatogonial transcripts also detected expression in all primary spermatocytes and even in spermatids. Second, certain gene products were detected initially in leptotene and zygotene spermatocytes and were abundant throughout the remainder of meiotic prophase and early spermiogenesis, with some becoming undetectable in condensing spermatids and others being absent from residual bodies (Fig. 13) (Thomas *et al.*, 1989). Third, a number of gene transcripts first appeared in pachytene spermatocytes and then increased in abundance during early to mid spermiogenesis (Fig. 14). These transcripts invariably were present at lower abundancies in condensing spermatids and/or residual bodies. Fourth, a variety of novel transcripts were expressed abundantly only during particular steps of spermiogenesis (Fig. 15). These gene products, all originating from haploid gene transcription, encoded for proteins not expressed by somatic cells. Finally, the various patterns of RNA abundance reflect a combination of

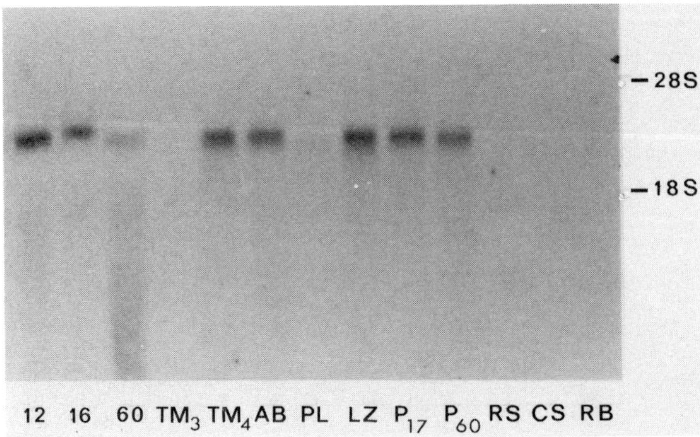

Fig.12 Northern analysis of RNA abundance during mouse spermatogenesis, as detected with a [^{32}P]-labelled, restriction fragment from cDNA clone C3D. Expression of this 3500-bp sequence was detected in types A and B spermatogonia, at much lower levels in preleptotene spermatocytes, and then at greater levels in more advanced spermatocytes. Samples included the following: seminiferous epithelium at postnatal days 12 (12), 16 (16) and 60 (60); TM_3 Leydig cells (TM_3) and TM_4 Sertoli cells (TM_4); and the germ cells: Types A and B spermatogonia (AB), preleptotene spermatocytes (PL), leptotene/zygotene spermatocytes (LZ), prepuberal pachytene spermatocytes (P_{17}), adult pachytene spermatocytes (P_{60}); round spermatids, steps 1 to 8 (RS), condensed spermatids, steps 12 to 16 (CS), and residual bodies (RB). [Data are from Thomas *et al.* (1989).]

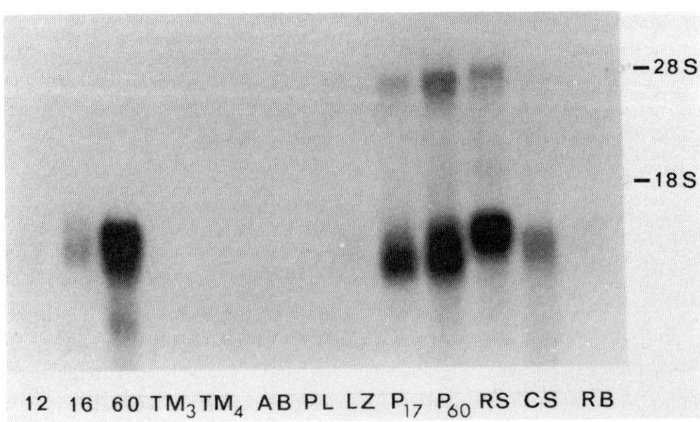

Fig. 13 Northern analysis of RNA abundance during mouse spermatogenesis on hybridizing with a [^{32}P]-labelled, restriction fragment from cDNA clone D20B. Expression of this 700-bp sequence is detected first in leptotene/zygotene spermatocytes and thereafter the transcripts

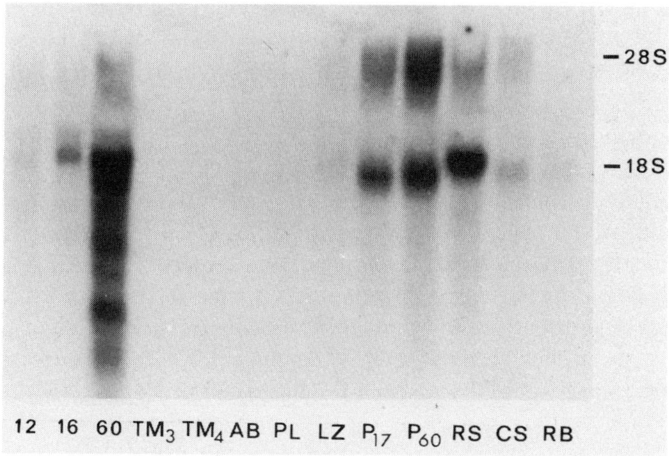

Fig. 14 Northern analysis of RNA abundance during mouse spermatogenesis on hybridizing with a [^{32}P]-labelled, restriction fragment from cDNA clone P3C. Expression of this 1000-bp RNA sequence was detected in pachytene spermatocytes and spermatids. Samples included the following: seminiferous epithelium at postnatal days 12 (12), 16 (16) and 60 (60); TM_3 Leydig cells (TM_3) and TM_4 Sertoli cells (TM_4); and the germ cells: Types A and B spermatogonia (AB), preleptotene spermatocytes (PL), leptotene/zygotene spermatocytes (LZ), prepuberal pachytene spermatocytes (P_{17}), adult pachytene spermatocytes (P_{60}); round spermatids, steps 1 to 8 (RS), condensed spermatids, steps 12 to 16 (CS), and residual bodies (RB). [Data are from Thomas *et al.* (1989).]

differential gene transcription and transcript degradation. Thus, defining the period of active gene transcription(s) will require more complex analyses. Regardless of this limitation, these studies have identified an array of gene products that are expressed preferentially or exclusively by germ cells. The libraries therefore will be of considerable value in elucidating the molecular mechanisms regulating mouse spermatogenesis.

Immuno-screening of the λgt11 cDNA expression libraries with mAbs PNT-1 and PNT-2 have not been successful in detecting the translational

Caption for Fig. 13 (cont.)

were abundant in pachytene spermatocytes and spermatids. Samples included the following: seminiferous epithelium at postnatal days 12 (12), 16 (16) and 60 (60); TM_3 Leydig cells (TM_3) and TM_4 Sertoli cells (TM_4); and the germ cells: Types A and B spermatogonia (AB), preleptotene spermatocytes (PL), leptotene/zygotene spermatocytes (LZ), prepuberal pachytene spermatocytes (P_{17}), adult pachytene spermatocytes (P_{60}); round spermatids, steps 1 to 8 (RS), condensed spermatids, steps 12 to 16 (CS), and residual bodies (RB). [Data are from Thomas *et al.* (1989).]

products. The difficulties encountered could have arisen from the rare PNT-1 and PNT-2 mRNAs not being represented in the cDNA libraries, or from the premature termination of translation so that the truncated polypeptides lacked the required epitopes. The former problem may be overcome by preparing unidirectional, full-length, λgt cDNA expression libraries from germ cell poly[A^+] RNA and then by undertaking reciprocal subtractive selection. The second problem may be circumvented by utilizing polyclonal Abs or an array of mAbs, directed against multiple epitopes located within the respective proteins. In the latter case, based on recent IEFimmunoblots, mAbs PNT-1 and PNT-2 are known to recognize different epitopes within the 50–48kD sperm proteins, and other mAbs directed against additional epitopes also appear to exist. It will be feasible, therefore, to screen the new undirectional λgt cDNA expression libraries with a cocktail of mAbs.

Conclusions

Mammalian spermiogenesis is accompanied by discrete patterns of transcriptional and translational events, each required to occur in a developmen-

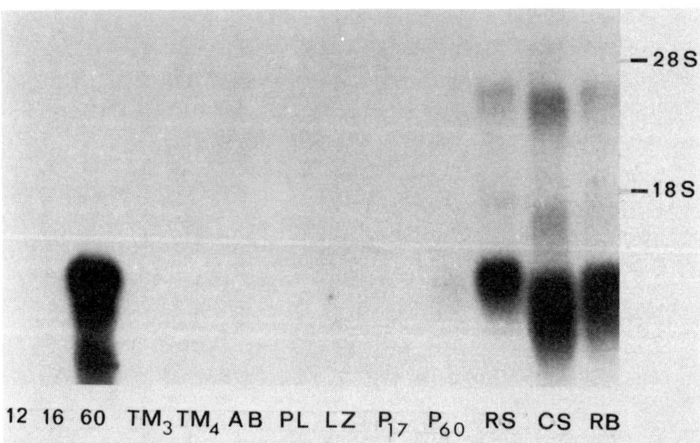

Fig.15 Northern analysis of RNA abundance during mouse spermatogenesis, undertaken with a [^{32}P]-labelled, restriction fragment from cDNA clone D28. This 600-bp sequence is expressed exclusively during spermiogenesis. Samples included the following: seminiferous epithelium at postnatal days 12 (12), 16 (16) and 60 (60); TM_3 Leydig cells (TM_3) and TM_4 Sertoli cells (TM_4); and the germ cells: Types A and B spermatogonia (AB), preleptotene spermatocytes (PL), leptotene/zygotene spermatocytes (LZ), prepuberal pachytene spermatocytes (P_{17}), adult pachytene spermatocytes (P_{60}); round spermatids, steps 1 to 8 (RS), condensed spermatids, steps 12 to 16 (CS), and residual bodies (RB). [Data are from Thomas *et al.* (1989).]

tally-regulated framework for assembling the spermatozoon. The production of germ cell-specific proteins is exemplified by the differential expression of the genes encoding for LDH-C, PGK-2, hsp 70.1, hsp 70.2, Hox 1.4, TCP-1, P1 and P2. In addition, an array of mAbs has enabled the identification of a diverse population of novel germ cell proteins, many of which are integrated into polar domains within the sperm cell. In this review, two of these mAbs have been used to demonstrate the ontogeny, polar distribution, and biochemical properties of proteins present in the perinuclear theca of haploid germ cells. These proteins, having M_rs of 83, 79, 76, 72, 50 and 48kD, each share at least two distinct epitopes and therefore are related immunologically.

The novel 83–72kD family appear coordinately during early spermiogenesis and assemble within a cap-like structure at the apical pole of the nucleus. This structure, the perinuclear theca, gradually expands peripherally and caudally during the remainder of spermiogenesis, eventually encompassing the entire anterior pole of the nucleus in mature sperm. These proteins therefore differ from calicin, a novel 60kD, 'cytoskeletal' protein that has been localized to the post-acrosomal dense lamina of bovine spermatozoa (Longo *et al.*, 1987). Significantly, mature spermatozoa do not express the 83–72kD proteins. Instead, the perinuclear theca contains two proteins at 50 and 48kD that share at least two epitopes with each member of the 83–72kD family. Based on this immunological relatedness, the existing evidence favors the interpretation that the 83–72kD family undergoes proteolytic processing, at some time during testicular and/or epididymal maturation, to form the 50–48kD constituents of the sperm perinuclear theca. This processing event may reflect autodigestion or involve a specific protease. It is speculated that the process may underly the transformation in the shape of the perinuclear theca that is known to occur during sperm maturation (Lalli & Clermont, 1981). Moreover, the biochemical changes may be prerequisites for the onset of capacitation and/or the acrosome reaction.

The temporal expression and relative abundance of mRNAs during spermatogenesis have been examined with Northern analyses, by using [^{32}P]-labelled cDNA probes selected on the basis of their exclusive or preferential expression in spermatogenic cells. These analyses have demonstrated that four principal periods of stage-specific gene expression occur during mouse spermatogenesis. The period of spermatogonial proliferation was characterized by several distinctive patterns of gene expression, but most of these transcripts were also in somatic cells, although at much reduced levels. Only a few of the genes in this class were transcribed subsequently during meiotic prophase, and fewer continued to be expressed during spermiogenesis. Concurrent with the gradual inactivation of genes during meiotic prophase, an increasing number of novel germ cell-specific transcripts were expressed. Some of these became abundant initially during leptonema/

zygonema, others during pachynema, and many at different steps of spermiogenesis. Clearly, the genome undergoes substantial changes in expression as some genes are repressed, while others are activated to provide the regulatory and structural proteins required for assembling the spermatozoon. Among the latter class presumably will be transcripts encoding those polypeptides recognized by mAbs PNT-1 and PNT-2. Availability of these cDNA probes will facilitate further studies on the ontogeny and function of these unusual proteins.

Acknowledgements

This review article, its presentation and the data, was funded by research grants from the National Institute of Child Health and Human Development (A.R.B.), Collagen Enterprises, S.A. (A.R.B.) and the American Cancer Society (M.I.S.). The authors are grateful to Drs Michael Griswold and Steven Sylvester, Washington State University, Pullman, WA for the cDNA probe to SGP-2, and to Dr Steven Li, National Institute of Environmental Health Sciences, Research Triangle Park, NC, for the probe to LDH-C.

References

Bellvé, A.R. (1979) The molecular biology of mammalian spermatogenesis. In: "Oxford Reviews of Reproductive Biology" (ed. C. A. Finn). Oxford University Clarendon Press, Oxford. Vol. 1, pp. 159–261.

Bellvé, A.R., Anderson, E. & Hanley-Bowdoin, L. (1975). Synthesis and amino acid composition of basic proteins in mammalian sperm nuclei. **Developmental Biology** 47, 349–365.

Bellvé, A.R., Cavicchia, J-. C., Millette, C.F., O'Brien, D.A., Bhatnagar, Y.M. & Dym, M. (1977a). Spermatogenic cells of the prepuberal mouse. Isolation and morphological characterization. **Journal of Cell Biology** 74, 68–85.

Bellvé, A.R., McKay, D.J., Renaux, B.S. & Dixon, G.H. (1988). Purification and chartacterization of mouse protamines P1 and P2. Amino acid sequence of P2. **Biochemistry** 27, 2890–2897.

Bellvé, A.R., Chandrika, R. & Hirtzer, P. (1989a). The perinuclear matrix: A structural element of the spermatozoon nucleus. (Submitted).

Bellvé, A.R., Chandrika, R. & Barth, A. (1989b). Temporal expression, polar distribution and processing of perinuclear theca proteins during mouse spermatogenesis. (Submitted).

Bellvé, A.R., Millette, C.F., Bhatnagar, Y.M. & O'Brien, D.A. (1977b). Dissociation of the mouse testis and characterization of isolated spermatogenic cells. In: Proceedings of the IV Engineering Foundation Conference on Automated Cytology. **Journal of Histochemistry & Cytochemistry** 25, 480–494.

Bellvé, A.R. & Moss, S.B. (1983). Monoclonal antibodies as probes of reproductive mechanisms. **Biology of Reproduction** 28, 1–26.

Bellvé, A.R. & O'Brien, D.A. (1983). The mammalian spermatozoon: Structure and temporal assembly. In: "Mechanism and Control of Animal Fertilization." (ed. J.F. Hartmann). Academic Press, Inc., New York and London, pp. 55–137.

Bragg, P.W. & Handel, M.A. (1979). Protein synthesis in mouse spermatozoa. **Biology of Reproduction** 20, 333–337.

Distel, R.J., Kleene, K.C. & Hecht, N.B. (1984). Haploid expression of a mouse testis α-tubulin gene. **Science** 224, 68–70.

Erickson, R.P., Kramer, J.M., Rittenhouse, J. & Salkeld, A. (1980). Quantitation of mRNAs during spermatogenesis: Protamine-like histone and phosphoglycerate kinase-2 mRNAs increase after meiosis. **Proceedings of the National Academy of Sciences, USA** 77, 6086–6090.

Gold, B., Fujimoto, H., Kramer, J.M., Erickson, R.P. & Hecht, N.B. (1983). Haploid accumulation and translational control of phosphoglycerate kinase-2 mRNA during mouse spermatogenesis. **Developmental Biology** 98, 392–399.

Goldman, D.B., Kiessling, A.A., Millette, C.F. & Cooper, G.M. (1987). Expression of c-Mos RNA in germ cells of male and female mice. **Proceedings of the National Academy of Sciences, USA** 84, 4509–4513.

Hecht, N.B., Kleene, K.C., Distel, R.J. & Silver, L.M. (1984). The differential expression of the actins and tubulins during spermatogenesis in the mouse. **Experimental Cell Research** 153, 275–280.

Ierardi, L.A., Moss, S.B. & Bellvé, A.R. (1983). Synaptonemal complexes are integral components of the isolated mouse spermatocyte nuclear matrix. **Journal of Cell Biology** 96, 1717–1726.

Ierardi, L.A., Moss, S.B. & Bellvé, A.R. (1989). The nuclear matrix of mouse spermatogenesis. Temporal appearance and synthesis of polypeptides coincident with the assembly of synaptonemal complexes. **Journal of Cell Science** (In Press).

Kleene, K.C., Distel, R.J. & Hecht, N.B. (1985). Nucleotide sequence of a cDNA clone encoding mouse protamine 1. **Biochemistry** 24, 719–722.

Lalli, M. & Clermont, Y. (1981). Structural changes in the head component of the rat spermatid during late spermiogenesis. **American Journal of Anatomy** 160, 419–434.

Longo, F.J., Krohne, G. & Franke, W.W. (1987). Basic proteins of the perinuclear theca of mammalian spermatozoa and spermatids: A class of novel cytoskeletal elements. **Journal of Cell Biology** 105, 1105–1120.

Mayer, J.F., Jr. & Zirkin, B.R. (1979). Spermatogenesis in the mouse., I. Autoradiographic studies of nuclear incorporation and loss of ^{3}H-amino acids.**Journal of Cell Biology** 81, 403–410.

Mayer, J.F., Jr., Chang, T.S.K. & Zirkin, B.R. (1981). Spermatogenesis in the mouse. 2. Amino acid incorporation into basic nucleoproteins of mouse spermatids and spermatozoa. **Biology of Reproduction** 25, 1041–1051.

Meistrich, M.L., Trostle, P.K., Frapart, M. & Erickson, R.P. (1977). Biosynthesis and localization of lactate dehydrogenase X in pachytene spermatocytes and spermatids of mouse testes. **Developmental Biology** 60, 428–441.

Millette, C.F. & Bellvé, A.R. (1980). Selective partitioning of plasma membrane antigens during mouse spermatogenesis. **Developmental Biology** 79, 309–324.

Monesi, V. (1965), Synthetic activities during spermatogenesis in the mouse: RNA and protein. **Experimental Cell Research** 39, 197–224.

Monesi, V. (1971). Chromosome activities during meiosis and spermiogenesis. **Journal of Reproduction and Fertility**, Supplement 13, 1–14.

Myles, D.G., Primakoff, P. & Bellvé, A.R. (1981). Surface domains of the guinea pig sperm defined with monoclonal antibodies. **Cell** 23, 433–439.

O'Brien, D.A. & Bellvé, A.R. (1980a). Protein constituents of the mouse spermatozoon. I. An electrophoretic characterization. **Developmental Biology** 75, 386–404.

O'Brien, D.A. & Bellvé, A.R. (1980b). Protein constituents of the mouse spermatozoon. II. Temporal synthesis during spermatogenesis. **Developmental Biology** 75, 405–418.

Oppi, C., Shore, S.K. & Reddy, E.P. (1987). Nucleotide sequence of a testis-derived c-abl cDNA: Implications for testis-specific transcription and abl oncogene activation. **Proceedings of the National Academy of Sciences, USA** 84, 8200–8204.

Ponzetto, C. & Wolgemuth, D.J. (1985). Haploid expression of a unique c-abl transcript in mouse male germ line. **Molecular and Cellular Biology** 5, 1791–1794.

Romrell, L.J., Bellvé, A.R. & Fawcett, D.W. (1976). Separation of mouse spermatogenic cells by sedimentation velocity. A morphological characterization. **Developmental Biology** 49, 119–131.

Rubin, M.R., Toth, L.E., Patel, M.D., D'Eustachio, P. & Ngyuyen-Huu, M.C. (1986). A mouse homeobox gene is expressed in spermatocytes and embryos. **Science** 233, 663–667.

Shackleford, G.M. & Varmus, H.E. (1987). Expression of the proto-oncogene int-1 is restricted to postmeiotic male germ cells and the neural tube of mid-gestational embryos. **Cell** 50, 89–95.

Thomas, K., Wilkie, T., Tomashefsky, P., Bellvé, A.R. & Simon, M.I. (1989). Differential gene transcription during mouse spermatogenesis. **Biology of Reproduction** (in press).

Villasante, A., Wang, D., Dobner, P., Dolph, P., Lewis, S.A. & Cowan, N.J. (1986). Six mouse α-tubulin mRNAs encode five distinct isotypes: Testis-specific expression of two sister genes. **Molecular and Cellular Biology** 6, 2409–2419.

Waters, S.H., Distel, R.J. & Hecht, N.B. (1985). Mouse testis contains two size classes of actin mRNA that are differentially expressed during spermatogenesis. **Molecular and Cellular Biology** 5, 1649–1654.

Weiben, E.D. (1981). Regulation of synthesis of lactate dehydrogenase X during spermatogenesis in mouse. **Journal of Cell Biology** 88, 492–498.

Willison, K.R., Dudley, K. & Potter, J. (1986). Molecular cloning and sequence analysis of a haploid expressed gene encoding t complex polypeptide 1. **Cell** 44, 727-738.

Wolgemuth, D.J., Engelmyer, E., Duggal, R.N., Gizang-Ginsberg, E., Mutter, G.L., Ponzetto, C., Viviano, C. & Zakeri, Z.F. (1986). Isolation of a mouse cDNA coding for a developmentally regulated, testis-specific transcript containing homeo box homology. **EMBO Journal** 5, 1229–1235.

Wolgemuth, D.J., Viviano, C.M., Gizang-Ginsberg, E., Frohman, M.A., Joyner, A.L. & Martin, G.R. (1987). Differential expression of the mouse homeobox-containing gene Hox-1.4 during male germ cell differentiation and embryonic development. **Proceedings of the National Academy of Sciences, USA** 84, 5813–5817.

Yelick, P.C., Balhorn, R., Johnson, P.A., Corzett, M., Mazrimas, J.A., Kleene, K.C. & Hecht, N.B. (1987). Mouse protamine 2 is synthesized as a precursor whereas mouse protamine 1 is not. **Molecular and Cellular Biology** 7, 2173–2179.

Zakeri, Z.F. & Wolgemuth, D.J. (1987). Developmental stage-specific expression of the hsp70 gene family during differentiation of the mammalian male germ line. **Molecular and Cellular Biology** 7, 1791–1796.

Zakeri, Z.F., Wolgemuth, D.J. & Hunt, C.R. (1988). Identification and sequence analysis of a new member of the mouse HSP70 gene family and characterization of its unique cellular and developmental pattern of expression in the male germ line. **Molecular and Cellular Biology** (in press).

Expression of the c-*abl* oncogene in mouse spermatids

D. MEIJER*, M. VON LINDERN*, T. VAN AGTHOVEN*, G. GROSVELD*, D. TALARICO**, G. DELLA VALLE**, P. MACKENBACH*** AND J.A. GROOTEGOED***

* *Department of Cell Biology and Genetics. Medical Faculty, Erasmus University Rotterdam, P.O. Box 1738, 3000 DR Rotterdam, The Netherlands.* ** *Dipartimento di Genetica e Microbiologia, Universita di Pavia, 27100 Pavia, Italy.* *** *Department of Biochemistry II, Medical Faculty, Erasmus University Rotterdam, P.O. Box 1738, 3000 DR Rotterdam, The Netherlands.*

Abstract

In mouse testes an abundant 4 kb c-*abl* mRNA is transcribed in addition to the two normal mRNAs of 5 and 6 kb expressed in all other tissues. Here we describe the differential expression of this specific transcript during mouse spermatogenesis, using isolated germinal cell populations and *in situ* hybridization on whole testis sections. The results confirm the findings of Ponzetto & Wolgemuth (1985) that the expression of the 4 kb mRNA is restricted to the haploid stage of spermatogenesis. Expression of this transcript is most abundant in elongating spermatids. Testis-specific c-*abl* cDNA molecules were cloned from a testis cDNA library and their structures were analyzed. The 4 kb c-*abl* mRNA was shown to arise from differential polyadenylation of a pre-RNA transcribed from the same promotor as the somatic 5 kb mRNA.

Polyadenylation is unusual as the transcript lacks a poly(A) addition sequence upstream of the poly(A) addition site. This differential polyadenylation does not affect the open reading frame of the transcript. Furthermore we show that protein-tyrosine kinase activity is present in spermatids and in cytoplasmic fragments detached from elongating spermatids.

Key words

c-*abl*, oncogene, testis, spermatogenesis, mRNA, cDNA, polyadenylation, differential expression, protein-tyrosine kinase.

Introduction

Gene expression in spermatogenic cells, passing through meiosis and the haploid stage of development, may not be controlled by direct action of hormones. There are indications that pachytene spermatocytes and round spermatids do not contain receptors for testosterone and FSH, the main hormonal regulators of spermatogenesis (Grootegoed *et al.*, 1977; Fritz, 1978). For a number of reasons, it seems likely that all effects of hormones and growth factors affecting spermatogenesis are mediated *via* Sertoli cells.

The differentiating germ cells contain adenylate cyclase, phosphodiesterase and cAMP-dependent protein kinase activities (Adamo *et al.*, 1980). In many cell types such enzyme activities play an important role in hormonal signal transduction. However, adenylate cyclase in spermatocytes and round spermatids is a soluble enzyme which does not interact with guanine nucleotide binding proteins (G-proteins). In spermatozoa, adenylate cyclase is membrane bound. Spermatozoa may contain an inhibitory guanine nucleotide binding protein (G_i) of the adenylate cyclase transducing system, but are lacking the stimulatory component (G_s) (Hildebrandt *et al.*, 1985; Kopf *et al.*, 1986). Adenylate cyclase activity in spermatozoa can be regulated by changes in Ca^{2+} levels or by other ion gradients across the sperm cell membrane (Tash & Means, 1983).

Two cAMP-dependent kinase isoenzymes are expressed during spermatogenesis (Conti *et al.*, 1983). Protein kinase 1 and 2 (PK1 and PK2) are present in mouse spermatocytes and spermatids. The activity of PK1 decreases during spermatid development; PK2 is the major isoenzyme present in elongating spermatids and epididymal spermatozoa (Conti *et al.*, 1983). The levels of cAMP-dependent protein kinase activity in sperm are relatively high (Tash & Means, 1983). Stimulation of sperm motility by cAMP may involve cAMP-dependent phosphorylation of sperm proteins, including the axonemal components tubulin and dynein (Tash & Means, 1983). Specific protein kinases do not phosphorylate at serine or threonine residues of proteins, but at tyrosine residues (Hunter & Cooper, 1985). Tyrosine phosphorylation has been implicated in the regulation of cell growth and differentiation. Phosphorylation of tyrosine residues in proteins is a relatively rare event, and rather small amounts of phosphotyrosine are found in cellular proteins. Two types of protein-tyrosine kinases have been described, although there is overlap between them. The first type includes the protein-tyrosine kinases associated with retroviral transforming gene products and their cellular homologs (oncogenes and proto-oncogenes), and the second type includes the protein-tyrosine kinases associated with growth factor receptors (Hunter & Cooper, 1985). Spermatogenic cells may not possess growth factor receptors with intrinsic protein-tyrosine kinase activity, such as for example IGF-I receptors. In rat testicular tubules IGF-I receptors are abundant on Sertoli cells, whereas these receptors were not detected on isolated spermatids (Oonk & Grootegoed, 1988). However, in the present experiments it was observed that round and elongating spermatids isolated from mouse testes contained protein-tyrosine kinase activity (see below). This protein-tyrosine kinase activity might reflect the expression of distinct proto-oncogenes which encode nonreceptor type cytoplasmic proteins. With respect to this, the present experiments concern the transcription of the c-*abl* oncogene during spermatogenesis.

The c-*abl* proto-oncogene

The cellular Abelson proto-oncogene (c-*abl*) encodes a protein-tyrosine kinase, and belongs to a family of tyrosine kinases involved in growth control (see Konopka & Witte, 1985a). The genome of the transforming Abelson murine leukemia virus (A-MuLV) contains a gene (v-*abl*) which appeared to be derived, *via* a process of recombination, from the cellular c-*abl* gene. As a consequence of this recombination event the cellular derived gene is expressed as part of a 160 kD fusion-protein consisting of N-terminal viral gag sequences and C-terminal *abl* sequences. This P160 gag-*abl* protein displays protein-tyrosine kinase activity and confers to A-MuLV the capacity to transform a wide range of hematopoietic cells as well as NIH 3T3 fibroblasts. That the transforming potential of this virus is tightly associated with the kinase activity is supported by the fact that mutants defective in phosphorylation do not transform (for review, see Witte, 1986). In humans, the c-*abl* gene is involved in Chronic Myelocytic Leukemia (CML). The disease is characterized by the presence of a shortened chromosome 22, called the Philadelphia (Ph) chromosome, which is the result of a reciprocal translocation between chromosomes 9 and 22. As a consequence of this translocation, part of the c-*abl* gene has moved from chromosome 9 to 22. This results in the formation of a new chimaeric gene on the Ph chromosome. This new gene encodes a 210 kD protein which has protein-tyrosine kinase activity. Also, the normal c-*abl* genes in mouse and human cells encode proteins (P150 and P145 respectively) that have been shown to be associated with protein-tyrosine kinase activity (Konopka & Witte, 1985b). However, the protein-tyrosine kinase activities of the viral P160 and the CML-associated P210 are much more robust than that of the normal cellular P150 and P145, which also do not show *in vivo* autophosphorylation.

The c-*abl* gene is conserved throughout the vertebrate world, which suggests an important role for the c-*abl* protein in normal cell activities. Although the malignancies in which aberrant c-*abl* expression is found are all of hematopoietic origin, expression of the normal c-*abl* gene is not restricted to these cell types. In all mouse and human tissues examined, the c-*abl* gene is transcribed mainly into two different mRNA molecules (Ben-Neriah *et al.*, 1986; Shtivelman *et al.*, 1986). However, in mouse testes a third specific mRNA of approx. 4 kb is produced in addition to the 5 and 6 kb mRNAs found in all other mouse tissues (Muller *et al.*, 1982). A role for a c-*abl*-encoded protein in spermiogenesis is suggested by the abundant expression of the testis-specific c-*abl* mRNA, which is correlated with the

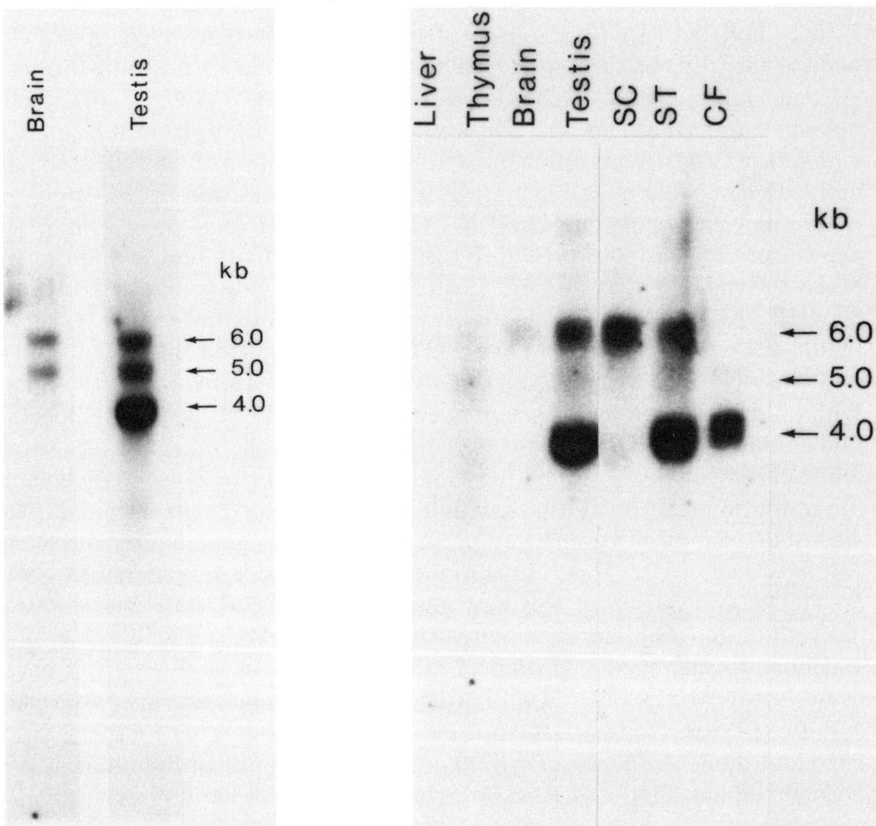

Fig. 1, left. c-*abl* expression in brain and testis.
Total RNA from mouse testes and brain was isolated using the guanidium-isothiocyanate method (Collins et al., 1984). Poly(A^+) RNA was selected twice on an oligo(dT) cellulose column. 10 μg of poly (A^+) RNA of both tissues were run on a denaturing formaldehyde 0.8% agarose gel and

appearance of spermatids in developing testes (Ponzetto & Wolgemuth, 1985).

As a first step to define a role for the *abl* protein(s) in spermatogenesis, we have cloned the testis-specific mRNA from a testis cDNA library and analyzed its structure. The results are discussed below. We further confirm the findings of Ponzetto & Wolgemuth (1985) that the 4 kb mRNA is specific for the haploid stage of spermatogenesis, using isolated populations of spermatogenic cells and *in situ* hybridization.

c-*abl* expression in spermatogenic cells

A cDNA probe, encompassing the c-*abl* protein-tyrosine kinase domain, was used to examine the expression of the c-*abl* mRNAs in different tissues. The c-*abl* expression in mouse brain and testis is illustrated in Fig. la. In testis, the most abundant mRNA is a transcript of 4 kb in length. This transcript was absent in all other tissues examined, as was shown before by Ben-Neriah *et al.*, (1986). As the mature testis contains all stages of developing germ cells, we have isolated different populations of spermatogenic cells to determine at which stage of spermatogenesis the shorter transcript is produced. It was observed that the testis-specific 4 kb transcript was completely absent in pachytene spermatocytes, but was abundant in round spermatids and cytoplasmic fragments detached from elongating spermatids (Fig. lb). On the other hand, the somatic transcripts of approximately 5 and 6 kb were present in spermatocytes and round spermatids, but virtually absent in the cytoplasmic fragments (Fig. lb). These results indicate that the 4 kb transcript appears after the meiotic divisions, and is present up to the later stages of spermiogenesis.

These results were further corroborated by *in situ* hybridization on whole testis sections. For these experiments an oligonucleotide was synthesized

Caption for Fig. 1, left. (cont.)

subsequently blotted onto nitrocellulose. The blot was probed with a mouse *abl* cDNA probe covering the entire kinase domain, washed to a final stringency of 0.3 × SSC, 65°C and exposed to Kodak X-AR5.

Fig. l, right. c-*abl* expression in different tissues and different stages of spermatogenesis. Total RNA from mouse tissues and isolated mouse spermatogenic cells was prepared following the LiCl/urea protocol. The different populations of spermatogenic cells were isolated according to methods described by Grootegoed *et al.* (1982, 1984, 1986) using sedimentation at unit gravity (Staput procedure) followed by density gradient centrifugation (Percoll gradients). Approximately 20 μg RNA were applied per lane. Further experimental procedures were as described above. Abbreviations: pachytene spermatocytes (SC), round spermatids (ST) and cytoplasmic fragments from elongating spermatids (CF).

that specifically recognizes the 4 kb c-*abl* mRNA under appropriate hybridization conditions (see below). These experiments indicate that the elongating spermatids are heavily labeled using this oligonucleotide (Fig. 2). As a control, a similar experiment was performed using an rRNA probe; cells close to the tubular wall were labeled, whereas the elongating spermatids were not.

The c-*abl* gene is present in the mouse genome as a single copy locus. The somatic *abl* transcripts of 5 and 6 kb arise from the use of different promotors, and different 5′ exons are spliced to a common set of 3′ exons (Ben-Neriah *et al.*, 1986; Shtivelman *et al.*, 1986). The testis-specific 4 kb mRNA might arise from an as yet unidentified promotor which is developmentally regulated. Other mechanisms which could give rise to a shorter mRNA include differential polyadenylation or alternative splicing of the

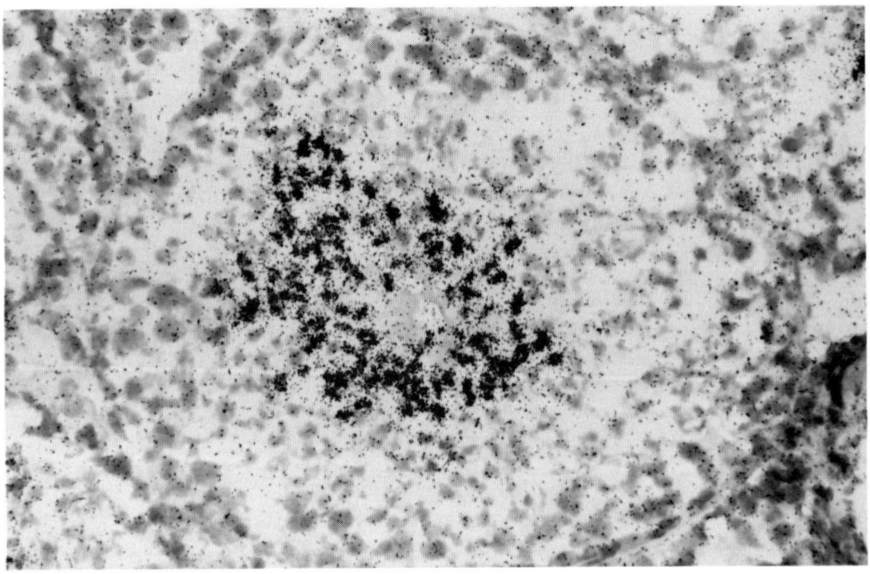

Fig. 2 *In situ* hybridisation of c-*abl* 4 kb mRNA. *In situ* hybridisation on whole testis sections was performed according to a protocol developed by two of us (D.T. and G.D.V.). This method is based on published methods used for *in situ* hybridization on *Drosophila* sections (Akam, 1983; Hafen *et al.*, 1983). In brief, testis cryostat sections were fixed onto microscope slides and hybridized with a ^{35}S-labeled oligonucleotide with a length of 30 nucleotides. This oligonucleotide covered the last 20 nucleotides of the testis-specific 4 kb transcript and the first 10 adenine nucleotides of the poly(A) tail (see Fig. 3). Hybridization and washing conditions were selective for a full 30-mer hybrid, and it was shown on Northern blots that the oligonucleotide specifically detected the 4 kb transcript.

pre-mRNAs generated by one or both of the known promotors. To address these questions and to characterize the 4 kb mRNA in more detail, we have cloned testis-specific *abl* cDNAs from a mouse testis cDNA library.

Cloning of a testis-specific c-*abl* cDNA

cDNA was synthesized from poly(A) RNA from testes of BCBA mice using oligo(dT) as a primer. The cDNA was tailed with dGTP and cloned into lambda phage gtl0 using a synthetic adaptor (Le Bouc *et al.*, 1986). The library was screened with a 3′ v-*abl* probe and yielded more than 50 positive clones. The positive clones with inserts longer than 1.2 kb were mapped for restriction enzyme cleavage sites. Based on their restriction maps the clones could be grouped into two classes. The first class contained inserts that were colinear with the previously described 1.2 kb 3′ cDNA clone isolated from a mouse L-cell cDNA library (Wang *et al.*, 1984). Most of the positive clones, however belonged to a second class, having inserts that were colinear with v-*abl* sequences, missing all sequences contained in the 3′ cDNA clone mentioned above. The restriction maps indicated that all clones in this class were primed at exactly the same position, making it unlikely that the clones were artifacts of cloning. To determine whether these clones arose from differential splicing or polyadenylation, we sequenced the 3′ part of three independently derived clones. It was found that all three clones were colinear with the published v-*abl* sequence up to nucleotide 4023 (Reddy *et al.*, 1983). The structure of the 3′ end of the 4 kb mRNA is depicted in Fig. 3. From the present results we conclude that the cDNAs were derived from 4 kb c-*abl* mRNAs, which have a poly(A) tail starting 1171 nucleotides upstream of the normal poly(A) addition site of the 5 and 6 kb mRNAs. S1 nuclease mapping showed that this alternative poly(A) addition site is the most abundantly used for testis c-*abl* mRNAs, whereas it is not used in brain (data not shown). S1 nuclease mapping and Northern blot hybridization experiments further showed that the testis-specific 4 kb mRNA is transcribed from the same promotor as the somatic 5 kb mRNA.

The results indicate that the difference in size between the 5 kb mRNA and the testis-specific 4 kb mRNA is fully accounted for by the use of a different poly(A^+) addition site within the last exon of the c-*abl* gene. This exon, which is more than 2.8 kb long (Wang *et al.*, 1984), is also present in the somatic 6 kb mRNA, and it is likely that this transcript will be differentially poly-adenylated as well. Indeed, Ben-Neriah *et al.* (1986) detected a faint RNA band of approx. 5 kb in testis, using a probe that specifically recognizes the somatic 6 kb mRNA. This band was not observed in any of the other tissues examined and may represent a testis-specific shortened variant of the somatic 6 kb transcripts. The 3′ exon of c-*abl* contains a very long (approx. 1600 nucleotides) untranslated region (UTR).

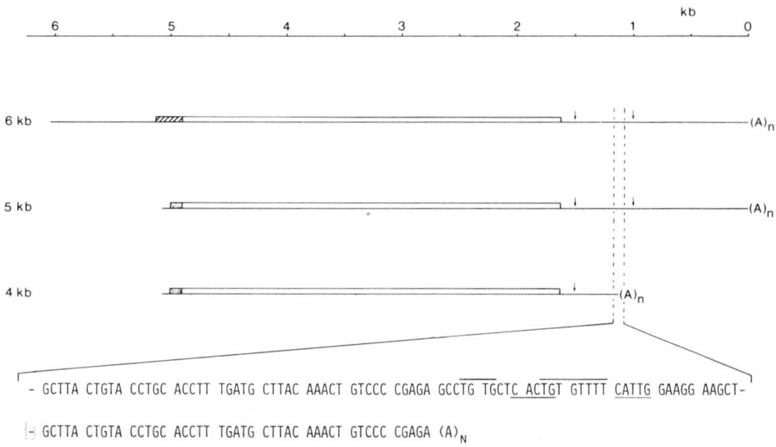

Fig. 3 Relationship between three different c-*abl* mRNAs. The schematic representation of the 5 and 6 kb *abl* mRNAs is based on the work of Ben-Neriah *et al.* (1986). Sequences were obtained using the chemical degradation method of Maxam & Gilbert (1977). The underlined and overlined sequences have been suggested to be involved in cleavage/polyadenylation (Berget, 1984; Birnstiel *et al.*, 1985).

Therefore the differential polyadenylation in spermatids does not affect the main open reading frame (ORF) of the c-*abl* mRNA. On the other hand, poly(A) addition in spermatids interrupts a hypothetical reading frame found in the 3′ UTR. This minor ORF potentially codes for a protein of 18 kD. A product from this minor ORF, however, has not been detected using antibodies raised against ORF sequences expressed in bacteria (Konopka & Witte, 1985a). Furthermore, preliminary results suggest that this ORF is not conserved in the human c-*abl* gene. Thus, it seems doubtful that this minor ORF is of any biological significance.

The present results indicate that the testis-specific 4 kb c-*abl* mRNA contains an intact main open reading frame. The 4 kb mRNA and the somatic 5 kb mRNA may therefore encode the same proteins. To obtain evidence for this, we have constructed full length cDNA clones out of independently derived cDNAs. The integrity of these cDNA clones was checked by Sl-nuclease mapping on testis RNA, using brain RNA as a control. From these cDNAs, representing the testis-specific 4 kb mRNA and the somatic 5 kb mRNA (see Fig. 3) RNA was transcribed *in vitro* using T7 RNA polymerase. This RNA was subsequently translated *in vitro* in a cell-free translation system (rabbit reticulocyte lysate) in the presence of ^{35}S-methionine. It was found that both the 4 kb and 5 kb c-*abl* mRNAs were translated into proteins with a MW of 150 kD, which were precipitated using the anti-*abl* antibodies described below. This result indicates that the

4 kb testis-specific mRNA encodes the ubiquitous protein c-*abl* P150, and raises the question as to what is the function of the alternative 3′ processing of the c-*abl* mRNA that generates the testis-specific shorter transcript. One possible explanation is that the 4 kb mRNA is more stable than the 5 kb mRNA, as discussed below.

Poly(A) addition is at an unusual site

The hexanucleotide AAUAAA is a crucial sequence in pre-mRNA cleavage/polyadenylation during processing of an immature transcript (Birnstiel *et al.*, 1985). The importance of this sequence is stressed by the finding that a form of α-thallassemia in man is caused by a mutation within this hexanucleotide (Higgs *et al.*, 1983). In all transcripts of other genes examined, polyadenylation takes place 10 to 30 nucleotides downstream of the hexanucleotide sequence. However, this sequence is not included in the testis-specific 4 kb c-*abl* mRNA, which contains a sequence with only a very weak homology to the hexanucleotide 11 nucleotides upstream of the poly(A) addition site (Fig. 3). This homologous sequence contains two mutations at positions that have been found to be crucial for its function (Orkin *et al.*, 1985).

The absence of the highly conserved hexanucleotide sequence upstream of the poly(A) addition site in the testis-specific transcript indicates that other, but as yet unidentified, sequences govern the cleavage/polyadenylation of the c-*abl* transcript in spermatids. Sequence motifs, usually found downstream of the poly(A) addition site, are indicated in Fig. 3. The importance of these sequences in 3′ processing is less well defined but might play a role in the formation of the 4 kb mRNA.

Transcripts of other genes which are expressed in spermatids might be processed *via* a similar mechanism. It has been suggested that small nuclear ribonucleoprotein particles (snRNPs) are involved in the 3′ processing of primary transcripts (Berget, 1984). These particles consist of a small molecule (U-RNA) complexed with a set of proteins. Recognition of conserved sequences within the primary transcript may be facilitated by base pairing between this U-RNA molecule and the conserved sequence. Alternative cleavage/polyadenylation of the testis-specific c-*abl* mRNA could therefore be accomplished by interaction with a U-RNA which is present during the haploid stage of spermatogenesis. If this indeed is the case, we may find other alternatively processed transcripts and new conserved sequences which are involved in mRNA processing. In this respect, it is of interest that the haploid expressed mouse protamine genes MP1 and MP2 contain a polyadenylation sequence (Hecht, 1986), indicating that such sequences play a role in 3′ processing in spermatids. Cloning and sequencing of other haploid expressed genes may clarify this issue.

Tyrosine phosphorylation in spermatids

Total soluble protein-tyrosine kinase activity was estimated in round spermatids and in cytoplasmic fragments from elongating spermatids, as described by Braun *et al.*, (1984) and Mackie *et al.*, (1986), using the random polyamino acid polymer poly(Glu, Ala, Tyr) (6:3:1) as a phosphate acceptor. The results, presented in Table 1, indicate that mouse spermatids (round and elongated) contain a low protein-tyrosine kinase activity. One should take into account that the cytoplasmic fragments (elongating spermatids) contained very little protein. The tyrosine kinase activity in the cytoplasmic fragments (0.47 pmol/10^6 cells per min) was not much lower than the total cAMP-dependent protein kinase activity in elongating spermatids (1.76 pmol/10^6 cells per min: Conti *et al.*, 1983).

It is not known if the 4 kb c-*abl* mRNA in spermatids is translated into an active protein-tyrosine kinase. Initially, it was believed that the normal cellular c-*abl* product in all cell types, in contrast to the v-*abl* gene product, was not an active kinase (Davis *et al.*, 1985). However, modified reaction conditions have allowed the detection of protein-tyrosine kinase activity associated with the c-*abl* products murine P150 and human P145 (Konopka & Witte, 1985a). The c-*abl* proteins have the ability to autophosphorylate, and to phosphorylate other proteins such as enolase, angiotensin, histones and synthetic peptides. The endogenous substrates *in vivo* may include glycolytic enzymes and cytoskeletal proteins (Hunter & Cooper, 1984).

Site-directed anti-*abl* antisera, kindly provided by Dr. O.N. Witte (Department of Microbiology and The Molecular Biology Institute, UCLA), were harvested from rabbits immunized with v-*abl* P160 sequences expressed as fusion proteins in bacteria, and were designated *abl* anti-pEX (Konopka

Table 1. *Phosphorylation of poly(Glu, Ala, Tyr) (GAT)*

	−GAT	+GAT	
Specimen	(*a*)	(*b*)	(*b* − *a*)
ST	1.13 ± 0.16	2.09 ± 0.16	0.96 ± 0.06
CF	0.73 ± 0.19	1.21 ± 0.22	0.48 ± 0.06

The round spermatids (ST) and cytoplasmic fragments from elongating spermatids (CF), isolated from mice, were suspended in buffer containing 20 mM $MnCl_2$ and sonicated. Supernatants were incubated with 50 μM [γ-^{32}P]ATP in the absence or presence of 1 mg/ml GAT (purchased from Sigma). Incorporated radioactivity is expressed as pmol/min per 10^6 cells or fragments (mean ± S.D. of 3 different experiments). Conti *et al.* (1983) have reported a total cytosol activity of cAMP-dependent kinase in mouse cells of 4.90 and 1.76 pmol/min per 10^6 cells, in round and elongating spermatids respectively.

et al., 1984). Immunoprecipitations of spermatid proteins were performed using anti-pEX 4, and the protein A-sepharose bound immunoprecipitates were subjected to *in vitro* phosphorylation (Konopka & Witte, 1985b). The results indicate that a protein with an apparent MWof 46 kD was phosphorylated on tyrosine residues. This phosphoprotein was abundant in immunoprecipitates of the cytoplasmic fragments from elongating spermatids (Fig. 4). A less intense phosphorylation of this protein was observed using cell lysates from pachytene spermatocytes, round spermatids and epididymal spermatozoa.

It has been reported that a MW 50 kD protein is associated with $pp60^{src}$, the transforming protein of Rous sarcoma virus with protein-tyrosine kinase activity, and is immunoprecipitated by antibody to $pp6O^{src}$ (Hunter & Sefton, 1980). It was suggested that this MW 50 kD protein, which contains phosphotyrosine residues, may be an endogenous substrate for the transforming protein of Rous sarcoma virus. Similarly, the MW 46 kD protein in spermatids may form an enzyme-substrate complex with c-*abl* P150 and may co-precipitate using anti-*abl* antiserum. Consequently, the MW 46 kD protein becomes phosphorylated on tyrosine during the *in vitro* phosphorylation assay. In the present experiments, no marked autophosphorylation of c-*abl P*150 was detected.

Concluding remarks

Spermiogenesis, the post-meiotic development of the haploid spermatids, involves the generation of complex structures (e.g., flagellum, acrosome) and the packaging of the chromosomes in a highly condensed and genetically inactive form in the nucleus. Haploid expression of specific protein-tyrosine kinases could play an important role in spermatid development. The testis-specific 4 kb c-*abl* mRNA may be much more stable than the somatic 5 kb and 6 kb c-*abl* mRNAs. Consequently, the 4 kb mRNA may accumulate in transcriptionally active round spermatids, resulting in high levels of the mRNA also after the genome is inactivated during late spermiogenesis. Furthermore, this stability may allow equilibration of mRNA pools between daughter cells *via* cytoplasmic bridges. Stability of specific mRNAs could very well be a prerequisite for haploid expression of the corresponding genes.

As described above, the testis-specific 4 kb c-*abl* mRNA is transcribed from the same promotor as the somatic 5 kb mRNA. The NH_2 termini of the c-*abl* proteins encoded by the somatic 5 and 6 kb mRNAs are different. Translation of the 6 kb mRNA results in a protein with a putative acceptor site for myristylation at the N-terminus (Ben-Neriah *et al.*, 1986). Myristylation is a hydrophobic modification which may result in membrane anchorage. The testis-specific 4 kb mRNA, like the 5 kb mRNA, encodes a protein lacking the putative myristylation site (Witte, 1986). Consequently, the

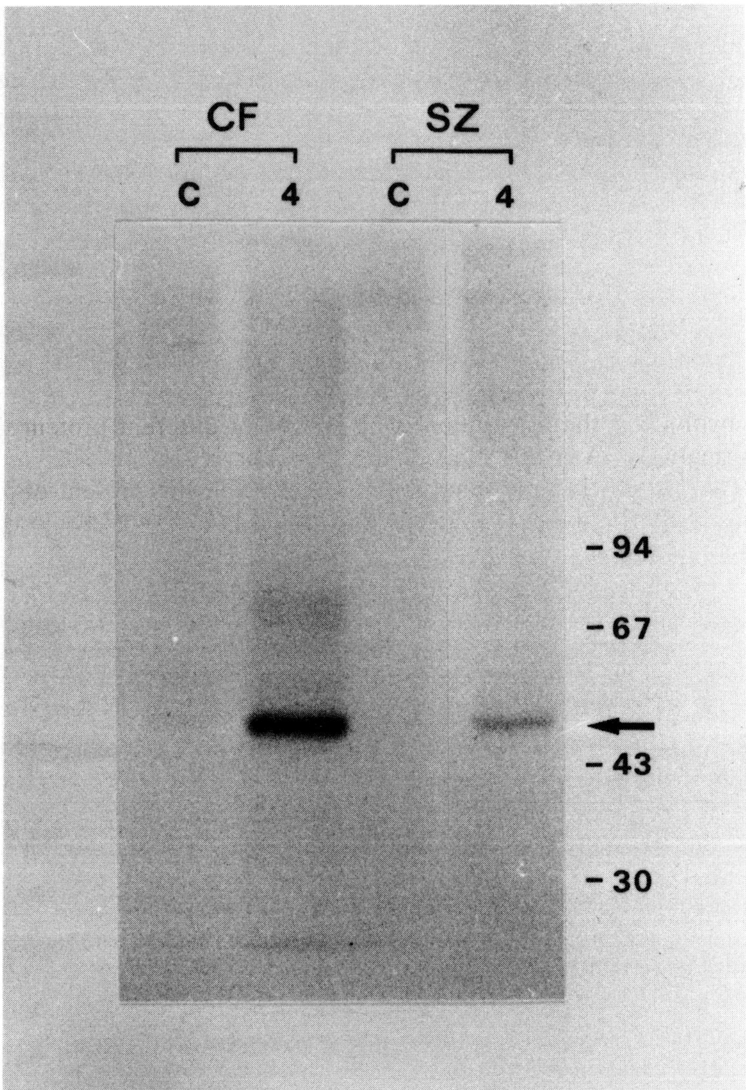

Fig. 4 Phosphorylation of proteins in immunoprecipitates. Cytoplasmic fragments from elongating spermatids (CF), and epididymal spermatozoa (SZ) were lysed in phosphate lysis buffer (Davis *et al.*, 1985). The lysates were clarified by centrifugation at 100 000 × g and incubated overnight with control serum (C) or anti-pEX 4 (4). Subsequently, the immune complexes were harvested using protein A-sepharose beads (Konopka & Witte, 1985b), suspended in phosphorylation buffer and incubated for 20 min at 30°C in the presence of [γ-^{32}P]ATP. The phosphoproteins were analyzed by SDS-polyacrylamide gel electrophoresis and autoradiography.

c-*abl* encoded kinase in spermatids may be a soluble cytoplasmic protein-tyrosine kinase.

Translation of the haploid-expressed 4 kb c-*abl* mRNA may take place specifically in elongating spermatids. Identification of the endogenous substrates is of prime importance to clarify the possible function of the c-*abl* encoded protein-tyrosine kinase. In this respect, it is of interest that phosphotyrosyl-proteins may have effects on gene transcription through specific binding to DNA. In murine fibroblasts transformed by the v-*abl* protein-tyrosine kinase, a phosphoprotein phosphorylated on tyrosine has been found associated with the nuclear fraction (Bell *et al.*, 1987). The protamines, the predominant proteins in the nucleus of vertebrate spermatozoa, are synthesized in the mouse between steps 12 and 15 of spermiogenesis. It can be postulated that phosphorylation on tyrosine residues is involved in the association of a number of different proteins with the spermatid DNA at this stage of spermiogenesis.

The MW 46 kD phosphoprotein observed in the present experiments may not represent such a nuclear protein. Rather, the results indicate that this phosphoprotein is abundant in cytoplasmic fragments from elongating spermatids. The glycolytic enzyme enolase is noticeably phosphorylated on tyrosine in many retroviral transformed cells (Hunter & Cooper, 1985). This phosphorylation may be adventitious and cannot change the activity of the soluble enzyme. On the other hand, phosphorylation could affect the spatial organization of the enzyme as it may exist in cells (Fulton, 1982). The flow of intermediates through the glycolytic pathway in round spermatids is impaired (Grootegoed *et al.*, 1986; Nakamura *et al.*, 1986), whereas this problem is largely or completely solved in mature spermatozoa. In this respect, it will be of interest to study if the MW 46 kD protein observed in the present experiments might represent enolase, also because a sperm-specific enolase isoenzyme is synthesized in elongating spermatids (Edwards & Grootegoed, 1983).

Much work has to be done to provide evidence for a specific function of tyrosine phosphorylation of endogenous substrates in spermiogenesis. Further investigations in this area will be of interest with respect to a better understanding of the molecular basis of spermiogenesis.

Acknowledgements

This work was partially supported by p.F. Ingegneria Genetica & Basi Molecolari delle Malattie Ereditarie of CNR to G.D.V.; D.T. is a fellow

Caption for Fig. 4 (cont.)

The gels were treated before autoradiography with 1 M KOH at 55°C for 2 h; phosphotyrosine has a higher stability to alkali than phosphoserine and phosphothreonine. The positions of LMW markers with MWs of 94, 67, 43 and 30 kD are indicated, and the arrow points to the MW 46 kD phosphoprotein.

of the Ass. Ital. Ric. Cancro. We wish to thank Dr O.N. Witte (Department of Microbiology and The Molecular Biology Institute, University of California at Los Angeles) for providing anti-*abl* antibodies.

References

Adamo, S., Conti, M., Geremia, R., Boitani, C. & Monesi, V. (1980). The cyclic nucleotide system in isolated male germ cells. In: Serio, M. & Martini, L., eds. **Animal Models in Human Reproduction**. New York, Raven Press, pp. 159–168.

Akam, M.E. (1983). The location of Ultrabithorax transcripts in Drosophila tissue sections. **The EMBO Journal** 2:2075–2084.

Bell, J.C., Mahadevan, L.C., Colledge, W.H., Frackelton Jr, A.R., Sargent, M.G. & Foulkes, J.G. (1987). Abelson-transformed fibroblasts contain nuclear phosphotyrosyl-proteins which preferentially bind to murine DNA. **Nature** 325:552–554.

Ben-Neriah, Y., Bernards, A., Paskind, M., Daley, G.Q. & Baltimore, D. (1986). Alternative 5′ exons in c-*abl* mRNA. **Cell** 44:577–586.

Berget, S.M. (1984). Are U4 small nuclear ribonucleoproteins involved in polyadenylation? **Nature** 309:179–182.

Birnstiel, M.L., Busslinger, M. & Strub, K. (1985). Transcription termination and 3′ processing: the end is in site! **Cell** 41:349–359.

Braun, S., Raymond, W.E. & Racker, E. (1984). Synthetic tyrosine polymers as substrates and inhibitors of tyrosine-specific protein kinases. **The Journal of Biological Chemistry** 259:2051–2054.

Collins, S.J., Kubonishi, I., Miyoshi, I. & Groudine, M.T. (1984). Altered transcription of the c-*abl* oncogene in K562 and other chronic myelogenous leukemia cells. **Science** 225:72–74.

Conti, M., Adama, S., Geremia, R. & Monesi, V. (1983). Developmental changes of cyclic adenosine monophosphate-dependent protein kinase activity during spermatogenesis in the mouse. **Biology of Reproduction** 28:860–869.

Davis, R.L., Konopka, J.B. & Witte, O.N. (1985). Activation of the c-*abl* oncogene by viral transduction or chromosomal translocation generates altered c-*abl* proteins with similar in vitro kinase properties. **Molecular and Cellular Biology** 5:204–213.

Edwards, Y.H. & Grootegoed, J.A. (1983). A sperm-specific enolase. **Journal of Reproduction and Fertility** 68:305–310.

Fritz, I.B. (1978). Sites of action of androgens and follicle stimulating hormone on cells of the seminiferous tubules. In: Litwack G, ed. **Biochemical Actions of Hormones**, vol. V. New York, Academic Press, pp. 249–281.

Fulton, A.B. (1982). How crowded is the cytoplasm? **Cell** 30:345–347.

Grootegoed, J.A., Peters, M.J., Mulder, E., Rommerts, F.F.G. & van der Molen, H.J. (1977). Absence of a nuclear androgen receptor in isolated germinal cells of rat testis. **Molecular and Cellular Endocrinology** 9:159–167.

Grootegoed, J.A., Kruger-Sewnarain, B.C., Jutte, N.H.P.M., Rommerts, F.F.G. & van der Molen, H.J. (1982). Fucosylation of glycoproteins in rat spermatocytes and spermatids. **Gamete Research** 5:303–315.

Grootegoed, J.A., Jansen, R. & van der Molen, H.J. (1984). The role of glucose, pyruvate and lactate in ATP production by rat spermatocytes and spermatids. **Biochimica et Biophysica Acta** 767:248–256.

Grootegoed, J.A., Jansen, R. & van der Molen, H.J. (1986). Effect of glucose on ATP dephosphorylation in rat spermatids. **Journal of Reproduction and Fertility** 77:99–107.

Hafen, E., Levine, M., Garber, R.L. & Gehring, W.J. (1983). An improved in situ hybridization method for the detection of cellular RNAs in Drosophila tissue sections and its application for localizing transcripts of the homeotic Antennapedia gene complex. **The EMBO Journal** 2:617–623.

Hecht, N.B. (1986). Evidence for haploid-specific genes in the mammalian testis. In: Stefanini, M., Conti, M., Geremia, R. & Ziparo, E., eds. **Molecular and Cellular Endocrinology of the Testis**. Amsterdam, Elsevier, pp. 199–213.

Higgs, D.R., Goodbourn, S.E.Y., Lamb, J., Clegg, J.B., Weatherall, D.J. & Proudfoot, N.J. (1983). a-Thalassaemia caused by a polyadenylation signal mutation. **Nature** 306:398–400.

Hildebrandt, J.D., Codina, J., Tash, J.S., Kirchick, H.J., Lipschultz, L., Sekura, R.D. & Birnbaumer, L. (1985). The membrane-bound spermatozoal adenylyl cyclase system does not share coupling characteristics with somatic adenylyl cyclases. **Endocrinology** 116:1357–1366.

Hunter, T. & Sefton, B.M. (1980). Transforming gene product of Rous sarcoma virus phosphorylates tyrosine. **Proceedings of the National Academy of Sciences USA** 77:1311–1315.

Hunter, T. & Cooper, J.A. (1984). Tyrosine protein kinases and their substrates: an overview. In: Greengard, P. *et al.*, eds. **Advances in Cyclic Nucleotide and Protein Phosphorylation Research** vol. 17, New York, Raven Press, pp. 443–455.

Hunter, T. & Cooper, J.A. (1985). Protein-tyrosine kinases. **Annual Review of Biochemistry**, 54:897–930.

Konopka, J.B., Davis, R.L., Watanabe, S.M., Ponticelli, A.S., Schiff-Maker, L., Rosenberg, N. & Witte, O.N. (1984). Only site-directed antibodies reactive with the highly conserved src-homologous region of the v-abl protein neutralize kinase activity. **Journal of Virology** 51:223–232.

Konopka, J.B. & Witte, O.N. (1985a). Activation of the abl oncogene in murine and human leukemias. **Biochimica et Biophysica Acta** 823:1–17.

Konopka, J.B. & Witte, O.N. (1985b). Detection of c-abl tyrosine kinase activity in vitro permits direct comparison of normal and altered abl gene products. **Molecular and Cellular Biology** 5:3116–3123.

Kopf, G.S., Woolkalis, M.J. & Gerton, G.L. (1986). Evidence for a guanine nucleotide-binding regulatory protein in invertebrate and mammalian sperm. **The Journal of Biological Chemistry** 261:7327–7331.

Le Bouc, Y., Dreyer, D., Jaeger, F., Binoux, M. & Sondermeyer, P. (1986) Complete characterization of the human IGF-I nucleotide sequence isolated from a newly constructed adult liver cDNA library. **FEBS Letters** 196:108–112.

Mackie, K., Lai, Y., Nairn, A.C., Greengard, P., Pitt, B.R. & Lazo, J.S. (1986). Protein phosphorylation in cultured endothelial cells. **Journal of Cellular Physiology** 128:367–374.

Maxam, A.M. & Gilbert, W. (1977). A new method for sequencing DNA. **Proceedings of the National Academy of Sciences USA** 74:560–564.

Muller, R., Slamon, D.J., Tremblay, J.M., Cline, M.J. & Verma, I.M. (1982). Differential expression of cellular oncogenes during pre- and postnatal development of the mouse. **Nature** 299:640–644.

Nakamura, M., Okinaga, S. & Arai, K. (1986). Studies on metabolism of round spermatids: glucose as unfavorable substrate. **Biology of Reproduction** 35:927–935.

Oonk, R.B. & Grootegoed, J.A. (1988). Insulin-like growth factor (IGF-I) receptors on Sertoli cells from immature rats and age-dependent binding of IGF-I and insulin. **Molecular and Cellular Endocrinology** 55:33–43.

Orkin, S.H., Cheng, T-C., Antonarakis, S.E. & Kazazian Jr, H.H. (1985). Thalassemia due to a mutation in the cleavage-polyadenylation signal of the human ß-globin gene. **The EMBO Journal** 4:453–456.

Ponzetto, C. & Wolgemuth, D.J. (1985). Haploid expression of a unique c-*abl* transcript in the mouse male germ line. **Molecular and Cellular Biology** 5:1791–1794.

Reddy, E.P., Smith, M.J. & Srinivasan, A. (1983). Nucleotide sequence of Abelson murine leukemia virus genome: structural similarity of its transforming gene product to other one gene products with tyrosine-specific kinase activity. **Proceedings of the National Academy of Sciences USA** 80:3623–3627.

Shtivelman, E., Lifshitz, B., Gale, R.P., Roe, B.A. & Canaani, E. (1986). Alternative splicing of RNAs transcribed from the human *abl* gene and from the bcr-*abl* fused gene. **Cell** 47:277–284.

Tash, J.S. & Means, A.R. (1983). Cyclic adenosine 3′, 5′ monophosphate, calcium and protein phosphorylation in flagellar motility. **Biology of Reproduction** 28:75–104.

Wang, J.Y.J., Ledley, F., Goff, S., Lee, R., Groner, Y. & Baltimore, D. (1984). The mouse c-*abl* locus: molecular cloning and characterization. **Cell** 36:349–356.

Witte, O.N. (1986). Functions of the *abl* oncogene. **Cancer Surveys** 5:183–197.

The chromatoid body in early spermatids: its dependence on haploid gene expression and cytoplasmic microtubules

MARTTI PARVINEN AND PEKKA MALI

Institute of Biomedicine, Department of Anatomy, University of Turku, Kiinamyllynkatu 10, SF-20520 Turku, Finland.

Abstract

The chromatoid body is a unique cytoplasmic structure of spermatogenic cells. It is most prominent in early spermatids, where it has been shown histochemically to contain RNA, basic proteins, polysaccharides, actin and possibly DNA. The chromatoid body is apparently dependent on haploid gene expression in early spermatids: radioactivity derived from [^{3}H]uridine is incorporated in this organelle, and it is rapidly deformed and remains non-radioactive if RNA synthesis is inhibited by actinomycin D. In living cells, the chromatoid body moves rapidly on the nuclear envelope having frequent interactions with nuclear contents and with the Golgi complex. The movements and structural integrity of the chromatoid body may involve storage and transport of presynthesized long-lived messenger RNAs that direct protein synthesis during late spermiogenesis, when chromatin becomes condensed and inactivated.

Key words

Rat, seminiferous tubules, spermatids, chromatoid body, haploid gene products, phase-contrast microscopy, vincristine, mitochondria, spermiogenesis, Golgi, Golgi complex, acrosome, flagellum, spermiation, residual body, RNA, spermatocytes, nuclear pore complexes, actin, microtubules.

Morphological observations

The germ determinants, called polar granules in insects, germinal plasm in amphibians and nuage in mammals, are morphologically demonstrable cytoplasmic substances common to all cells of the germ line, including the period of embryonic development (Beams & Kessel, 1974). The chromatoid body is a special form of nuage, characteristic in spermatocytes and spermatids (Eddy, 1974; Söderstrom, 1981a; for review see Sud, 1961). In early round spermatids, it is a lobulated cytoplasmic structure 1–2 μm in diameter (Fig. 1).

The chromatoid body was first decribed by Benda (1891) who suggested that it was derived from the nucleus or the nucleolus. This origin has also been proposed by electron microscopic observations (Comings & Okada, 1972; Reger *et al.*, 1977), but a cytoplasmic origin of the chromatoid body has also been considered possible (Sud, 1961; Fawcett *et al.*, 1970; Fawcett, 1972). During the first appearance of the chromatoid material in pachytene spermatocytes at stage VIII of the seminiferous epithelial cycle (Russell & Frank, 1978), it is often seen to be intimately associated with mitochondria (Fawcett *et al.*, 1970; Fawcett, 1972). During late pachytene and diplotene stages of meiotic prophase, the amount of the chromatoid material increases and specific intranuclear membrane modifications may be involved in its formation (Russell, 1977). During meiotic divisions, the chromatoid body is dispersed in the cytoplasm as small 30 nm granules and as larger 0.5 μm spherical particles (Russell & Frank, 1978). Its condensation to the typical

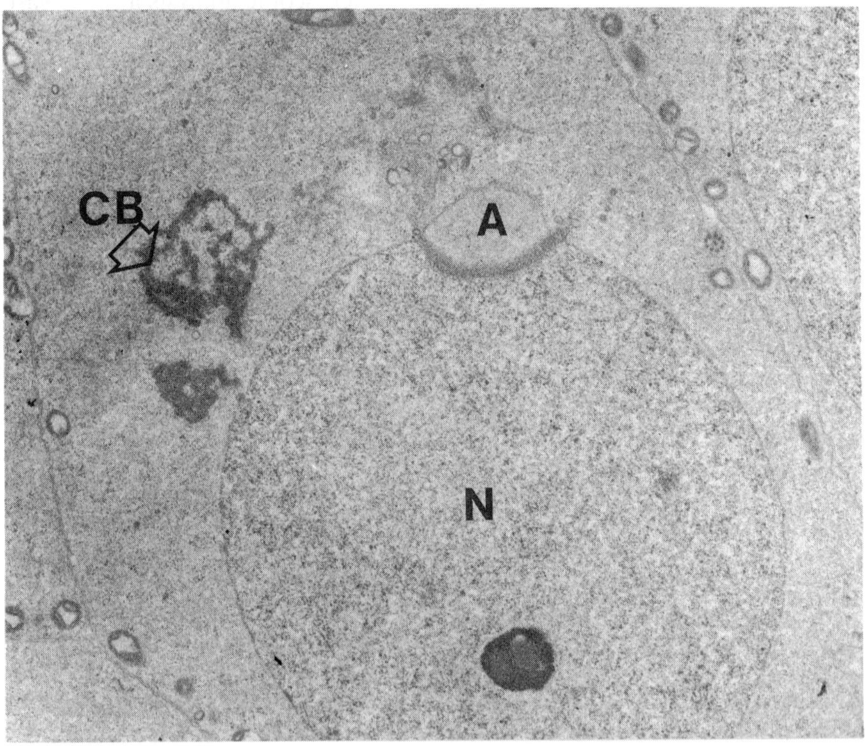

Fig. 1 The normal lobulated ultrastructure of the chromatoid body (CB) of a spermatid from a stage IV seminiferous tubule segment cultured for 24 h in Ham's F 12/Dulbecco's MEM (1:1), supplemented with Hepes, gentamycin, bovine serum albumin and sodium bicarbonate in 5% CO_2 in air at 32°C. A, acrosomic system; N, nucleus. × 15 000.

lobulated form occurs during step 1 of spermiogenesis (Söderstrom & Parvinen, 1978). During the Golgi and early cap phases of spermiogenesis, the chromatoid body is located on the surface of the nucleus close to the Golgi complex and the developing acrosome (Susi & Clermont, 1970). Later in spermiogenesis, the chromatoid body moves to the area of the attachment of the flagellum to the spermatid nucleus (at step 7), where it forms a ring-like structure (Fawcett *et al.*, 1970). Towards the end of spermiogenesis, the chromatoid body diminishes in size and is found in the region of the developing flagellum. It is removed from the spermatozoon at spermiation in the residual body.

Chemical composition

Histochemical pyronin staining in young spermatids showed an intense positive reaction at the chromatoid body suggesting the presence of RNA (Daoust & Clermont, 1955), but electron microscopic analysis after RNase digestion did not support this finding (Eddy, 1970). More recently, specific localization of RNase-gold particles suggests RNA is present in the chromatoid body (Walt & Armbruster, 1984). RNA content of the chromatoid body has been strongly suggested by *in vitro* labelling studies, where radioactivity derived from [^{3}H]uridine was found to incorporate in the chromatoid body after the nuclear labelling both in spermatocytes (Söderstrom & Parvinen, 1976a; Söderstrom, 1981b) and in spermatids (Söderstrom & Parvinen, 1976a). This agrees with observations concerning an increased amount of nuclear pore complexes in the vicinity of the chromatoid body (Fawcett *et al.*, 1970) and material continuities to intranuclear ribonucleoprotein particles through these pores (Söderstrom & Parvinen, 1976b; Parvinen & Parvinen, 1979). The dependence of the chromatoid body on transcription of the haploid genome has been suggested, since after inhibition of transcription by actinomycin D, it becomes morphologically altered (Parvinen *et al.*, 1978) and remains unlabelled after [^{3}H]uridine pulse (Söderstrom, 1977).

More recent high-resolution histochemical analyses have demonstrated that basic proteins are present in the chromatoid body itself, with polysaccharides present in surrounding vesicles (Krimer & Esponda, 1980; Paniagua *et al.*, 1985). DNase-gold complex has been shown to be specifically incorporated on the chromatoid body, but since this marker also binds to actin, the presence of DNA in the chromatoid body remains to be confirmed (Dadoune *et al.*, 1987).

The chromatoid body in living cells

In living cells, the chromatoid body moves rapidly on the nuclear surface in a non-random saltatory fashion (Parvinen & Jokelainen, 1974). Two different patterns of the movement have been described (Parvinen & Parvinen, 1979). During step 1, the chromatoid body moves parallel to the nuclear envelope around the Golgi complex and has frequent transient contacts with the complex (Fig. 2). During step 3, the chromatoid body has been shown to move perpendicularly in relation to the nuclear envelope with material continuities through nuclear pore complexes to intranuclear particles. Both types of movements have been suggested to play a role in the incorporation and transport of haploid gene products during Golgi and early cap phases (steps 1–5) of spermiogenesis (Parvinen & Parvinen, 1979).

Dependence on cytoskeleton

The movements of the chromatoid body may be dependent on the components of the cytoskeleton. When incubated with gold-labeled anti-actin, chromatoid body displays high concentrations of gold particles in electron microscopic immunohistochemical analysis (Walt and Armbruster, 1984), suggesting the presence of actin in the electron dense material of the chromatoid body. Although not analyzed in detail, some electron micrographs show several microtubules in the vicinity of the chromatoid body (Parvinen & Parvinen, 1979).

Vincristine, an inhibitor of microtubule polymerization, has been shown to influence both the movements and structural integrity of the chromatoid body (Mali *et al.*, 1987; Fig. 3). After 16 h incubation *in vitro*, 0.01 and 0.1 μg/ml of vincristine had a slight inhibitory effect on the movements of the chromatoid body , whereas 1 μg/ml slowed the movements to a very low level and 10 μg/ml completely inhibited them. In addition, the concentrations of 10 and 20 μg/ml of vincristine caused marked morphological alterations to the chromatoid body in early spermatids. The earliest changes could be observed after 16 h incubation in the presence of 1 μg/ml of vincristine (Fig. 3). These observations suggest that not only the movements, but also the structural integrity of the chromatoid body are dependent on cytoplasmic microtubules of early spermatids.

Possible functions

The chromatoid body must have several functions during spermiogenesis. The interaction with Golgi complex (Söderstrom & Parvinen, 1976b) suggests that it may be involved in the formation of the acrosomic system. Perhaps the most interesting of the proposed functions is its participation in the RNA metabolism of the spermatogenic cells. Although this is far from clearly understood, there is evidence that a considerable proportion of the RNA synthesized in pachytene spermatocytes is preserved through spermatid development until late spermiogenesis (Geremia *et al.*, 1978). The chromatoid body may be a storing organelle for this long-lived mRNA which is suggested to direct protein synthesis during late spermiogenesis when the genome of the spermatid is inactive. It is also possible that presynthesized proteins are stored in the chromatoid body, to be assembled in spermatid structures at appropriate times. Studies on these problems await techniques such as labelling with specific antibodies and cDNA probes at ultrastructural level, together with its isolation for biochemical analyses.

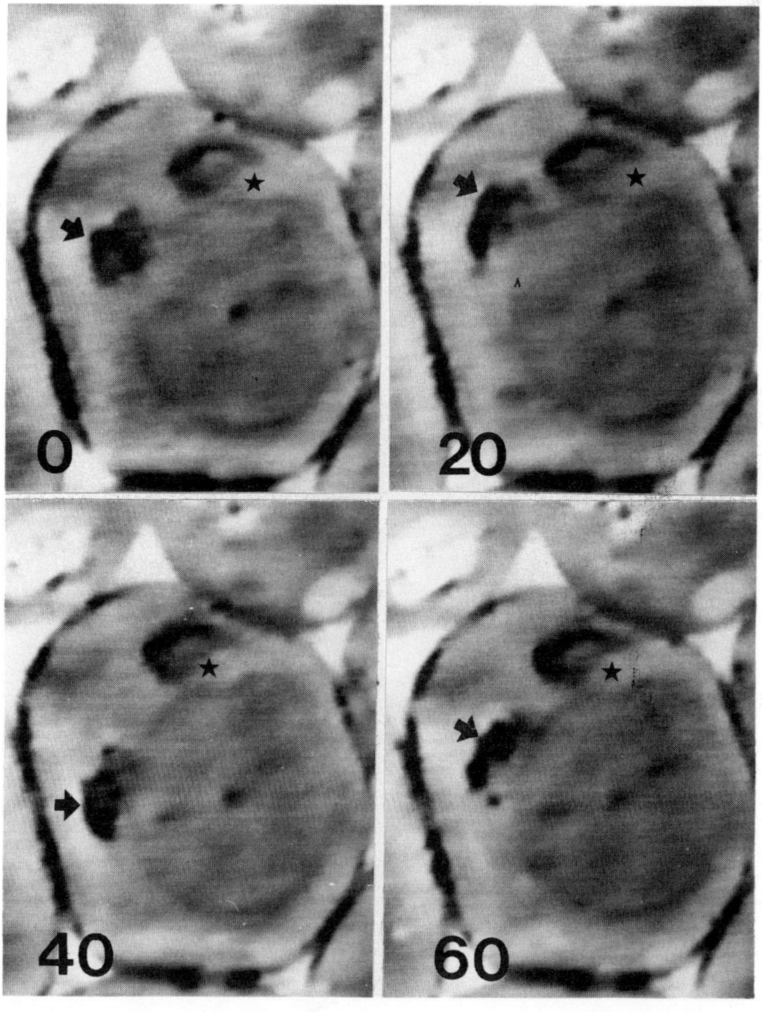

Fig. 2 Phase-contrast video micrographs of a step 4 spermatid cultured for 24 h in control conditions as above. A frozen TV-screen was photographed at 20 s intervals. The movements of the chromatoid body (arrow) on the nuclear envelope are evident when compared with a stationary Golgi complex and acrosomic system (star). × 3 500.

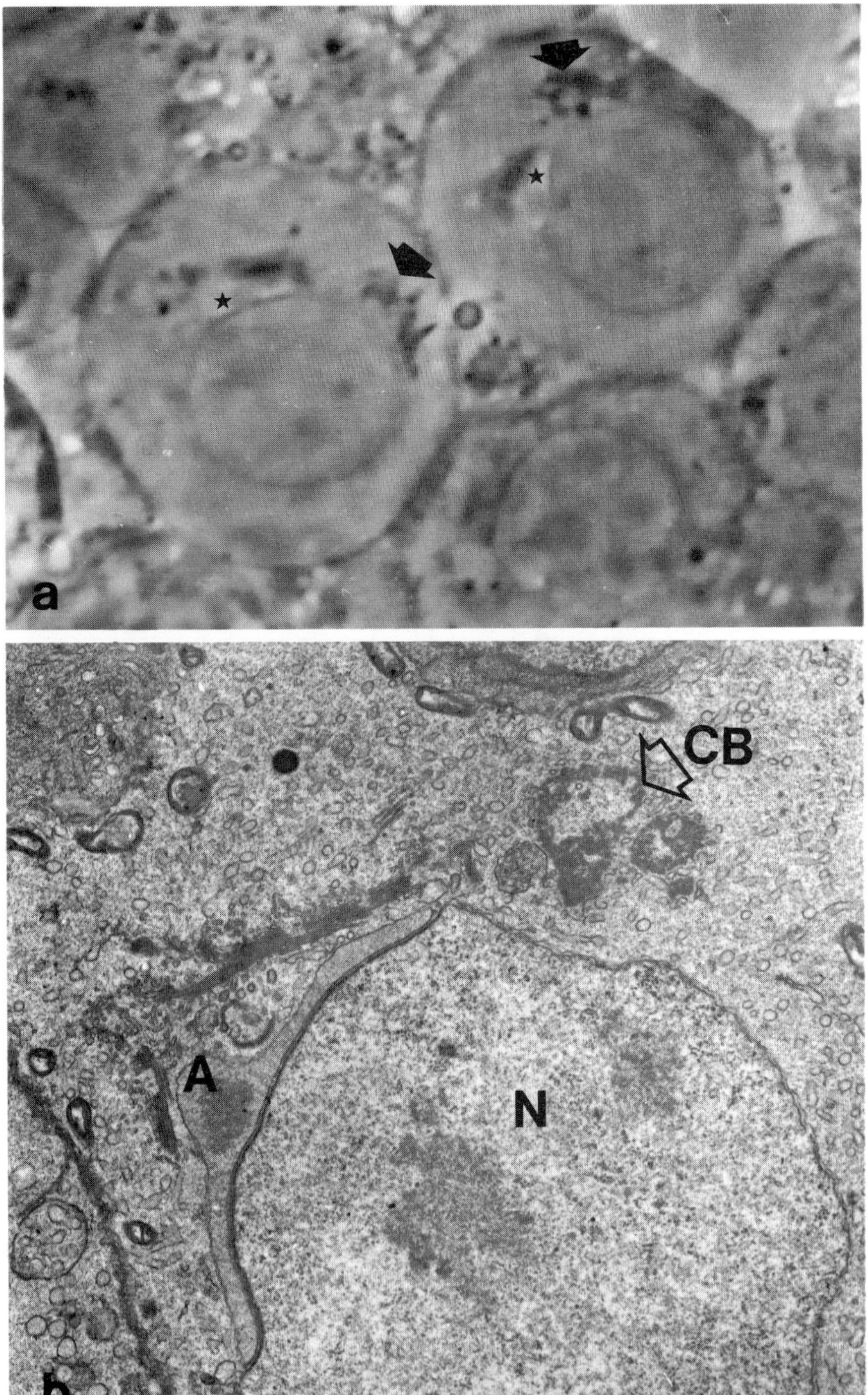
a
CB
A
N
b

Acknowledgements

This project was supported by The Academy of Finland (Project no. 200 at the Medical Research Council).

References

Beams, H.W. & Kessel, R.G. (1974). The problem of germ cell determinants. **International Review of Cytology** 39:413–479.

Benda, C. (1891). Neue Mitteilungen über die Entwickelung der Genitaldrüsen und die Metamorphose der Samenzellen (Histogenese der Spermatozoen). Verhandlungen der Berliner Physiologischen Gesellschaft. **Archiv für Anatomie und Physiologie** 1891:549–552.

Comings, D.E. & Okada, T.A. (1972). The chromatoid body in mouse spermatogenesis: Evidence that it may be formed by the extrusion of nucleolar components. **Journal of Ultrastructure Research** 39:15–23.

Dadoune, J-P., Alfonsi, M-F. & Fain-Maurel, M-A. (1987). Marquage ultrastructural du corps chromatoide et du corps associé au centriole à l'aide du complexe DNase-or colloidal dans les spermatides de singe. **Comptes Rendus de l'Academie des Sciences Serie III. Sciences de la Vie** 305:135–140.

Daoust, R. & Clermont, Y. (1955). Distribution of nucleic acids in germ cells during the cycle of the seminiferous epithelium of the rat. **American Journal of Anatomy** 96:255–283.

Eddy, E.M. (1970) Cytochemical observations of the chromatoid body of male germ cells. **Biology of Reproduction** 2:114–128.

Eddy, E.M. (1974). Fine structural observations on the form and distribution of nuage in germ cells of the rat. **The Anatomical Record** 178:731–758.

Fawcett, D.W. (1972). Observations on cell differentiation and organelle continuity in spermatogenesis. In Beatty, R.A. & Gluecksohn-Waelsch, S., eds. **The Genetics of Spermatozoon**. Edinburgh, pp. 37–67.

Fawcett, D.W., Eddy, E.M. & Phillips, D.M. (1970). Observations on the fine structure and relationships of the chromatoid body in mammalian spermatogenesis. **Biology of Reproduction** 2:129–153.

Geremia, R., d'Agostino, A. & Monesi, V. (1978), Biochemical evidence of haploid gene activity in spermatogenesis of the mouse. **Experimental Cell Research** 111:23–30.

Krimer, D.B. & Esponda, P. (1980). Presence of polysaccharides and proteins in the chromatoid body of mouse spermatids. **Cell Biology International Reports** 4:265–270.

Mali, P., Fagerhed, R-M. & Parvinen, M. (1987). Cytoskeletal regulation of the chromatoid body. In Parvinen, M., Huhtaniemi, I. & Pelliniemi, L.J., eds.

Caption for Figs. 3a,b

Fig. 3a, b. The movements of the chromatoid body slowed down when seminiferous tubule segments from stage V of the cycle were cultured for 24 h in the presence of 1 μg/ml of vincristine in the medium. Large vacuoles inside the chromatoid body and fragmentation was frequently seen in phase-contrast videomicrographs (solid arrow in a) and in electron micrographs (open arrow in b). A, acrosome; N, nucleus; CB, chromatoid body . a, × 3 500; b, × 15 000.

Development and Function of the Reproductive Organs. Serono Symposia Review, New York, Raven Press, (in press).
Paniagua, R., Nistal, M., Amat, P. & Rodriguez, M.C. (1985). Presence of ribonucleoproteins and basic proteins in the nuage and intermitochondrial bars of human spermatogonia. **Journal of Anatomy** 143:201–206.
Parvinen, L-M., Jokelainen, P.T. & Parvinen, M. (1978). Chromatoid body and haploid gene activity: Actinomycin D induced morphological alterations. **Hereditas** 88:75–82.
Parvinen, M. & Jokelainen, P.T. (1974). Rapid movements of the chromatoid body in living early spermatids of the rat. **Biology of Reproduction** 11:85–92.
Parvinen, M. & Parvinen, L-M. (1979), Active movements of the chromatoid body. A possible transport mechanism for haploid gene products. **Journal of Cell Biology** 80:621–628.
Reger, J.F., Fain-Maurel, M.A. & Cassier, P. (1977). The origin, distribution and fate of the chromatoid body (germ plasm) during spermatogenesis and spermiogenesis in two peracaridae. **Journal of Ultrastructure Research** 60:84–94.
Russell, L. (1977). Membranous structures within nuclei of spermatocytes in the adult male rat. **Biology of Reproduction** 17:184–191.
Russel, L. & Frank, B. (1978). Ultrastructural characterization of nuage in spermatocytes of the rat testis. **Anatomical Record** 190:79–98.
Söderstrom, K-O. (1977). Effect of actinomycin D on the structure of the chromatoid body in the rat spermatids. **Cell and Tissue Research** 184:411–421.
Söderstrom K-O. (1981a). The relationship between the nuage and the chromatoid body during spermatogenesis in the rat. **Cell and Tissue Research** 215:425–430.
Söderstrom, K-O. (1981b). Labelling of the chromatoid body with 3H-uridine in the pachytene spermatocytes of the rat. **Experimental Cell Research** 131:488–491.
Söderstrom, K-O. & Parvinen, M. (1976a). Incorporation of 3H-uridine by the chromatoid body during rat spermatogenesis. **Journal of Cell Biology** 70:239–246.
Söderstrom, K-O. & Parvinen, M. (1976b). Transport of material between the nucleus, the chromatoid body and the Golgi complex in the early spermatids of the rat. **Cell and Tissue Research** 168:335–342.
Söderstrom, K-O. & Parvinen, M. (1978). Observations on living rat spermatogenic cells in different developmental stages. **Acta Anatomica** 100:557–572.
Sud, B.N. (1961). The "chromatoid body" in spermatogenesis. **Quarterly Journal of Microscopical Science** 102:273–292.
Susi, F.R. & Clermont, Y. (1970). Fine structural modifications of the rat chromatoid body during spermatogenesis. **American Journal of Anatomy** 129:177–192.
Walt, H. & Armbruster, B.L. (1984). Actin and RNA are components of the chromatoid bodies in spermatids of the rat. **Cell and Tissue Research** 236:487–490.

Gene expression during spermiogenesis

NORMAN B. HECHT

Department of Biology, Tufts University, Medford, MA 02155, USA.

Abstract

The high levels of RNA synthesized during meiosis in the mammalian testis and the 'long lived' meiotic mRNAs utilized during spermiogenesis, have led many investigators to conclude that the meiotic nucleus is the sole source of mRNA transcripts needed for male germ cell differentiation to reach completion. However, recent cell-free translations of mRNAs isolated from meiotic and post-meiotic cell types and differential screening of mouse testicular cDNA libraries have demonstrated that there is a temporal regulation of gene products programmed from *both* the meiotic and post-meiotic genomes (see Hecht, 1986b).

In this chapter the identification of nuclear and cytoplasmic proteins that represent promising spermiogenic targets for fertility control by the use of recombinant DNA cloning approaches will be discussed. Using stage-specific testicular cDNAs to differentially screen a testis cDNA library, it has been determined that 3 nuclear DNA-binding proteins involved in chromatin alterations during spermiogenesis, a transiently appearing testis-protein (TP) and the two protamine variants, are first expressed during spermiogenesis. DNA sequence analysis of the protamine cDNAs (mP1 and mP2) reveals that although mP1 and mP2 are co-ordinately expressed and chromosomally linked, they differ markedly in sequence, size of protein, and their requirement for post-translational processing.

Two additional spermiogenic targets, isotypes of actin and α-tubulin have also been identified. In addition to the β- and γ-actins that are differentially expressed throughout spermatogenesis, a novel γ-like actin is first detectable in post-meiotic testicular cells. Screening testis cDNA libraries with α-tubulin cDNAs from somatic cell types, cDNAs coding for at least 6 different testicular α-tubulin isotypes have been isolated and characterized. Using antisera prepared against a peptide whose sequence was predicted from one of these mouse testis α-tubulin cDNAs, an α-tubulin epitope that is not detectable in the brain has been found in the testis. Immuno-blotting and indirect immunofluorescence of testis sections have been used to localize this α-tubulin to two germ cell specific structures, the manchette and the meiotic spindle.

Key words

Spermatogenesis, spermiogenesis, testis, gene expression, protamine, testis-protein (TP), actin, tubulin, manchette, fertility markers.

Introduction

The ordered sequence of events leading to the production of the spermatozoon offers many opportunities for regulatory intervention. During meiotic prophase, chromosomes pair and genetic recombination occurs. These events are followed by the haploid phase of spermatogenesis, spermiogenesis, an interval of cytodifferentiation culminating in the transformation of the spermatozoon from the haploid spermatid (Figs. 1 and 2). During this lengthy continuum of cell differentiation the nucleus condenses and the acrosome and flagellum develop. These are unique post-meiotic events dependent upon temporal and spatial alterations in gene expression.

In the mammalian testis, autoradiographic and biochemical studies from many laboratories have demonstrated that RNA and protein synthesis levels are maximum during meiosis and then decrease abruptly during early spermiogenesis to an undetectable level (reviewed in Monesi *et al.*, 1978). These observations have led many to conclude that the transcripts from the meiotic nucleus code for all the proteins needed for spermatogenesis to reach

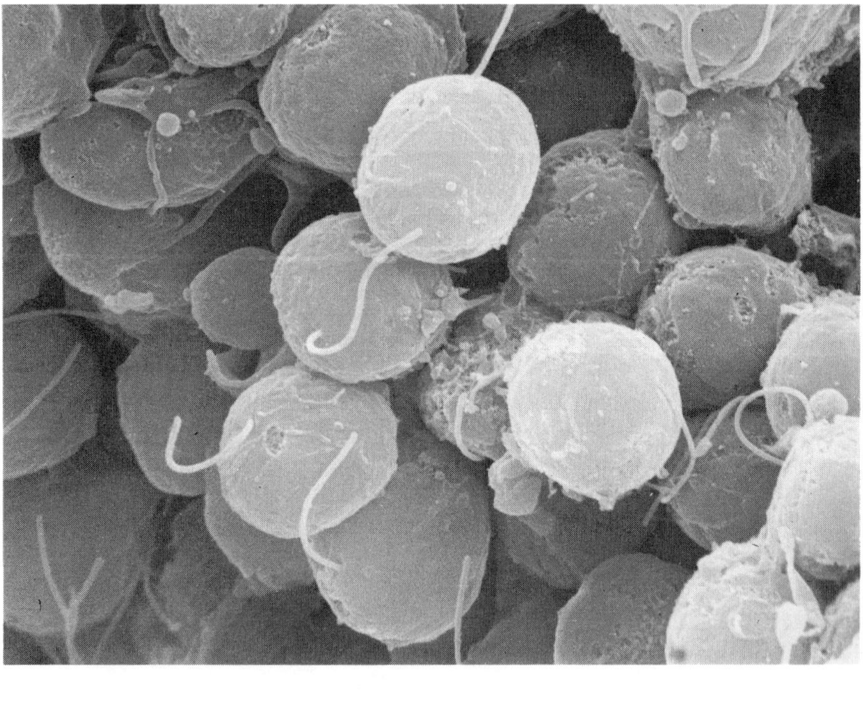

Fig. 1 Scanning electron micrograph of human round spermatids. The cells are viewed from the lumen of the seminiferous tubule. Short, slightly bent flagella protrude from the cells (from Holstein & Roosen-Runge, 1981).

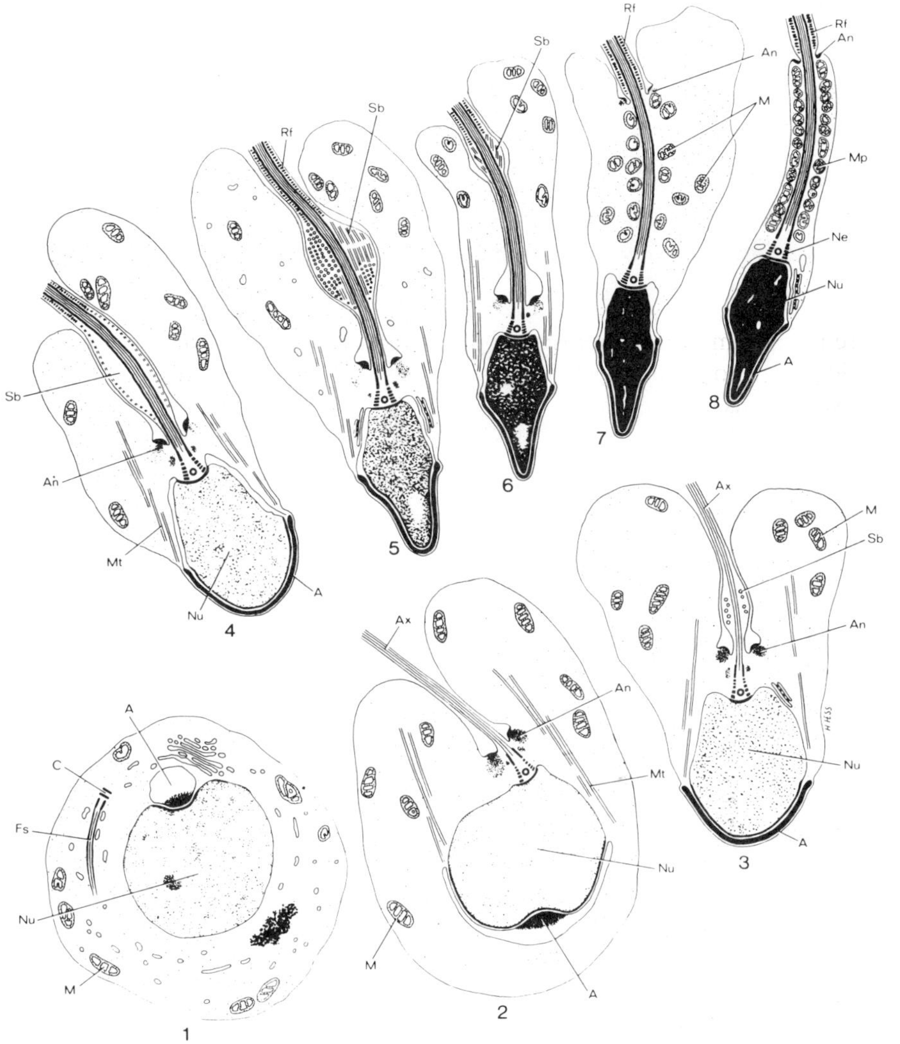

Fig. 2 Differentiation of spermatids in the human. This diagram presents the three major developmental processes of spermiogenesis: (1) condensation of the nucleus, (2) the formation of the acrosome, and (3) the development of the flagellum. A, acrosome; An, annulus; Ax, axoneme; C, centriole; F, flower-like structures; Fs, flagellar substructures; M, mitochondria; Mp, middle piece; Mt, manchette; Ne, neck; N, nucleus; PP, principal piece; Rf, ring fibers; Sb, spindle-shaped body (from Holstein & Roosen-Runge, 1981).

completion. Although meiotic 'long lived' mRNAs are utilized during spermiogenesis, *in vivo* and cell-free translation studies argue strongly that many genes are expressed for the first time in the testis during spermiogenesis (Geremia *et al.*, 1978; Boitani *et al.*, 1980; Kramer & Erickson, 1982; Fujimoto & Erickson, 1982; Gold *et al.*, 1983; Stern *et al.*, 1983a, 1983b). Because of their tissue-specific expression and their unique temporal appearance, many of these gene products represent promising targets for male fertility regulation.

The temporal expression of a few representative testicular proteins is shown in Fig. 3, which is not comprehensive but is meant only to provide an overview of gene expression for a few primarily post-meiotic proteins of the mammalian testis. A more complete examination of molecular events during spermiogenesis can be found in several recent discussions of this topic (Bellvé, 1979; Bellvé & O'Brien, 1983; Hecht 1986a, b, 1987a, b, c, d).

In this chapter several approaches using recombinant DNA techniques that have been successfully used to identify spermiogenic genes will be presented. These studies demonstrate that the post-meiotic interval of early spermiogenesis should be seriously considered for male fertility regulation. Among a large number of possible spermiogenic targets, the discussion will be focused on 5 structural proteins that are expressed during spermiogenesis: 3 DNA-binding proteins involved in the chromatin transitions during spermiogenesis and isotypes of actin and tubulin. Special attention will be given to a novel testicular α-tubulin that is not detectable in the brain and is shown to be a microtubular component of the manchette.

Differential screening of a testis cDNA library identifies testis-protein (TP) and the protamines as spermiogenic gene products

To identify genes expressed during spermiogenesis, we have differentially screened a testis cDNA library with meiotic and post-meiotic cDNAs. Utilizing a combination of two cell separation procedures, Staput unit gravity sedimentation in a bovine serum albumin gradient, which separates cells by sedimentation rate, and Percoll density gradient centrifugation, which separates cells by the criterion of density, highly enriched populations of meiotic pachytene spermatocytes and post-meiotic round spermatids were isolated (Stern *et al.*, 1983a; Kleene *et al.*, 1983). Polyadenylated RNA was prepared from these two cell types and used as template to make stage-specific radiolabeled cDNAs. Using the radiolabeled cDNAs from the meiotic and post-meiotic cell types to differentially screen a testicular cDNA library, we reasoned that although the majority of the cDNAs representing expressed genes would be present in both meiotic and post-meiotic cells we would expect a small number of genes to show cell type specificity of expression. Following an initial screening of the mouse testis cDNA library, 17 of 750 cDNA clones hybridized solely or much more strongly with the

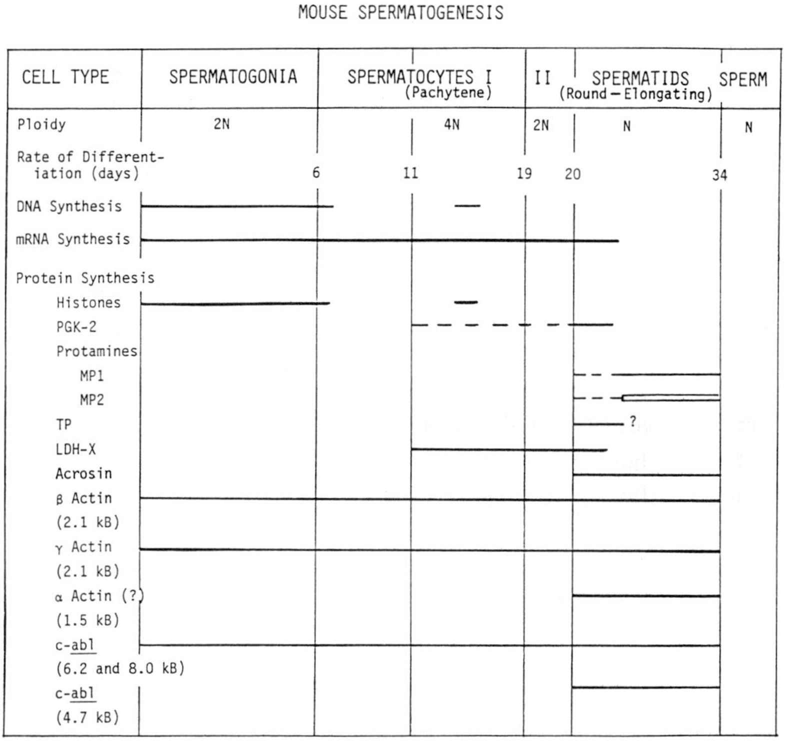

Fig. 3 Summary of the expression of selected proteins during spermiogenesis (data derived from several reviews and chapters (Bellvé, 1979; Bellvé & O'Brien, 1983; Hecht, 1986a, b, 1987a, b, c, d). A solid line represents the approximate time period when the protein is synthesized. A dashed line indicates the interval when mRNA for the protein has been detected, but the protein has not yet been detected i.e. the interval when the mRNA is probably stored as a ribonucleoprotein. An open box indicates the time period when mP2 is synthesized as a precursor and subsequently processed to its mature form. Question marks represent uncertainty as to the exact time when TP synthesis terminates and to the identity of the ~1.5 kB actin. PGK-2, phosphoglycerate kinase 2; MP, mouse protamine; TP, testis-specific protein; LDH-X, lactate dehydrogenase C4.

round spermatid cDNAs (Kleene *et al.*, 1983). Additional analysis demonstrated that these 17 clones represented several different sequences which were not expressed in liver, brain, cultured Sertoli cells, testes from 16–17-day-old animals (where spermatogenesis has only proceeded to meiosis) or in mouse mitochondria. DNA sequence analysis of three groups of cDNA clones that did not cross-hybridize identified cDNA probes that code for a prominent mouse testis-protein (TP) and the two protamine variants. A discussion of the properties and confirmatory evidence that these three DNA-binding proteins are indeed post-meiotic gene products follows.

Testis-protein (TP)

Striking changes occur in the composition of nuclear proteins during spermatogenesis (Fig. 4). In addition to the usual complement of somatic histones, several germ cell-specific histone variants, Hlt, TH2A, and TH2B and a high mobility group protein HMG2, accumulate in the meiotic nucleus. As germ cell differentiation proceeds into spermiogenesis, an additional unique class of basic protein, designated testis-protein (TP), transiently appears. In turn the TPs are displaced from the spermatid chromatin towards the end of spermiogenesis by the protamines, the predominant nuclear protein of the spermatozoon.

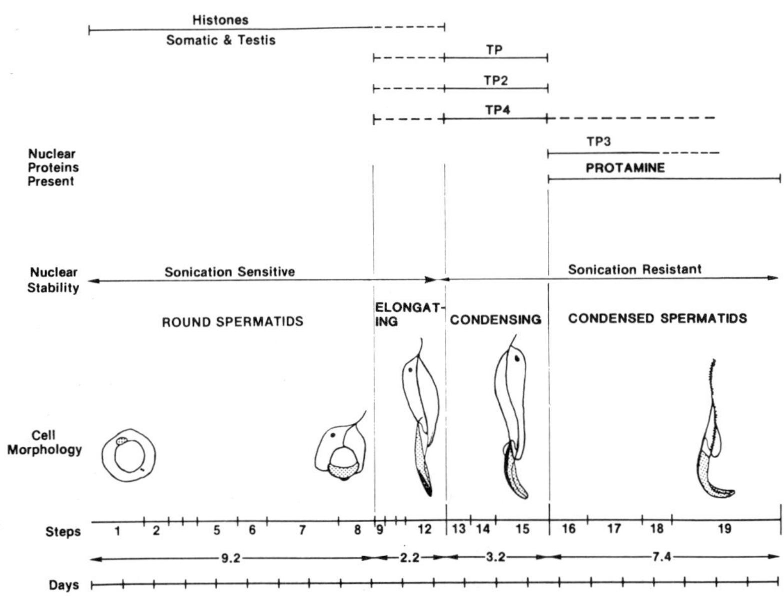

Fig. 4 Transition proteins during spermiogenesis. Outline of the basic nuclear proteins present during various stages of spermatid development in the rat correlated with the cell morphology and kinetics of the process. The definite presence of a nuclear protein at the indicated stages is represented by the solid line. The dashed line indicates uncertainty or gradual accumulation or loss of the protein. Each tic mark represents one day and the duration (in days) of the 4 phases of spermatid development is indicated by the arrows (from Meistrich, 1987).

TPs are a heterogeneous group of DNA-binding proteins that have been found in several mammals (for a recent review, see Meistrich, 1987). The proteins are synthesized during spermiogenesis and are associated with the condensing spermatid nucleus as it transforms from a nucleosome type structure to the smooth branching fibril structure found in the spermatozoon. From the sequence of one rat and mouse TP, we conclude that TPs have a lower arginine and higher lysine content than the mammalian protamines; in rodents at least one TP lacks cysteine, an amino acid present in all eutherian mammalian protamines (Kistler *et al.*, 1975; Meistrich, 1987). It has not yet been established how many distinct TP sequences are encoded in the mammalian genome, i.e., whether the multiple TPs detected by gel electrophoresis reflect post-translational modifications of one or more TP sequences.

DNA sequence analysis of one group of cDNAs, previously called 11A (Hecht *et al.*, 1986a), obtained by the differential screening of a mouse testis cDNA library described above, reveals that it codes for a mouse TP (Kleene *et al.*, 1983). Knowing that in the mouse TP is synthesized in the early part of spermiogenesis before the protamines, we have utilized the TP cDNA to determine when TP mRNA can first be detected in the testis (Hecht *et al.*, 1986a). Hybridization of Northern blots of total testicular RNA from the testes of prepuberal mice ranging from 16 to 26 days of age reveals that TP mRNA is first detectable in testis extracts 22 days after birth, a time when the first wave of spermatogenesis of the prepuberal testis has reached early spermiogenesis. The absence of TP mRNA in testicular extracts of 18- and 20-day-old animals suggests that TP mRNA is not present at a detectable level in primary or secondary spermatocytes. The post-meiotic appearance of mouse TP is confirmed by Northern blot analysis of RNA isolated from pachytene spermatocytes, round spermatids, and residual bodies and by *in situ* hybridization studies (Hecht & Penshow, 1987). High levels of TP mRNA in the ribonucleoprotein fraction of testicular extracts suggest that TP synthesis may also be translationally regulated during spermiogenesis as has been found for the mouse protamine genes (see below). The similar temporal appearance of the mRNAs coding for TP and the protamine mRNAs raises a fascinating post-transcriptional regulatory problem, because the TP and protamine proteins are sequentially synthesized during spermiogenesis.

The protamines

Protamines are the predominant nuclear proteins of the spermatozoa of fish, some birds and all mammals (for a recent review, see Hecht, 1987a). In the mouse and man, two protamine variants (P1 and P2) that differ markedly in both size and sequence have been described (Kleene *et al.*, 1985; Yelick *et al.*, 1987; Ammer *et al.*, 1986; McKay *et al.*, 1985, 1986). It has generally been thought that, with the exception of mouse and man, most mammalian spermatozoa contain only the P1 protamine variant. Recent hybridization studies with protamine cDNAs indicate that DNA sequences showing homology to both mouse protamine variants are more widespread in mammals than predicted by gel electrophoresis of sperm nuclear proteins. For instance, species such as the rat, where only the P1 protamine has been detected in spermatozoa, contain a P2 gene and synthesize both P1 and P2 protamines (Bower *et al.*, 1987). In the mouse, both protamine variants, mP1 and mP2, are synthesized towards the end of spermatogenesis in elongated spermatids (Fig. 4) (Lam & Bruce, 1971; Bellvé *et al.*, 1975; Mayer & Zirkin, 1979; Mayer *et al.*, 1981; Balhorn *et al.*, 1984). The synthesis of mP1 is detected in late-stage spermatids one day earlier than mP2, because mP2 is synthesized as a precursor that is subsequently processed to the mature form of P2 found in spermatozoa (see below) (Balhorn *et al.*, 1984; Yelick *et al.*, 1987).

Properties of the P1 protamines

P1 protamines isolated from spermatozoa, have been sequenced from a diverse group of mammals including rat, boar, bull, ram, and human (reviewed in Hecht, 1987a). The sequence of the mouse P1 protamine has been derived from its cDNA (Fig. 5) (Kleene *et al.*, 1985). All P1 protamines sequenced to date are molecules 50 amino acids long with clusters of arginine in the core of the sequences and less basic regions at the amino and carboxyl termini. Although the 6 amino acids at the N terminal and the arginine-rich core of the molecule appear to be strongly conserved, P1 sequences differ substantially among mammals. For example, the mouse P1 protamine differs in 9, 16, and 18 amino acids (out of 50) from boar, bull, ram, and human, respectively. Based upon the reported inability to synthesize mouse protamine in cell-free systems (Elsevier, 1982) and the presence of transiently appearing cysteine-containing proteins during spermiogenesis (Balhorn *et al.*, 1984), it has been proposed that the protamines are synthesized as precursor molecules. The sequence of P1 cDNAs for mouse and bull demonstrates that this is not true (Kleene *et al.*, 1985; Lee *et al.*, 1987).

Properties of the P2 protamines

Although both P1 and P2 protamines contain clusters of arginine, the two protamine variants of the mouse differ in sequence, size, and post-translational processing (Kleene *et al.*, 1985; Yelick *et al.*, 1987). The P2 protamine is substantially longer than the P1 protamine and contains 13 residues of histidine, an amino acid absent in mP1. Moreover, in contrast to the P1 protamine which does not require post-translational cleavage, the predicted

```
A.  mP1                                                                                                  1
                                                                                                     Ala Arg
5'   GACCCCTGCTCACAGGTTGGCTGGCTCGACCCAGGTGGTGTCCCCTGCTCTGAGCCAGCTCCCGGCCAAGCCAGCACC ATG GCC AGA
     -78                -60                                  -30                                  1
                               10                                      20
    Tyr Arg Cys Cys Arg Ser Lys Ser Arg Ser Arg Cys Arg Arg Arg Arg Arg Arg Cys Arg Arg Arg Arg
    TAC CGA TGC TGC CGC AGC AAA AGC AGG AGC AGA TGC CGC CGT CGC AGA CGA AGA TGT CGC AGA CGG AGG
                               30                                      60
                30                                      40
    Arg Arg Cys Cys Arg Arg Arg Arg Arg Arg Cys Cys Arg Arg Arg Arg Ser Tyr Thr Ile Arg Cys Lys
    AGG CGA TGC TGC CGG CGG AGG AGG CGA AGA TGC TGC CGT CGC CGC CGC TCA TAC ACC ATA AGG TGT AAA
                  90                                      120
      50
    Lys Tyr ***
    AAA TAC TAG ATGCACAGAATAGCAAGTCCATCAAAACTCCTGCGTGAGAATTTTACCAGACTTCAAGAGCATCTCGCCACATCTTGAA
     150                              180                            210

    AAATGCCACCGCCGATGAAAAACAGGAGCCTGCTAAGGAACAATGCCACCTGCCAATAAATGTTGAAAACTCAAAAAAAAAAAAAAAAAA  3'
       240                            270                            300                          326

B.  mP2        1                                    10
           Met Val Arg Tyr Arg Met Arg Ser Pro Ser Glu Gly Pro His Gln Gly Pro Gly Gln Asp
5'   CGACCC ATG GTT CGC TAC CGA ATG AGG AGC CCC AGT GAG GGT CCG CAC CAG GGG CCT GGA CAA GAC

     20                                      30                                      40
     His Glu Arg Glu Glu Gln Gly Gln Gly Gln Gly Leu Ser Pro Glu Arg Val Glu Asp Tyr Gly Arg
     CAT GAA CGC GAG GAG CAG GGG CAG GGG CAA GGG CTG AGC CCA GAG CGC GTA GAG GAC TAT GGG AGG

             ↓                           50                                  60
     Thr His Arg Gly His His His His Arg His Arg Arg Cys Ser Arg Lys Arg Leu His Arg Ile His
     ACA CAC AGG GGC CAC CAC CAC CAC AGA CAC AGG CGC TGC TCT CGT AAG AGG CTA CAT AGG ATC CAC

                             70                                      80
     Lys Arg Arg Arg Ser Cys Arg Arg Arg Arg Arg His Ser Cys Arg His Arg Arg Arg His Arg Arg
     AAG AGG CGT CGG TCA TGC AGA AGG CGG AGG AGA CAC TCC TGC CGC CAC AGG AGG CGG CAT CGC AGA

                         90                                      100
     Gly Cys Arg Arg Ser Arg Arg Arg Arg Arg Cys Arg Cys Arg Lys Cys Arg Arg His His His ***
     GGC TGC AGA AGA TCC CGA AGG AGG AGG AGA TGC AGG TGC AGG AAA TGT AGG AGG CAC CAT CAC TAA

     GCCTCCCCAGGCCTGTCCATTCTGCCTGGAGCCAAGGAAGTCACTTGCCCAAGGAATAGTCACCTGCCCAAGCAACATCATGTGAGGC

     CACACCACCATTCCATGTCGATGTCTGAGCCCTGAGCTGCCAAGGAGCCACGAGATCTGAGTACTGAGCAAAGCCACCTGCCAAATAAA

     GCTTGACACGAGAAAAAAAAAAAAAAAAA  3'
```

Fig. 5 Sequences of cDNAs coding for mouse protamines 1 and 2. The predicted amino acid sequences of the cDNAs are presented above the nucleotide triplets (Kleene *et al.*, 1985; Yelick *et al.*, 1987). Amino acids are numbered beginning with alanine for mP1 and valine for mP2 because methionine does not appear at the amino termini of protamines. The arrow in the mP2 sequence indicates the first amino acid of the processed form of mP2 isolated from spermatozoa (Yelick *et al.*, 1987). The termination codons, TAG for mP1 and the TAA for mP2, are denoted by asterisks. The canonical hexanucleotide, AATAAA, is underlined in each cDNA sequence. The common repeat sequence in the 3′ untranslated regions of the two cDNAs is denoted by a line over it.

amino acid sequence from the mP2 cDNA suggests the P2 variant is synthesized as a precursor of 106 amino acids which is subsequently processed to its 63 amino acid mature form (Fig. 5) (Yelick *et al.*, 1987). Partial sequence determinations of purified P2 mouse protamine and a group of cysteine-rich proteins, whose synthesis is temporally coincident with that of protamine, confirms that the mouse P2 protamine is synthesized as a precursor (Yelick *et al.*, 1987). The requirement for cleavage of the P2 protamine offers a promising intervention site to perturb a critical step of spermiogenesis.

The reason that one but not both of the protamine variants is synthesized as a precursor can only be the subject of speculation. One possibility may reflect different protein stabilities of the P1 and P2 protamines. Protein turnover rates have been proposed to be determined by the amino terminal residue of a polypeptide, with half lives ranging from 3 min to over 20 h (Bachmair *et al.*, 1986). These protein turnover studies would predict that a protein with an arginine N- terminal, such as the mature form of mP2, would have a half life of 2–3 min, whereas valine, the N- terminal amino acid of the mP2 precursor and alanine, the N- terminal amino acid of MP1, would confer half lives of more than 20 h.

In the mouse the P1 and P2 protamine variants are single copy genes that have been mapped to chromosome 16 by Southern blot analysis of DNA from hamster-mouse cell lines (Hecht *et al.*, 1986b). Further mapping efforts using mouse restriction fragment length polymorphisms have demonstrated that the P1 and P2 protamines are closely linked about 7–10 map units from the centromere of mouse chromosome 16 (Reeves and Hecht, unpublished observations). The close linkage of P1 and P2 protamines in mouse and hamster suggest that the human P1 and P2 protamines are also likely to be closely linked.

Although the mP1 and mP2 cDNAs do not cross-hybridize, their cDNAs share a common nucleotide sequence in the 3′ untranslated region. Fifteen of 17 nucleotides in the sequence GA(A or G)CAA(Tor A)GCCACCTGCC directly upstream and adjacent to the polyadenylation signal, AA(possibly A)TAAA, are identical in both mouse protamine variants (Fig. 5). In the mP2 cDNA, a modified form of this sequence is repeated in the 3′ untranslated region. The conservation of this sequence in two co-ordinately expressed genes argues for an essential regulatory role for this region of the protamine mRNAs. DNA sequence analysis of 5′ flanking regions of genomic protamine DNA have identified additional shared sequence regions of the P1 and P2 protamine genes (Johnson *et al.*, 1988).

Temporal appearance of the mouse protamines

In the mouse, protamines are synthesized between steps 12 and 15 of spermiogenesis. Northern blot analysis with the cDNAs for mP1 and mP2 has allowed us to determine when mP1 and mP2 mRNAs are first detectable during testicular germ cell development. Analysis of RNA extracts from the testes of prepuberal animals (Hecht *et al.*, 1986a), from isolated populations of meiotic and post-meiotic cell types (Kleene *et al.*, 1984), and by *in situ* hybridization (Hecht & Penshow, 1987) has established that, as found for TP, protamine mRNAs are first detectable during spermiogenesis. Moreover, the protamine mRNAs are present in the testis approximately one week before the protamine proteins are synthesized. Hybridization analysis of polysomal and ribonucleoprotein fractions from whole testis and from individual populations of round spermatids, elongating spermatids and residual bodies has demonstrated that mP1 and mP2 are translationally regulated. Approximately 85% of total protamine mRNA is found in the ribonucleoprotein fraction. A more detailed discussion of protamine mRNA metabolism during spermatogenesis and the use of protamine mRNAs as biomarkers of spermatogenesis is reviewed elsewhere (Hecht *et al.*, 1985; Hecht 1987a, b). From these temporal studies, we conclude that the two mouse protamine variants and TP are three nuclear structural proteins that are solely expressed during spermiogenesis. They represent inviting organ-specific targets for fertility regulation in the haploid male germ cell.

Identification of a novel post-meiotic actin isotype

Actins are ubiquitous structural proteins encoded by a multigene family in mammals. They play roles in many cellular functions including cell division and motility, organelle movement, and in the maintenance of cell structure. In the testis, actins are also involved in chromosome pairing during meiosis, in shaping and anchoring the acrosome, and they are present in mature spermatozoa.

We have initiated a study of actin metabolism during spermatogenesis to determine whether the unique structural demands of spermiogenesis require novel testicular actin isotypes. Analysis of *in vivo* labeled testicular proteins by two-dimensional gel electrophoresis has documented that the predominant cytoplasmic isotypes of actin, β- and γ-actin are differentially synthesized in meiotic and post-meiotic cells (Hecht *et al.*, 1984). Using a series of isotype specific actin cDNA probes, Waters *et al.* (1985) have extended these protein studies to the mRNA level showing that β- and γ-actin are encoded by a population of testicular 2.1 kb mRNAs, presumably identical to the β- and γ-actin mRNAs found in somatic cells. In addition, a second size class of actin mRNA of 1.5 kb not coding for β- or γ-actin was detected in spermiogenic cells. The 1.5 kb actin mRNA was especially abundant in elongating spermatids and in residual bodies.

To identify the actin coded for by the 1.5 kb mRNA, we have isolated mRNAs from a population of residual bodies and elongating spermatid cytoplasts and used them to prepare a small cDNA library. After screening this cDNA library with a mouse actin coding-region probe, a group of cDNAs were isolated. DNA sequence analysis was used to identify several cDNAs coding for a smooth muscle γ-actin (Kim, Waters, and Hecht, unpublished observations). Thus, in addition to the continually expressed β- and γ-actins, the testis expresses an additional γ-actin in round and elongating spermatids. We do not know whether this actin is testis-specific or not, but its unique temporal appearance provides an additional target for intervention during spermiogenesis. Determination of the organ and cellular distribution of this post-meiotic actin isotype may help to define its function(s) in the testis.

Identification of a novel alpha – tubulin of the manchette

Microtubules are filamentous heterodimers of α- and β-tubulin which play essential roles in cell shape and cytoarchitecture, in intracellular transport, and in cell division (Cleveland & Sullivan, 1985; see chapters by Parvinen & Mali and Kuriyama & Sellitto in this book). The meiotic spindle and the manchette are two germ cell-specific structures known to be especially rich in microtubules. In the testis, the meiotic spindle is essential for proper chromosome disjunction in the primary spermatocyte. Although the function of the unique manchette remains to be elucidated, it is known that it is made up of a parallel arrangement of up to about 1100 microtubules attached to a circum-nuclear dense protein plate. The manchette is transiently present during spermiogenesis, forming at the time the round spermatids begin to lose their spherical shape and disappearing as the spermatid elongates. The structure and temporal appearance of the manchette have led to widespread speculation regarding its function, including involvement in shaping the sperm nucleus, transportation of cytoplasmic components from the anterior to posterior regions of the elongating spermatid and structural partitioning of the spermatid cytoplasm. Aside from containing somatic tubulin (Cherry & Hsu, 1984), the structural components of the manchette are unknown.

In the mammalian genome, both α- and β-tubulin are coded for by multigene families. Although many genomic tubulin sequences represent pseudogenes, a growing number of functional tubulin genes have been identified, leading to speculation that different tubulin genes exist to fulfill specific cellular requirements. In support of this, a testis-specific β-tubulin has been identified in *Drosophila* (Kemphues *et al.*, 1982) and testis-specific α-tubulin cDNAs have been characterized in the mouse (Distel *et al.*, 1984; Villasante *et al.*, 1986) and chicken (Pratt *et al.*, 1987). Differential expression of the α-and β-tubulins in the testis is documented by two-dimensional gel electrophoresis of [^{35}S]methionine labeled *in vivo* testicular cell extracts, which have suggested that several isoforms of mouse testicular α-and β-tubulin exist in spermatids (Hecht *et al.*, 1984). Although multiple electrophoretic forms of tubulin could represent post-translational modification of a single tubulin sequence, to date, cDNAs that code for at least 7 different tubulin mRNAs in the mouse testis have been isolated (Distel *et al.*, 1984; Villasante *et al.*, 1986).

To determine which α-tubulin genes are expressed during spermatogenesis, somatic cDNA tubulin probes have been utilized to screen mouse testis cDNA libraries (Distel *et al.*, 1984; Villasante *et al.*, 1986). This approach is based upon the assumption that in the coding regions of genes homologous sequences will be conserved between species and between isotypic variants of proteins such as the tubulins. In the following section, the discussion will be focused on how the amino acid sequence predicted from one testicular cDNA has been used to define and localize a novel α-tubulin in the mammalian testis (Hecht *et al.*, 1987).

Using a 1650 nucleotide cDNA that codes for part of a rat brain α-tubulin, a mouse testis cDNA, pRDαTT1, was isolated. Sequence analysis reveals it codes for the carboxyl half and part of the 3′ untranslated region of an α-tubulin (Fig. 6). To aid in interpretation we have maximized homology between the somatic rat brain α-tubulin sequence and pRDαTT1, by creating two interruptions in the α-tubulin sequence. One proline residue was deleted from the mouse testis cDNA at amino acid 360 and a 12 amino acid region was deleted between amino acids 410 and 422. 0n both sides of the larger deletion there is a direct repeat of 12 nucleotides, suggesting a similar deletion mechanism as found in human globin deficiencies (Efstradiatis *et al.*, 1980). When the carboxyl terminal region of 196 amino acids of the rat brain α-tubulin was compared to the equivalent region of the mouse testis cDNA, 39 amino acids (about 20%) differed. This marked difference in sequence suggested to us that the testis α-tubulin could serve a specialized function during spermatogenesis. To relate molecular structure and protein function, we reasoned that a unique peptide predicted from the cDNA could be synthesized and used to produce antibody to localize the tubulin protein.

Antisera were raised in rabbits to a peptide containing most of the sequence of amino acids 355 to 370 (hereafter called αTT1) (Hecht *et al.*, 1988). The resultant antisera recognized one band of about 58 kD in an immunoblot of a mouse testis extract or from a reticulocyte cell-free translation directed by mouse testis poly(A)$^+$ RNA. The serum was not immunoreactive, however, by Western blotting with a mouse brain extract. This difference in specificity of recognition between testis and brain extracts has been confirmed by immunoblot and indirect immunofluorescence staining of mouse testis and cerebellum sections. Antisera to the αTT1 reacted with the manchette of elongated spermatids and with meiotic spindles. In contrast, the αTT1 antisera did not react with any brain antigen. Control studies with a monoclonal antibody to chick brain α-tubulin reacted strongly with neurons in the mouse brain as well as the microtubule structures of the testis. Similar results were obtained with testis cells from guinea pig and hamster suggesting that the epitope recognized by the αTT1 antisera is not solely restricted to the mouse.

```
              260                                                                     270
Rat Brain     ...Gln Thr Asn Leu Val Pro Tyr Pro Arg Ile His Phe Pro Leu Ala Thr
                 CAG ACC AAC CTG GTG CCC TAC CCT CGC ATC CAC TTC CCT CTG GCC ACT
                          *           *           *                   *
Mouse Testis...CAG ACC AGC CTG GTG CTC TAC CCT CAC ATC CAC TTC CCT CTC GCC ACT
                          Ser         Leu         His

                                            280
   Tyr Ala Pro Val Ile Ser Ala Glu Lys Ala Tyr His Glu Gln Leu Ser Val Ala
   TAT GCC CCT GTC ATC TCT GCT GAG AAA GCC TAC CAT GAA CAG CTT TCT GTA GCA
        *                                       *   *               *   *
   TAT GCT CCT GTC ATC TCT GCT GAG AAA GCC TAC CAG GAG CAG CTT TCT ATA GAA
                                               Gln                 Ile Glu

   290                                     300
   Glu Ile Thr Asn Ala Cys Phe Glu Pro Ala Asn Gln Met Val Lys Cys Asp Pro
   GAG ATC ACC AAT GCC TGC TTT GAG CCA GCC AAC CAG ATG GTG AAA TGT GAC CCT
   *    *   *   *                               *                   * *
   AAG ATT ACT AGT GCC TGC TTT GAG CCA GCC AAC CAG AAG GTG AAA TGT GAT TCT
   Lys         Ser                             Lys                 Ser

       310                                     320
   Arg His Gly Lys Tyr Met Ala Cys Cys Leu Leu Tyr Arg Gly Asp Val Val Pro
   CGC CAT GGT AAA TAC ATG GCT TGC TGC CTG CTG TAC CGT GGT GAT GTG GTC CCC
   * *                                 * *       ** *    *   *   * *
   TGT CAT GGT AAA TAC ATG GCT TGC TGC CTA TTG TAC CAC AGT GGT ATG ATT CCC
   Cys                                           His Ser Gly Met Ile

           330                                     340
   Lys Asp Val Asn Ala Ala Ile Ala Thr Ile Lys Thr Lys Arg Thr Ile Gln Phe
   AAA GAT GTC AAT GCT GCC ATT GCC ACC ATC AAG ACC AAG CGT ACC ATC CAG TTT
            *   *       *           *           *  *  *  *          *
   AAA GAT ATC CAT GCT GCT ATT GCC ACC ACC AAG ACC AGG TGC AGC ATC CAG CTT
           Ile His                     Thr         Arg Cys Ser         Leu

               350                                     360
   Val Asp Trp Cys Pro Thr Gly Phe Lys Val Gly Ile Asn Tyr Gln Pro Pro Thr
   GTG GAC TGG TGC CCC ACT GGC TTC AAG GTT GGC ATT AAT TAC CAG CCT CCC ACT
                *                                                   ***
   GTG GAC TGG TAC CCC ACT GGC TTC AAG GTT GGC ATT AAT TAC CAG CCT --- CAC
               Tyr                                                 His

                                   370
   Val Val Pro Gly Gly Asp Leu Ala Lys Val Gln Arg Ala Val Cys Met Leu Ser
   GTG GTC CCT GGT GGC GAC CTG GCC AAG GTC CAG AGA GCT GTG TGT ATG CTG AGC
   *** **   *  *  *  *                                 *
   TGT GGT CCC TGG AGC AAC CTG GCC AAG GTC CAG AGA GCT GTG TGC ATG CTG AGC
   Cys Gly     Trp Ser Asn

   380                                     390
   Asn Thr Thr Ala Ile Ala Glu Ala Trp Ala Arg Leu Asp His Lys Phe Asp Leu
   AAC ACC ACA GCC ATT GCT GAG GCC TGG GCT CGC CTG GAT CAC AAG TTT GAT CTG
                                           ** *  *   *
   AAC ACC ACA GCC ATT GCT GAG GCC TGG GCT CCT GTA GAC CAC AAG TTT GAT CTG
                                           Pro Val

       400                                     410
   Met Tyr Ala Lys Arg Ala Phe Val His Trp Tyr Val Gly Glu Gly Met Glu Glu
   ATG TAT GCC AAG CGT GCC TTT GTG CAC TGG TAC GTG GGT GAG GGC ATG GAG GAG
   *   *                                   **
   TTG CAT GCC AAG CGT GCC TTT GTG CAC TGG TGT GTG GGT ------------------
   Leu His                                 Cys

           420                                     430
   Gly Glu Phe Ser Glu Ala Arg Glu Asp Met Ala Ala Leu Glu Lys Asp Tyr Glu
   GGA GAG TTC TCT GAG GCC CGT GAG GAC ATG GCT GCC CTA GAG AAG GAT TAT GAG
                                                                *
   -------------------------- GAG GAC ATG GCT GCC CTA GAG AAG GGT TAT GAG
                                                               Gly

               440                                     450
   Glu Val Gly Val Asp Ser Val Glu Gly Glu Gly Glu Glu Glu Gly Glu Glu Tyr END
   GAG GTT GGT GTG GAT TCT GTG GAG GGT GAG GGT GAG GAA GAA GGA GAG GAA TAC TAA
                *   *       *   *  *               *   *
   GAG GTT GGT ATG GGT TCT GTT GAA GCT GAG GGT GAG GAG GAG ...............
               Met Gly             Ala
```

These studies demonstrate that the manchette, a structure unique to the testis, and the meiotic spindle, contain at least one α-tubulin with an epitope not detectable in any brain α-tubulin. The immunofluorescent data confirm Northern blot analyses which demonstrated that pRDαTT1 mRNAs were present in testis extracts but were not detectable in brain extracts (Distel *et al.*, 1984). Thus, at least one of the many α-tubulin genes that are expressed during spermatogenesis can be related to a synthesized protein and more importantly, to two defined testicular structures. The finding that at least one α-tubulin is not ubiquitously expressed in the mouse argues that microtubule differentiation may be partially controlled by variations in tubulin subunits themselves. The growing evidence that unique testicular structures such as the manchette contain novel and perhaps unique protein variants suggests the manchette offers a most promising target for male fertility intervention.

Conclusions

The 'spermiogenic' genes discussed represent only a small first sample of 'haploid' expressed gene products in the mammalian testis. By nature of the screening approaches used, the post-meiotic genes described code for highly abundant structural proteins of the testis. In light of the many testicular isozyme variants and the unique events and structures of spermiogenesis, there will be many other less abundant post-meiotic targets for male fertility regulation.

The α-tubulin observations are especially noteworthy because they establish that novel variants of structural multigene families are expressed in the testis and that testicular structures such as the manchette contain novel, possibly unique, subsets of these structural genes. The α-tubulin studies also confirm that combining DNA cloning and immunology can be a valu-

Caption for Fig. 6

Fig 6 Nucleotide and predicted amino acid sequences of part of the coding regions of rat brain and mouse testis α-tubulin cDNAs. The part of the mouse cDNA pRDαTT1 is shown below that of a rat brain sequence pILαT1 (Distel *et al.*, 1984; Hecht *et al.*, 1987). Aligning the mouse sequence to the rat sequence, the mouse testis cDNA begins at amino acid 256. Two gaps in the mouse sequence are required to best align the two sequences. The direct repeat sequences adjacent to the 12 amino acid deletion are underlined. The coding region of the rat α-tubulin terminates with TAG after residue 450, while the coding region of the mouse α-tubulin continues beyond amino acid 448. Stars (*) designate nucleotide substitutions, and amino acid difference are written beneath the mouse α-tubulin sequence. The two deletions introduced in this section of the mouse cDNA to maximize homology are at amino acid 360 and between amino acids 410 to 422.

able approach to relate sequence and function and to define molecular targets. Moreover, the availability of both DNA and immunological probes that recognize novel products will aid in the isolation and composition analysis of structures such as the manchette. Such knowledge will lead to strategies that directly perturb its normal assembly. Efforts to localize additional spermiogenic gene products to specific cellular sites such as the chromatoid body or the surface of spermiogenic cells should be encouraged.

The many unique processes in the post-meiotic male germ cell offer many exciting options for intervention. Perhaps because of its presumed quiescent role in testicular gene expression, little attention has been paid to spermiogenic function at the molecular level. Clearly, factors that influence the integrity of gene expression in post-meiotic cells at either the transcriptional or translational level have the profound ability to directly and specifically impair the outcome of spermatogenesis. A concerted effort to understand the molecular mechanisms for the synthesis and regulation of proteins unique to post-meiotic germ cells should lead to new fertility control methodologies for the male.

Acknowledgements

This work was supported by a grant from the National Institutes of Health (GM 29224).

References

Ammer, H., Henschem, A. & Lee, C.H. (1986). Isolation and amino acid sequence analysis of human sperm protamines P1 and P2. **Biol. Chem. Hoppe-Seyler** 357:515–522.

Bachmair, A., Finley, D. & Varchavsky, A. (1986). In vivo half-life of a protein is a function of its amino-terminal residue. **Science** 234:179–186.

Balhorn, R., Weston, S., Thomas, C. & Wyrobek, A. (1984). DNA packaging in mouse spermatids. Synthesis of protamine variants and four transition proteins. **Experimental Cell Research** 150:298–308.

Bellvé, A.R. (1979). The molecular biology of mammalian spermatogenesis. In: Finn, C.A., ed. **Oxford Reviews of Reproductive Biology**. London and New York, Oxford, pp.159–261.

Bellvé, A.R., Anderson, E. & Hanley-Bowdoin, L. (1975). Synthesis and amino acid composition of basic proteins in mammalian sperm nuclei. **Developmental Biology** 47:349–365

Bellvé, A.R. & O'Brien, D.A. (1983). The mammalian spermatozoon: structure and temporal assembly. In: Hartman, J.F., ed. **Mechanisms and Control of Fertilization**. Academic Press, New York, pp.55–137.

Boitani, C., Geremia, R., Rossi, R. & Monesi, V. (1980). Electrophoretic pattern of polypeptide synthesis in spermatocytes and spermatids of the mouse. **Cell Differentiation** 9:41–49.

Bower, P.A., Yelick, P.C. & Hecht, N.B. (1987). Both protamine 1 and 2 genes are expressed in the mouse, hamster, and rat. **Biology of Reproduction** 37:479–488.

Cleveland, D.W. & Sullivan, K.F. (1985). Molecular biology and genetics of tubulin. **Annual Review of Biochemistry** 54:331–365.

Cherry, L.M. & Hsu, T.C. (1984). Antitubulin immuno-fluorescence studies of spermatogenesis in the mouse. **Chromosoma** 90:265–274.

Distel, R.J., Kleene, K.C. & Hecht, N.B. (1984). Haploid expression of a mouse testis tubulin gene. **Science** 224:68–70.

Efstradiatis, A., Posakony, J.W., Maniatis, T., Lawn, R.M., O'Connell, C., Spritz, R.A., Dekiel, J.K., Forget, B.G., Weissman, S.M., Slightom, J.L., Blechl, A.E., Smithies, O., Baralle, F.E., Shoulders, C.C. & Proudfoot, N.J. (1980). The structure and evolution of the human β-globin gene family. **Cell** 21: 653–668.

Elsevier, S. (1982). Messenger RNA encoding basic chromosomal protein of mouse testis. **Developmental Biology** 90:1–12.

Fujimoto, H. & Erickson, R.P. (1982). Functional assays for mRNA detect many new messages after male meiosis in mice. **Biochemical Biophysical Research Communication** 108:1369–1375.

Geremia, R., D'Agostino, A. & Monesi, V. (1978). Biochemical evidence of haploid gene activity in spermatogenesis of the mouse. **Experimental Cell Research** 111:23–30.

Gold, B., Stern, L. & Hecht, N.B. (1983). Gene expression during spermatogenesis II. Evidence for stage-specific differences in mRNA populations. **Journal of Experimental Zoology** 225:123–134.

Hecht, N.B. (1986a). Evidence for haploid-specific genes in the mammalian testis. In: Stefanini, M., Conti, M., Geremia, R. & Ziparo, E., eds. **Proceedings of the IVth European Workshop on Molecular and Cellular Endocrinology of the Testis**. Elsevier, Amsterdam, pp.199–213.

Hecht, N.B. (1986b). Regulation of gene expression during mammalian spermatogenesis. In: Rossant, J. & Pedersen, R.A., eds. **Experimental Approaches to Mammalian Embryonic Development**. Cambridge University Press, New York, pp.151–193.

Hecht, N.B. (1987a). Mammalian protamines and their expression. In: Stein, G. & Stein, J., eds. **Histones and Other Basic Nuclear Proteins**. CRC Press, Boca Raton, in press.

Hecht, N.B. (1987b). Gene expression during spermatogenesis. In: Orgebin-Crist MC, ed. **Cell Biology of the Testis and Epididymis**. **Ann. New York Academy Sci.**, 513:90–101.

Hecht, N.B. (1987c). Haploid gene expression in the mammalian testis. In: Mohri H, ed. Recent Advances in Spermatology. **Japan Scientific Societies Press**, Tokyo, pp. 451–467.

Hecht, N.B. (1987d). Post-meiotic gene expression during spermatogenesis. In: Haseltine, F., ed. **Meiotic Inhibition: Molecular Control of Meiosis**, 291–313.

Hecht, N.B., Kleene, K.C., Distel, R.J. & Silver, L.M. (1984). The differential expression of the actins and tubulins during spermatogenesis in the mouse. **Experimental Cell Research** 153:275–280.

Hecht, N.B., Bower, P.A., Kleene, K.C. & Distel, R.J. (1985). Size changes of protamine mRNA provide a molecular marker to monitor spermatogenesis in wild type and mutant mice. **Differentiation** 29:189–193.

Hecht, N.B., Bower, P.A., Waters, S.H., Yelick, P.C. & Distel, R.J. (1986a). Evidence for haploid expression of mouse testicular genes. **Experimental Cell Research** 164:183–190.

Hecht, N.B., Kleene, K.C., Yelick, P.C., Johnson, P.A., Pravtcheva, D.D. & Ruddle, F.H. (1986b). Haploid gene mapping: The genes for both mouse protamines are located on chromosome 16. **Somatic Cell and Molecular Genetics** 12:191–196.

Hecht, N.B. & Penchow, J. (1987). *In situ* localization of mRNAs coding for mouse testicular structural genes. **Experimental Cell Research** 173:274–281.

Hecht, N.B., Distel, R.J., Yelick, P.C., Driscoll, C.E., Goldberg, E. & Tung, K.S.K. (1988). Localization of a highly divergent mammalian testicular α-tubulin that is not detectable in brain. **Molecular and Cellular Biology** 8:996–1000.

Holstein, A.E. & Roosen-Runge, E.C. (1981). **Atlas of Human Spermatogenesis**, Grosse Verlag, Berlin.

Johnson, P.A., Peschon J.J., Yelick, P.C., Palmiter, R.D. & Hecht, N.B. (1988). Sequence homologies in the protamine 1 and 2 genes in the mouse. **Biochim. Biophys. Acta** 950:45–53.

Kemphues, K.J., Kaufman,T.C., Raff, R.A. & Raff, E.C. (1982).The testis-specific tubulin subunit in Drosophila melanogaster has multiple functions in spermatogenesis. **Cell** 31:655–670.

Kistler, W.S., Geroch, M.E.& Williams-Ashman, H.G. (1975). The amino acid sequence of a testis-specific basic protein that is associated with spermatogenesis. **Journal of Biological Chemistry** 250:1847–1853.

Kleene, K.C., Distel, R.J. & Hecht, N.B. (1983). cDNA clones encoding cytoplasmic poly $(A)^+$ RNAs which first appear at detectable levels in haploid phases of spermatogenesis in the mouse. **Developmental Biology** 98:455–464.

Kleene, K.C., Distel, R.J. & Hecht, N.B. (1984). Translational regulation and co-ordinate deadenylation of a haploid mRNA during spermiogenesis in the mouse. **Developmental Biology** 105:71–79.

Kleene, K.C., Distel, R.J. & Hecht, N.B. (1985). The nucleotide sequence of a cDNA clone encoding mouse protamine 1. **Biochemistry** 24:719–722.

Kramer, J.M. & Erickson, R.P. (1982). Analysis of stage-specific protein synthesis during spermatogenesis of the mouse by two-dimensional gel electrophoresis. **Journal of Reproduction and Fertility** 64:139–144.

Lam, D.M.K. & Bruce, W.R. (1971). The biosynthesis of protamine during spermatogenesis of the mouse: extraction, partial characterization, and site of synthesis. **Journal of Cell Physiology** 78:13–24.

Lee, C.H., Mansouri, A., Hecht, N.B. & Engel, W. (1987). Nucleotide sequence of a bovine protamine cDNA. **Biol. Chem. Hoppe-Seyler** 368:131–135.

Mayer, J.F. & Zirkin, B.R. (1979). Spermatogenesis in the mouse 1. Autoradiographic studies of nuclear incorporation and loss of $[^3H]$ amino acids. **Journal of Cell Biology** 81:403–410.

Mayer, J.E., Chang, T.S.K. & Zirkin, B.R. (1981) Spermatogenesis in the mouse 2. Amino acid incorporation into basic nuclear proteins of mouse spermatids and spermatozoa. **Biology of Reproduction** 25:1041–1051.

Meistrich, M. (1987). Histone and basic nuclear protein transitions in mammalian spermatogenesis. In: Stein, G. & Stein, J., ed. **Histone and Other Basic Nuclear Proteins**, CRC Press, Boca Raton, in press.

McKay, D.J., Renaux, B.S. & Dixon, G.H. (1985). The amino acid sequence of human sperm protamine P1. **Bioscience Reports** 5:383–391.

McKay, D.J., Renaux, B.S. & Dixon, G.H. (1986). Human sperm protamines. Amino acid sequences of two forms of P2. **European Journal of Biochemistry** 156:5–8.

Monesi, V., Geremia, R., D'Agostino, A. & Boitani, C. (1978). Biochemistry of male germ cell differentiation in mammals: RNA synthesis in meiotic and postmeiotic cells. In: Moscona, A. & Monroy, A., eds. **Current Topics in Developmental Biology**. Academic Press, New York, pp.11–36.

Pratt, L.F., Okamura, S. & Cleveland, D.W. (1987). A divergent testis-specific tubulin isotype that does not contain a coded C-terminal tyrosine. **Molecular and Cell Biology** 7:552–555.

Stern, L., Gold, B. & Hecht, N.B. (1983a). Gene expression during mammalian spermatogenesis I. Evidence for stage-specific synthesis of polypeptides *in vivo*. **Biology of Reproduction** 28:483–496.
Stern, L., Kleene, K.C., Gold, B. & Hecht, N.B. (1983b). Gene expression during mammalian spermatogenesis III. Changes in populations of mRNA during spermiogenesis. **Experimental Cell Research** 143:247–255.
Villasante, A., Wang, D., Dobner, P., Dolph, P., Lewis, S.A. & Cowan, N.J. (1986). Six mouse tubulin mRNAs encode five distinct isotypes: Testis-specific expression of two sister genes. **Molecular and Cell Biology** 6:2409–2419.
Waters, S.H., Distel, R.J. & Hecht, N.B. (1985). Mouse testes contain two size classes of actin mRNA that are differentially expressed during spermatogenesis. **Molecular and Cell Biology** 5:1649–1654.
Yelick, P.C., Balhorn, R., Johnson, P.A., Corzett, M., Mazrimas, J.A., Kleene, K.C. & Hecht, N.B. (1987). Mouse protamine 2 is synthesized as a precursor whereas mouse protamine 1 is not. **Molecular and Cell Biology** 7:2173–2179.

Male contraception and chemotherapy for trypanosomiasis and leishmaniasis: an analysis of biochemical similarities and considerations for future research

ROBERTO DOCAMPO*

Instituto de Microbiologia, Universidade Federal do Rio de Janeiro, Bloco I, Cidade Universitaria, Rio de Janeiro, RJ21.944, Brazil.

Abstract

A number of peculiar aspects common to spermatozoa and flagellated protozoa which are not present in other eukaryotic cells are reviewed. These aspects include: surface changes during differentiation, the involvement of calcium ions and cyclic AMP in motility, the sensitivity to microtubule and mitochondrial inhibitors and the deficiency in defense mechanisms against oxygen toxicity. At this time, these characteristics appear to explain best the similar sensitivity of flagellated protozoa and spermatozoa to various chemical agents. Compounds of current experimental interest are also discussed. These are agents that have shown promise in the laboratory and for which data exist regarding probable mechanisms of action. The activities of these agents are correlated, in so far as possible, with those structures or metabolic pathways in the spermatozoa and the flagellated protozoa that are affected by their actions. The compounds are of two general groups: drugs capable of inducing an oxidative stress, and membrane-active agents.

Key words

Calcium, cyclic AMP, differentiation, free radicals, microtubule, motility, oxygen toxicity, spermatozoa, trypanosomatids.

Introduction

Most studies concerning the effects of chemicals on the male reproductive system have focused on two areas:

(1) the therapeutic use of drugs as possible male contraceptives or as stimulators of sperm production in subfertile men, and

*Present address Department of Molecular Parasitology, The Rockefeller University, 1230 York Avenue, New York, N.Y. 10021–6399, USA.

(2) potential toxic side effects of drugs administered for therapeutic purposes or of environmental agents (Whorton, 1982).

It is clear that virtually all aspects of physiological control of the male reproductive system are possible targets for primary drug effects and side effects. The hypothalamic–pituitary axis could be one target; direct action on the gonadal functions of androgen production and spermatogenesis may be a second; and drug effects on the functioning of the sex accessory organs is a third possibility.

From the point of view of the use of drugs as possible male contraceptives, the more appropriate target for drug effects is the rapid cell division and specialized processes involved in spermatogenesis. Because spermatogenesis in man takes approximately 70 days, a drug that has an effect on the early stages of sperm production will not produce detectable effects for many weeks. However, because spermatogenesis is a renewable process, it may be possible to restore full sperm production even after a virtually complete loss of developing sperm cells. Additionally, the later stages of spermatogenesis are relatively protected from exogenous toxins by the blood–testis barrier. In recent years it has been apparent that several drugs active against flagellated protozoa may interfere with spermatogenesis causing male infertility. The mode of action of the drugs has been studied and some similarities in response by protozoa and sperm have been detected. This chapter is an attempt to review aspects of the biology of some protozoa which are demonstrably similar to those of sperm, and different from those of other mammalian cells and which, for this reason, may be susceptible to similar chemical agents. The chemical agents with antiprotozoal and antispermatogenic actions will be correlated, in so far as possible, with those structures or metabolic pathways in the cells known to be related to or affected by the drugs. For a broader listing of agents that are active against either protozoa (Gutteridge & Coombs, 1977) or spermatozoa (Duncan & Pharris, 1970) the reader is referred to previous reviews.

The biology of flagellated protozoa as it may relate to chemotherapy: similarities with sperm

Some biologists have seen a sort of phylogenetic repetition in the many similarities between flagellated protozoa and spermatozoa. The unicellular flagellate stage through which all the metazoans pass at the initial step of their development would be a repetition of the first stage of evolution of multicellular animals. Although this interpretation should not be carried too far, flagellated protozoa and spermatozoa do share several common characteristics.

Information about flagellated protozoa of medical importance is summarized in Table 1. In addition to the parasites mentioned in the table, several species of parasites that are used as models for the human pathogenic

Table 1. *Flagellated protozoa of medical importance*

Protozoan	Habitat	Disease
Trypanosoma cruzi	Blood, heart, brain, muscles	Chagas' disease
Trypanosoma brucei gambiense	Blood, lymphatics, brain	Sleeping sickness (chronic form)
Trypanosoma brucei rhodesiense	Blood, lymphatics, brain	Sleeping sickness (acute form)
Leishmania brasiliensis	Skin, mucous membranes of mouth and nose	Leishmaniasis (mucocutaneous)
Leishmania tropica	Skin	Leishmaniasis (cutaneous)
Leishmania donovani	Liver, spleen, skin	Leishmaniasis (visceral, Kala-azar)
Giardia lamblia	small intestine, gall bladder	Giardiasis
Trichomonas vaginalis	vagina	Trichomoniasis

species will be discussed (such as *Crithidia fasciculata*, used as a model for trypanosomatids, and *Trypanosoma brucei brucei* used as a model for African trypanosomes).

The protozoan haemoflagellates belonging to the order Kinetoplastida are perhaps some of the organisms most closely related to spermatozoa amongst the unicellular eukaryotes of medical importance. The most important biochemical similarities that have been found in the genuses *Trypanosoma* and *Leishmania* and that may have relevance for chemotherapy are described below.

Surface changes during differentiation of intracellular protozoa

The mammalian spermatozoon leaves the testis as a highly differentiated, but functionally immature cell. Spermatozoa undergo a series of post-testicular developmental steps involving alterations in their metabolism, composition, and structure as they pass through the excurrent duct system of the male tract. These changes, collectively termed maturation, yield a motile gamete capable of fertilizing an egg. However, if sperm are injected into the oviduct, they are not immediately capable of penetrating the egg. The necessary changes that occur in the female tract are termed capacitation and are mainly a membrane phenomenon consisting of several distinct but related steps. Similarly, intracellular parasites undergo a series of developmental stages involving alterations in their metabolism, composition, and structure, as they pass through their invertebrate or vertebrate hosts. As occurs with spermatozoa these developmental stages are not immediately capable of penetrating a host cell and surface changes occur that 'capacitate' them for that process. These processes have been better studied in *Trypanosoma cruzi*, although they also are important in the *Leishmania sp*. that also have an intracellular stage of development.

Trypanosoma cruzi has at least three well-defined differentiation stages (Fig. 1): the amastigote, which lives and reproduces inside vertebrate cells; the epimastigote, a dividing form that inhabits the midgut of reduviid insects; and the trypomastigote, a nondividing cell which, in nature, is the link between the vertebrate and the invertebrate phases of the protozoan biological cycle. These forms are distinct not only in terms of their host environments but also with respect to their morphologies and biological behavior (De Souza, 1984). Epimastigotes, which are noninfective to vertebrates, are unable to invade cultured mammalian cells, except for macrophages where they are destroyed by lysosomes. In contrast, trypomastigotes are fully infective to vertebrate hosts, possessing the capacity of becoming intracellular and of resisting lysosomal digestion (Andrews *et al*., 1984).

Table 2 shows the functional and structural changes common to spermatozoa during epididymal maturation and capacitation and to *Trypanosoma cruzi* during differentiation of epimastigotes to infective trypomastigotes. Structural changes in spermatozoa undergoing maturation include migration of the cytoplasmic droplet, reduction in the amount of

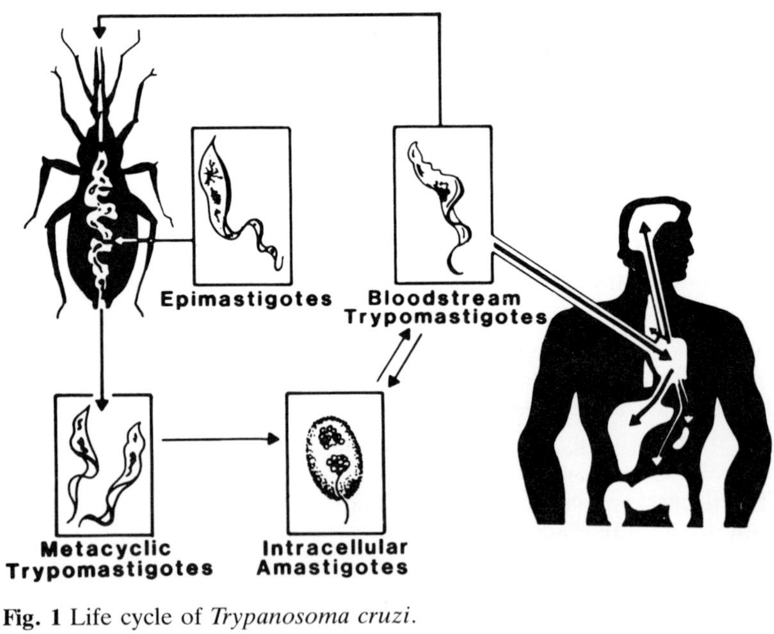

Fig. 1 Life cycle of *Trypanosoma cruzi*.

cytoplasm, and alterations in the size, shape and contents of organelles such as in the acrosome, the mid-piece and the sperm membranes. Structural changes also occur in epimastigotes as they transform into trypomastigotes, such as migration of the kinetoplast, reduction in the amount of cytoplasm, and alterations in the size, shape and contents of organelles such as the mitochondrion and the trypanosome membranes. There is an increase in the content of cAMP during sperm passage in the epididymis (Hoskins *et al.*, 1976) and during differentiation of epimastigotes into trypomastigotes (Rangel-Aldao *et al.*, 1987). The increase in the content of cAMP in spermatozoa combined with the action of a special 'forward motility protein' is held to be responsible in part for initiation and gradual full development of motility (Hoskins *et al.*, 1976). In *T. cruzi* motility is present in both epimastigotes and trypomastigotes, although it is faster in trypomastigotes (De Souza, 1984).

Whole cell electrophoresis studies have demonstrated that intact epimastigotes possess a net negative surface charge and that there is a marked increase in the negative surface charge as the epimastigotes mature into trypomastigotes (De Souza, 1984). This is similar to what occurs during epididymal maturation of spermatozoa (see Olson & Orgebin-Crist, 1982). In both cases the increase in negative surface charge has been attributed to the addition of sialic acid groups to the cell surface. This increase of the negative surface charge of both spermatozoa and differentiated trypomastigotes might have an important role with respect to their increased sensitivity to cationic drugs and deserves further study.

The presence of sialic acid on the surface of *T. cruzi* has been shown by interaction with lectins (Pereira *et al.*, 1980), by changes in the electrophoretic mobility upon neuraminidase treatment of the parasites (Souto-Padron & De Souza, 1985), and by biochemical analysis (Schauer *et al.*, 1983; Couto et al., 1985). This sialic acid does not arise from intracellular

Table 2. *Functional and structural changes common to spermatozoa during maturation and capacitation and to* Trypanosoma cruzi *during differentiation of epimastigotes into infective trypomastigotes*

1.	Reduction in cell volume and alterations in the size, shape, and content of organelles.
2.	Increased content of cyclic AMP
3.	Increased motility
4.	Modifications of the plasma membrane:
	(a) Increased negative surface charge
	(b) Sialic acid requirement
	(c) Different lectin–mediated agglutination
	(d) Redistribution of intramembranous particles
	(e) Specific glycoproteins involved in capacitation for fertilization or penetration into the host cell, respectively.

precursors, as *T. cruzi* is unable to synthesize them (Schauer *et al.*, 1983), but from exogenous sialoconjugates such as fetuin or other serum proteins *via* a sialyl transferase activity present on the surface of *T. cruzi* (Previato *et al.*, 1985). In addition, parasite infectivity is enhanced in the presence of serum sialoglycoproteins (Piras *et al.*, 1987), thus indicating the importance of the sialyl transferase activity on the trypomastigote surface in parasite penetration. It is interesting to recall that optimal levels of sialic acid are also regarded as essential for sperm maturation and survival in the epididymis (Hinrichsen-Kohane *et al.*, 1984). Differences in lectin-mediated agglutination of spermatozoa (Olson & Orgebin-Crist, 1982) and trypanosomes (De Souza, 1984) also indicate qualitative and quantitative differences in the carbohydrates exposed on the cell surface of the different cells. A redistribution of intramembranous particles occurs in both sperm cells and trypanosomes.

Finally, as occurs with the process of capacitation of the sperm cells (O'Rand, 1982) it has been shown that trypomastigotes recently released by the infected host cell are immature forms (Dvorak & Hyde, 1973) which undergo a maturation or activation process, dependent on protein synthesis in order to be able to invade the host cell efficiently (Zingales & Colli, 1985). Parasite adhesion and infectivity can be altered by glycosylation inhibitors like tunicamycin, thus indicating the involvement of glycoproteins containing N-linked oligosaccharides in this interaction (Piras *et al.*, 1983). Trypsinization of the parasite (Piras *et al.*, 1985) or incubation in the presence of monosaccharides like N-acetylglucosamine or mannose (Crane & Dvorak, 1982; Piras *et al.*, 1983; Zingales & Colli, 1985) were also found to alter the host-parasite interaction. All of these observations suggest the presence on the surface of mature parasites of a specific lectin-like protein(s) that facilitates adhesion to host cells, probably through interaction with host cell glycoprotein(s). Similarly, specific glycoproteins necessary for fertilization are present in mature spermatozoa (Lopo *et al.*, 1982), at least in invertebrates.

Several groups have succeeded in the identification of *T. cruzi* surface molecules involved in the host-parasite interaction. A polypeptide of 83 kD was detected in extracts of mature trypomastigotes that may facilitate parasite adhesion to surface receptors of Vero fibroblasts. This polypeptide probably acts as a lectin-like protein which interacts with mannose residues of a 33–38 kD glycoprotein present in the host cell membrane (Piras *et al.*, 1986). On the other hand, monoclonal antibodies that recognize an 85 kD glycoprotein which contains N-acetyl-D-glucosamine and/or sialic acid on the trypomastigote surface, partially prevented *T. cruzi* invasion of LLC-MK_2 cell monolayers (Alves *et al.*, 1986). In this regard, the molecular cloning of a genomic DNA fragment that encodes antigenic determinants present in the 85 kD trypomastigote surface protein was recently described (Peterson *et al.*, 1986). The polypeptide fragment encoded by the cloned

DNA is recognized by serum from a *T. cruzi* infected host and is inferred by DNA sequence analysis to contain a nonapeptide that is tandemly repeated 5 times (Peterson *et al.*, 1986).

Receptors for fibronectin have been detected on the surface of infective trypomastigotes (Ouaissi & Capron, 1986). The involvement of fibronectin receptors in the host cell-*T. cruzi* interaction, was suggested by the observations that both addition of anti-fibronectin and antifibronectin receptor antibodies in the culture medium, or saturation by an excess of fibronectin, led to a significant inhibition of the *in vitro* infection of host cells (Ouassi & Capron, 1986). Fibronectin consists of highly structured domains containing the binding sites for a variety of macromolecules that interact with it. A cell attachment site has been characterized and a particular amino acid sequence, Arg-Gly-Asp-Ser (RGDS), was shown to be involved as the recognition site for cells. Several synthetic peptides modelled from the sequence of this domain were constructed and it was demonstrated that the hydrophilic sequence RGDS binds to the parasite fibronectin receptors and inhibits the invasion of host cells *in vitro*. In addition, the immunization of Balb/c mice with tetanous toxoid-conjugated peptide, induced a significant protection against *T. cruzi* infection (Ouassi & Capron, 1986). This particular example illustrates that beyond the classical approach towards immunization with specific parasite surface antigens, parasite infection might be controlled by immunization against versatile membrane structures having a key role in parasite–cell interactions (Ouassi & Capron, 1986). On the other hand, the inhibition of epimastigote to trypomastigote differentiation *in vitro* by a monoclonal antibody specific for a 72 kD cell-surface glycoprotein of *T. cruzi* has been described and it has been proposed that this molecule may be a receptor controlling parasite differentiation in the vector (Sher & Snary, 1983).

In conclusion, an understanding of the mechanisms that control the capacity of parasites to penetrate a cell or the capacity of sperm to fertilize an egg might facilitate the formulation of strategies for blocking infection and fertilization through immunotherapy or chemotherapy.

Motility, cAMP and calcium

The cytoskeletal arrangements of spermatozoa and flagellated protozoa are of crucial importance for motility. The 9+2 configuration of the sperm tail axoneme has long been a model system for the study of ciliary motion in general (Schatten, 1982). The sliding of adjacent microtubule pairs mediated by dynein ATPase results in the lateral translocation of one ciliary doublet past another. Presumably, elastic spoke proteins, referred to as nexin, convert this lateral movement into bending (Schatten, 1982). The flagellum of trypanosomatids has a basic structure similar to that of sperm showing a 9+2 pattern of axonemal microtubules (De Souza, 1984). In trypanosomatids, the length of the flagellum varies according to their developmental stage. For example, amastigotes of *T. cruzi* have a short flagellum, 1 μm in length, whereas at the end of the intracellular cycle the parasite elongates and the flagellum also grows to about 20 μm (De Souza, 1984).

Although more than 70 years have passed since Tournade (1913) first showed that sperm acquire the ability to move progressively forward as they travel through the mammalian epididymis, the chemical nature of this process remains largely unknown. The acquisition of full motility by epididymal spermatozoa is a gradual process, subject to considerable species variation. In the bull, immotile caput epididymal spermatozoa can be activated *in vitro* by treating them with diluents containing phosphodiesterase inhibitors, thereby raising the intracellular cAMP level (Hoskins *et al.*, 1976). Under *in vivo* conditions, an increase in the content of cAMP during sperm passage in the epididymis, combined with the action of a special 'forward motility protein' is held to be responsible in part for the initiation and gradual full development of motility (Hoskins *et al.*, 1976). Calcium ions and the supply of androgens also influence the development of motility (Schatten, 1982). Calcium ions are also required for the initiation of the acrosome reaction, characterized by extensive vesicularization of the plasma and outer acrosomal membranes. This process results in the eventual loss of these membranes and the release of the hydrolytic contents of the acrosome which aid in the penetration of the ovum vestments. It is now generally believed that a transient increase in free cytosolic Ca^{2+} in response to a stimulus serves as a trigger for many Ca^{2+}-dependent regulatory processes that occur in animal cells (Lenz & Cormier, 1982). Free Ca^{2+} apparently exerts its effects by binding to low molecular weight Ca^{2+}-binding proteins (Lenz & Cormier, 1982). The Ca^{2+}-binding protein, calmodulin, appears to be particularly important since it has been found in all eukaryotic cells thus far examined and since it is apparently involved in the regulation of numerous cellular processes and enzyme activation. For example, calmodulin activates, in a Ca^{2+}-dependent manner, cyclic nucleotide phosphodiesterase, adenylate cyclase, the Ca^{2+}-transport ATPase, several protein kinases, and myosin light chain kinase, and it is apparently involved in regulating cellular processes such as the control of microtubule assembly/disassembly. It seems likely that calmodulin is a primary target for Ca^{2+} during the Ca^{2+}-dependent regulatory processes involved in the acrosome reaction (Lenz & Cormier, 1982). The Ca^{2+}-dependent binding of a number of phenothiazine drugs to calmodulin is known to inhibit the activation of a number of calmodulin-dependent enzymes (Lenz & Cormier, 1982). These drugs inhibit the acrosome reaction and the motility of spermatozoa and have been postulated as vaginal contraceptives (Lenz & Cormier, 1982).

Cyclic nucleotides and Ca^{2+} have also been related to regulation of the growth and differentiation transitions of *T. cruzi* as well as African trypanosomes. In African trypanosomes, Ca^{2+} might be a factor controlling the release of coat variable antigen (Voorheis *et al.*, 1982) and activation of the adenylate cyclase (Voorheis & Martin, 1981). Cyclic AMP, in turn, might be involved in the control of parasite growth and differentiation (Mancini & Patton, 1981). *T. cruzi* calmodulin activates, in a Ca^{2+}-dependent manner the homologous cAMP phosphodiesterase. Interestingly, several phenothiazine drugs inhibit this effect and the motility of the trypanosomes (Tellez-Iñon *et al.*, 1985). Phenothiazine derivatives also exert a clear and rapid cytotoxic effect in *Leishmania* (Pearson *et al.*, 1982, 1984) and African trypanosomes (Ruben *et al.*, 1985). Electron microscopic analysis of *T. brucei brucei*-treated cells shows that the pellicular layer of microtubules is selectively disintegrated by phenothiazine treatment, while the flagellar axoneme is entirely resistant to the drugs (Seebeck & Gehr, 1983). This evidence might open new therapeutic possibilities for the treatment of trypanosomiasis with this type of compound.

Microtubule inhibitors

Microtubule inhibitors have been shown to inhibit both fertilization and the replication of trypanosomatids. Taxol, a drug that has been shown to inhibit microtubule depolymerization (Schiff & Horwitz, 1980), inhibits the replication of *T. cruzi* (Baum *et al.*, 1981). The agent is active in micromolar concentrations and prevents the completion of cell division, although it permits the multiplication of cell organelles, such as the nucleus, kinetoplast, and flagellum (Baum *et al.*, 1981). It has been postulated that taxol stabilizes a portion of the microtubular cytoskeleton of the parasite and prevents cell division. The earliest sign of division in *T. cruzi* is the replication of the basal body and flagellum; microtubules subsequently form in the nucleus (De Souza & Meyer, 1974). After the daughter nuclei have separated, these intranuclear microtubules disappear. As the new cell membrane is duplicated and surrounds the daughter cells, new subpellicular microtubules must form. The microtubule assembly appears to originate at the basal bodies in the dividing daughter cells (Meyer & De Souza, 1976) and continues posteriorly as cytokinesis proceeds. In the presence of taxol, cytokinesis proceeds along the length of the organism to a fixed region; this suggested that some critical structure is present at that location. Whether this structure is the basal body is not yet clear. There appear to be two classes of microtubules in *T. cruzi* as defined by sensitivity to taxol. Those comprising the flagellar apparatus are unsusceptible to taxol, while those microtubules that make up in the subpellicular cytoskeleton are stabilized by taxol and cannot disassemble to allow duplication. Taxol also induces the formation of intermediate filaments in the nonpathogenic trypanosomatid *Crithida fasciculata* (Solari & Duschak, 1985) probably through the disruption of the normal equilibrium between microtubular assembly and disassembly.

Taxol does not affect sperm tail motility either, but inhibits the cytoplasmic migrations of the sperm and egg nucleus, thus inhibiting syngamy. Other studies with drugs that affect microtubules e.g. colchicine, vinblastine, and griseofulvin have shown that replication of several protozoa as well as syngamy (Juliano *et al.*, 1985; Schatten, 1982) are inhibited by these drugs. Taxol and these other drugs are toxic to mammalian cells (Schiff & Horwitz, 1980; Schatten, 1982) and may not be of therapeutic use. Nevertheless, the fact that compounds dissassembling microtubules may have the requisite specificity needed for male contraception or chemotherapy is encouraging, and agents with this type of activity deserve further study.

Mitochondrial inhibitors

Some mitochondrial inhibitors have been shown to be cytotoxic to both spermatozoa and trypanosomes. Lonidamine (1-(2,4-dichlorobenzyl)-lH-indazol-3-carboxylic acid), first developed as a reversible antispermatogenic agent (Corsi *et al.*, 1976) and later used as an antitumor agent (Floridi & Lehninger, 1983) inhibits *T. cruzi* and *T. brucei* growth in culture by interfering with electron transport through the respiratory chain (Turrens, 1986a). Lonidamine affects tumor energy metabolism at two different levels: glucose oxidation and mitochondrial electron transport (Floridi *et al.*, 1981; Floridi & Lehninger, 1983). The effect of lonidamine on mitochondrial respiration is unusual because it inhibits the rate of oxygen consumption only when mitochondria are actively phosphorylating or when they are uncoupled, but there is no inhibition of the respiration in the absence of phosphate acceptor. The two mechanisms that have been proposed for lonidamine action are inhibition of mitochondrial dehydrogenases (Floridi & Lehninger, 1983) and mitochondrial swelling associated with changes in the permeability to different ions (Floridi *et al.*, 1981). The selectivity of its action on spermatozoa and trypanosomes is apparently explained by a selective permeability in those cells (Turrens, 1986b).

Defenses against oxygen toxicity

Most of the oxygen consumed in biological systems is reduced to water by cytochrome oxidase. The coupling of oxidation to phosphorylation of ADP by the respiratory chain forms the energetic basis for life in eukaryotic cells (Kappus & Sies, 1981). However, there are other aspects of O_2 metabolism, such as its univalent reduction to superoxide anion (O_2^-) or bivalent reduction to hydrogen peroxide (H_2O_2), that are associated with toxic processes and are dealt with by powerful defense systems.

Several oxidative enzymes have been shown to produce substantial amounts of O_2^- and H_2O_2 in both sperm (Holland & Storey, 1981, 1982) and trypanosomatids (Docampo & Moreno, 1984a, b). However, in contrast to other cells, sperm and trypanosomatids are partially deficient in defense mechanisms against these oxygen metabolites. Superoxide dismutases, which catalyze the dismutation of superoxide anion to form H_2O_2 and O_2 are present in *T. cruzi, T. brucei* and *Leishmania* (Docampo & Moreno, 1984a) as well as in sperm (Alvarez & Storey, 1983). However, catalase is absent in both sperm and pathogenic trypanosomatids (Holland & Storey, 1981, 1982; Buhrley & Le Grande, 1973; Docampo & Moreno, 1984a). Glutathione peroxidase, which catalyzes the reaction of hydroperoxides with reduced glutathione to form oxidized glutathione and reduced products of the hydroperoxide, is present in sperm (Gorus *et al.*, 1986).

Trypanosomes do not have glutathione peroxidase (Docampo & Moreno, 1984a). Instead, they have a peroxidase activity dependent on a glutathione-spermidine cofactor termed trypanothione (Penketh & Klein, 1986; Fairlamb *et al.*, 1985). In addition, the amount of reduced glutathione is very low in trypanosomatids such as *T. cruzi* (Boveris *et al.*, 1980) and is almost undetectable in certain mammalian sperm (Holland & Storey, 1981).

There are no reports of the presence of α-tocopherol, β-carotene, or ascorbic acid in *T. cruzi*. In addition, since trypanosomatids lack xanthine oxidase (Marr & Berens, 1983), uric acid, another biochemical defense against oxygen reduction byproducts (Ames *et al.*, 1981) is not formed in these organisms.

These enzymatic and biochemical deficiencies have been correlated with the pronounced sensitivity of spermatozoa to peroxidative attack (Jones & Mann, 1977; Gorus *et al.*, 1986) and with the sensitivity of trypanosomatids to both intracellularly generated and phagocyte-derived by-products of oxygen reduction (Docampo & Moreno, 1984a).

The chemotherapeutic potential for trypanosomiasis of these enzymatic and biochemical deficiencies was first recognized during work on the mode of action of the trypanocidal o-naphthoquinone β-lapachone and its derivatives (Docampo & Moreno, 1984a, 1984b). These studies showed that the metabolism of these compounds by the parasites involved, at least in part, the generation of oxygen radicals, and hence, *via* superoxide dismutase, of H_2O_2 which accumulated in the cells to cytotoxic levels and was also excreted (Docampo *et al.*, 1978). Unfortunately, these compounds are inactive in experimental animal infections (Lopes *et al.*, 1978) though one of them might be useful for sterilization of blood used for transfusion (Goncalves *et al.*, 1980). Nevertheless, the importance of these enzymatic and biochemical deficiencies are clear in the case of the nitrofuran used in the treatment of Chagas' disease and several anticancer compounds which are effective trypanocidal drugs with antispermatogenic activity, and will be discussed later.

Perspective

The biochemical similarities discussed appear at this time to better explain the similar sensitivity of trypanosomatids and spermatozoa to different chemical agents. These structures and compounds hold promise for potential exploitation for chemotherapeutic purpose. Of course, there are other structures and compounds that could be addressed but they are not common to trypanosomatids and spermatozoa and are thus beyond the scope of this article.

Two main groups of compounds have been identified as active against both spermatozoa and flagellated protozoa: compounds capable of producing an oxidative stress, and membrane-active agents. Many other agents tested experimentally or clinically have been found to be ineffective or too toxic to warrant their further development. These agents have been carefully reviewed over several years (Brener, 1979, Marr & Docampo, 1986, Gutteridge & Coombs, 1977). The discussion that follows will concentrate on those two groups of compounds which appear to hold the greatest promise at this time.

Compounds of current experimental interest against trypanosomatids, possessing antispermatogenic action

Agents capable of producing oxidative stress

A wide variety of chemicals with vastly different structures may induce an oxidative stress in eukaryotic cells (Kappus & Sies, 1981). Some of them are active against both spermatozoa and trypanosomatids. One-electron reduction of several compounds results in the formation of free radicals (compounds with an odd or spin-unpaired electron in the outer orbital, Fig. 2). Because of the thermodynamic potential of the unpaired electron to form an electron pair, free radical intermediates are extremely reactive. They can undergo a variety of reactions, i.e., add across double bonds, abstract hydrogen from other molecules, or combine with themselves to form dimers (Pryor, 1976). In addition, free radical intermediates can activate molecular oxygen by univalent reduction (redox cycling) to the superoxide anion (O_2^-), which, in turn, can produce H_2O_2, either spontaneously or by the action of the enzyme superoxide dismutase (SOD). The interaction of O_2^- and H_2O_2 can lead to the production of the hydroxyl radical ($OH^\cdot$), which is one of the most damaging radical species. The formation of free radicals (drug-derived and oxygen derived radicals) may lead to extensive cellular damage through the production of secondary and tertiary free radicals from lipids, amino acids, and components of nucleic acids (oxidative stress). The selectivity of this action depends on several factors, including cellular uptake and subcellular localization of the agent, levels of protective (e.g. catalase, superoxide dismutase, glutathione

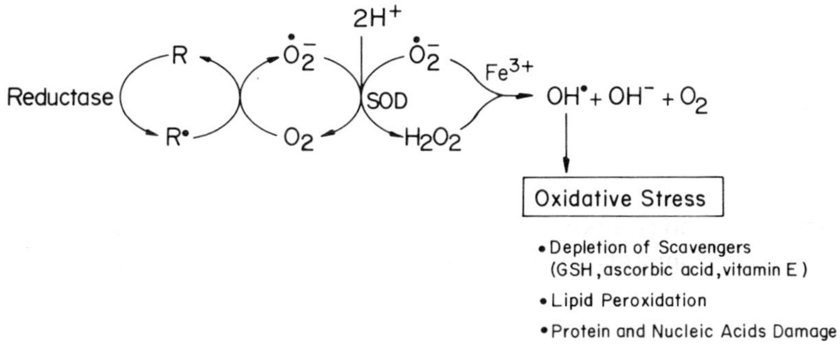

Fig. 2 Scheme of redox cycling of different pharmacologic agents leading to lipid peroxidation, depletion of scavengers, and protein and nucleic acid damage (oxidative stress). SOD, superoxide dismutase.

peroxidase) or amplifying (oxygen) reagents, and ultimately, the availability of enzymes competent to initiate the reductive process. In general, enzymes catalyzing redox cycling of foreign compounds are flavoproteins with relatively low substrate specificity. Therefore, the number of enzyme systems able to catalyze such a reduction step is limited.

Nitrofurans Nitrofurans have been widely used in the treatment of several infectious diseases (Docampo & Moreno, 1985). There is general agreement that metabolic reduction of the nitro group is of central importance in the toxicity of nitrofurans. The first step in the reduction of nitrocompounds is the formation of the nitro anion radical ($R\text{-}NO_2^{\cdot -}$). This intermediate is the first in a sequence of reduction products, any of which (except the amine product) may be responsible for the toxicity:

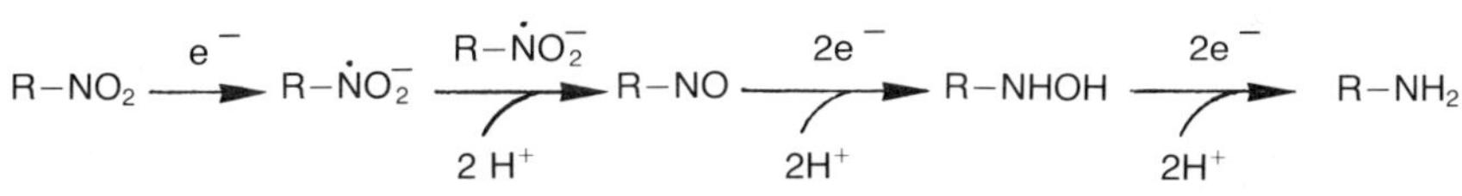

In addition, under aerobic conditions, this radical ($R\text{-}NO_2^{\cdot -}$) is rapidly oxidized by molecular oxygen, with regeneration of the nitro group and production of superoxide anion, whose dismutation yields H_2O_2 (Mason & Holtzman, 1975):

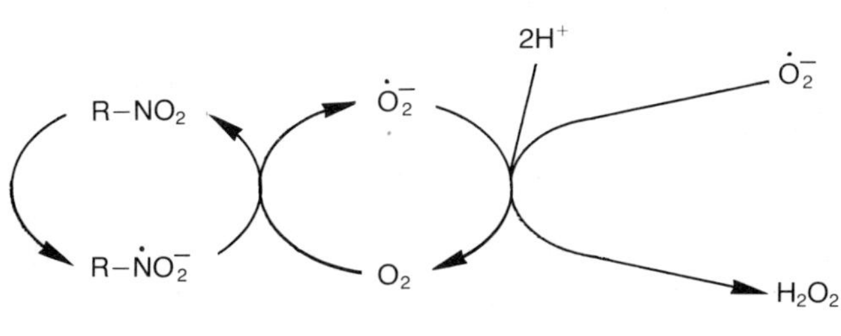

It seems likely that all of these molecular species, both the short-lived alkylating nitro-reduction products (which can covalently bind to cell macromolecules) and the oxygen–reduction byproducts (which can cause an oxidative stress) participate in the mechanism of toxicity of nitrocompounds. However, oxidative stress has been postulated as the main mechanism of toxicity of nitrofurans (Docampo & Moreno, 1984a, b). In this regard, it has been demonstrated that vitamin E deficiency potentiates the acute lung toxicity of nitrofurantoin in the rat model (Boyd *et al.*, 1979), and selenium deficiency increases the toxicity of nitrofurantoin to chicks (Peterson *et al.*, 1982). Oxidative stress has also been implicated in the toxicity of the nitrofuran nifurtimox to *T. cruzi* (Docampo & Moreno, 1984b).

The evidence supporting oxidative damage of *T. cruzi* induced by nifurtimox is as follows (reviewed by Marr & Docampo, 1986):

(1) The addition of nifurtimox to homogenates of *T. cruzi* that contain NAD(P)H leads to the appearance of a characteristic ESR spectrum corresponding to the nitro anion radical. The incubation of intact forms of *T. cruzi* with nifurtimox also leads to the appearance of the anion radical ESR signal.

(2) The nitro anion radical reacts with oxygen under aerobic conditions, as demonstrated by the increase in the cyanide-insensitive respiration of the intact cells and the release of H_2O_2 from the whole cells into the medium; the increase in the rate of O_2^- and H_2O_2 production after addition of nifurtimox to homogenates of *T. cruzi* in the presence of NAD(P)H; the stimulation of O_2^- production by the *T. cruzi* mitochondrial fraction supplemented with NADH (or succinate) and the enhancement of O_2^- generation by the microsomal fraction with NADPH as reductant.

(3) After administration to a human of a single oral dose of 15 mg of nifurtimox/kg, the serum concentration of the drug is 10–20 μM. This concentration of nifurtimox produces maximal stimulation of production of O_2^- by the *T. cruzi* mitochondrial fraction, initiates diffusion of H_2O_2 outside the cells, produces significant breakage of the DNA and inhibition of protein and nucleic acid synthesis, and completely inhibits growth of *T. cruzi*.

(4) The *in vitro* toxic effects of nifurtimox on *T. cruzi* (breakage of DNA, inhibition of protein and nucleic acid synthesis, alterations in ultrastructure, and inhibition of growth) have been demonstrated under aerobic conditions.

The involvement of other intermediates of nifurtimox reduction (e.g. nitroso or hydroxylamine) in its *in vitro* toxicity could be ruled out for the following reasons: the nitro anion radical is the first intermediate on the reduction of nifurtimox; under aerobic conditions redox cycling is the dominant pathway of metabolism of the nifurtimox anion radical; and this reaction results in regeneration of the parent drug without its net reduction. The evidence supporting oxidative stress as a mechanism for the toxicity of nifurtimox in animal tissues is also extensive but is circumstantial, and it rests on the observations (see Marr & Docampo, 1986) that:

(1) the target tissues have the capacity for nifurtimox-induced radical generation, and the most susceptible tissues (testis and brain) are deficient in defenses against oxygen-derived radicals;
(2) nifurtimox induces lipid peroxidation in these susceptible tissues;
(3) depletion of molecules of defense against oxygen-derived radicals (such as glutathione) enhances the nifurtimox toxicity, and
(4) the ultrastructural alterations observed in testes after treatment of rats with toxic doses of nifurtimox are similar to those observed after X-irradiation of animals.

Nifurtimox, nitrofurantoin, and other nitrocompounds like the anti-schistosomal agent niridazole, inhibit spermatogenesis in man and animals (Beeley, 1984). Although their mode of action on sperm cells has not been studied in detail, it is supposed that oxidative stress may play an important role.

Adriamycin. Anthracycline anticancer drugs have been extensively studied in respect to redox cycling, because it has been assumed that the antitumor effects of anthracyclines like adriamycin are due to oxygen radical formation occurring during redox cycling in tumor cells (Kappus, 1986). Also cardiotoxicity, a serious side effect of adriamycin has been related to oxygen radical formation (Kappus, 1986). Adriamycin produces long-term infertility and dominant lethal mutations in male mice (Meistrich *et al.*, 1985) and is cytotoxic against *T. cruzi* (Hammond *et al.*, 1984), although its mode of action in these cells has not been investigated.

Bleomycin. The bleomycins, a family of glycopeptide antibiotics, effective in the chemotherapy of several solid tumors and malignant lymphomas (Kappus, 1986) inhibit nuclear division and cause malformations of the nuclear microtubules in *T. gambiense* (Ono & Nakabayashi, 1980). Administration of bleomycin alone or in combination with DL-α-difluoromethylornithine, an irreversible inhibitor of polyamine synthesis, cures mice infected with *T. brucei* (Bacchi *et al.*, 1982). Bleomycin, used in combination therapy regimens with vinblastine or cisplatin also causes reversible male infertility (Beeley, 1984). The cytostatic activity of this compound is apparently due to DNA strand breaks, which have been shown to correlate with the reduction of cell proliferation (Kappus, 1986). Several reports (reviewed by Docampo & Moreno, 1984a) suggest that DNA breakage is produced *via* a tertiary oxygen–ferrous ion–bleomycin complex that is similar to heme-containing oxygenases. Whether this complex produces DNA breakage through the formation of O_2^- and $OH^{\cdot}$ or whether a ferryl ion intermediate is involved is not known. However, in either case free radicals are involved in its mechanism of action.

Membrane-active agents

The second group of trypanocidal compounds with antispermatogenic activity is formed by those compounds that exert their actions at the level of cell membranes. These drugs have been shown to affect the activities of some membrane-bound enzymes, to uncouple mitochondrial oxidative phosphorylation, and to inhibit ion transport in intact cells.

These actions can be attributed either to specific interactions with membrane proteins or to binding and modification of the properties of the lipid bilayer matrix. Some of these compounds may have a toxic action on cellular membranes by inducing an oxidative stress.

Tricyclic compounds. Amphiphilic cationic compounds such as chlorpromazine, imipramine, and clomipramine have been shown to inhibit *in vitro* the motility of sperm (Beeley, 1984) and trypanosomes (Tellez-Iñon *et al.*, 1985; Pearson *et al.*, 1982, 1984; Ruben *et al.*, 1983; Hammond *et al.*, 1984). Chlorpromazine and several of its derivatives kill *Leishmania donovani* promastigotes, extracellular amastigotes, and amastigotes within human macrophages (Pearson *et al.*, 1982). Furthermore, chlorpromazine reduces the parasite burden of hamsters infected with *L. donovani* (Pearson *et al.*, 1984). Similar *in vitro* effects of these compounds on *T. brucei* (Seebeck & Gehr, 1983) and *T. cruzi* (Hammond *et al.*, 1984) have been described. In an uncontrolled trial healing was observed following topical chlorpromazine ointment (2%) in 3 patients with diffuse cutaneous leishmaniasis caused by *L. aethiopica*, a condition usually resistant to conventional therapy with pentavalent antimonials (Hewlett *et al.*, 1985). As mentioned before, phenothiazines like chlorpromazine and trifluoperazine are calmodulin antagonists and inhibit the acrosome reaction and motility of spermatozoa and thus have been postulated as vaginal contraceptives (Lenz & Cormier, 1982). Phenothiazines like chlorpromazine and fluphenazine also block the activation of *T. cruzi* cyclic nucleotide phosphodiesterase by Ca^{2+}-calmodulin and impair *T. cruzi* motility (Tellez-Iñon *et al.*, 1985). Previous studies in mammalian cells demonstrated that both antipsychotic and antidepressant tricyclic compounds are also able to interact with proton-transporting ATPases. For example, the antipsychotic compound trifluoperazine was found to inhibit H^+, K^+-ATPase, and H^+ accumulation in porcine gastric cells (Hewlett *et al.*, 1985). Furthermore, this compound and other phenothiazine and thioxanthene derivatives efficiently inhibited the proton-translocating ATPase of mammalian mitochondria (Hewlett *et al.*, 1985). Tricyclic antidepressants also have a striking effect on energy-linked reactions in mitochondria. For example, both imipramine and clomipramine inhibit mitochondrial F_0,F_1-ATPase and uncouple oxidative phosphorylation (Weinbach *et al.*, 1986). The cumulative data suggest that the uncoupling of oxidative phosphorylation is due to specific interactions of the tricyclic antidepressants with the F_1-ATPase rather than to an ionophoric-like activity of these drugs. The tricyclic compounds appear to disrupt the leishmanial plasma membrane ATPase. As a result, membrane potential is reduced, active transport is inhibited, and intracellular pH homeostasis is presumptively disrupted, resulting in rapid cell death (Hewlett *et al.*, 1985).

Gossypol. Gossypol, a phenolic compound isolated from the cotton plant, is a powerful inhibitor of NAD-linked enzymes of *T. cruzi* and is toxic *in vitro* against *T. cruzi* epimastigotes (Montamat *et al.*, 1982). Because of its antispermatogenic activity gossypol has been tested as a male contraceptive. A recent report (Depyster *et al.*, 1984) has demonstrated that in addition to its multiple cellular effects, gossypol promotes the formation of oxygen radicals when incubated with human sperm. In addition, the fertilizing capacity of sea urchin sperm *in vitro* is destroyed when gossypol is added to the medium; superoxide dismutase or catalase inhibit this effect (Coburn *et al.*, 1980). The possibility of oxygen radical generation induced by gossypol in *T. cruzi* or other flagellated protozoa has not been investigated.

In addition to its capacity to generate oxygen radicals, gossypol has been shown to exert multiple effects on cellular membranes that cannot be explained by its redox properties. Gossypol has been shown to affect the activities of some membrane-bound mitochondrial enzymes, to uncouple mitochondrial oxidative phosphorylation, to inhibit the Na^+-K^+-ATPase, and to inhibit anion transport in red blood cells (reviewed in Reyes *et al.*, 1984). These actions of gossypol presumably are exerted at the level of cell membranes. In this regard, it has been shown that the charged form of gossypol binds to monolayers of different lipid compositions with apparent dissociation constants ranging from 0.7 to 2 μM. Gossypol also induces an increase in bilayer conductance accompanied by an increase in proton permeability. These changes induced in model membranes can account for the mitochondrial uncoupling effects of this molecule and may be responsible for the inhibitory effects of gossypol on several membrane transport systems (Reyes *et al.*, 1984). Interestingly, damage of the mitochondria of both spermatozoa (Nadavukaren *et al.*, 1979) and *T. cruzi* epimastigotes (Blanco *et al.*, 1983) is an early event in the toxicity of this compound.

Amphotericin B and Monensin. When the polyene macrolide antibiotic amphotericin B binds to sterols in cell membranes, several dose-dependent effects are observed. Increased membrane permeability to small ions occurs at intermediate concentrations in yeast, and higher concentrations of the drug in yeast cause cell lysis and death (Brajtburg *et al.*, 1980).

Despite the large amount of information that has been gathered about the interactions of amphotericin B with cellular and artificial membranes, the actual mechanism(s) that follow the binding to membrane sterols remains obscure, although oxidative damage has been recently implicated in the lethal and lytic effects of amphotericin B on *Candida albicans* (Sokol-Anderson *et al.*, 1986).

Amphotericin B has been widely used in several forms of dermal and subcutaneous leishmaniasis with considerable success (Sampaio *et al.*, 1971), and it has been shown to bind to membrane sterols of *L. donovani* resulting in loss of the permeability barrier to small metabolites (Saha *et al.*, 1986). Similar permeability changes occur in rabbit spermatozoa treated with this drug (Holland & Storey, 1981). Amphotericin B is also toxic against *T. cruzi* trypomastigotes (Cruz *et al.*, 1980; Hammond *et al.*, 1984) and has been suggested as a blood additive to prevent blood transmission of Chagas' disease (Cruz *et al.*, 1980). It is possible that monensin, a related anti-coccidial ionophoric antibiotic acts in a similar way (Gutteridge & Coombs, 1977). Monensin also causes infertility in male rats (Atef *et al.*, 1986).

Conclusions

Over the last several years our knowledge of the biochemistry of flagellated protozoa has expanded enormously. Gradually, it has become apparent that these parasites have a number of features similar to those in spermatozoa and different from other eukaryotic cells. From the point of view of obtaining a chemotherapeutic agent against pathogenic protozoa, this fact constitutes a drawback. However, compounds active against pathogenic protozoa may prove useful as male contraceptives. The two main groups of chemotherapeutic agents against pathogenic protozoa active on spermatozoa are those compounds capable of undergoing a reductive metabolism with generation of free radicals and membrane-active agents. The former lacks the specificity that is so important if one is to avoid toxic reactions in the organism. If this problem could be solved, such compounds could be quite effective because of the many mechanisms by which the active metabolites can damage biologic systems. On the other hand, membrane-active agents could prove useful as male contraceptives or as antiparasitic compounds because of the special structures of the spermatozoal and protozoal membranes and their role in differentiation or intracellular penetration, phenomena not observed in other mammalian cells.

Acknowledgements

The preparation of this publication was supported in part by grants from the UNDP/World Bank/World Health Organization Special Programme for

Research and Training in Tropical Diseases, and from CNPq and FINEP, Brazil.

References

Alvarez, J.G. & Storey, B.T. (1983). Role of superoxide dismutase in protecting rabbit spermatozoa from O_2 toxicity due to lipid peroxidation. **Biology of Reproduction** 28:1129–1136.

Alves, M.J.M., Abuin, G., Kuwajima, V.Y. & Colli, W. (1986). Partial inhibition of trypomastigote entry into cultured mammalian cells by monoclonal antibodies against a surface glycoproteins of *Trypanosoma cruzi*. **Molecular and Biochemical Parasitology** 21:75–82.

Ames, B.N., Catchcart, R., Schwiers, E. & Hochstein, P. (1981). Uric acid provides an antioxidant defense in humans against oxidant- and radical-caused aging and cancer: a hypothesis. **Proceedings of the National Academy of Sciences, USA** 78:6858–6862.

Andrews, N.W., Katzin, A.M. & Colli, W. (1984). Mapping of surface glycoproteins of *Trypanosoma cruzi* by two-dimensional electrophoresis. A correlation with the cell invasion capacity. **European Journal of Biochemistry** 140:599–604.

Atef, M., Shalaby, M.A., El-Sayed, M.G.A., El-Din, S., Youssef, A.H. & El-Sayed, M.A.I. (1986). Influence of monensin on fertility in rats. **Clinical and Experimental Pharmacology and Physiology** 13:113–121.

Bacchi, C.J., Nathan, H.C., Hutner, S.H., McAnn, P.P. & Sjoerdsma, A. (1982). New combination chemotherapy of experimental trypanosomiasis by using bleomycin and DL-α-difluoromethylornithine, reversal by polyamines. **Biochemical Pharmacology** 31:2833–2835.

Baum, S.G., Wittner, M., Nadler, J.P., Horwitz, S.B., Dennis, J.E., Schiff, P.B. & Tanowitz, H.B. (1981). Taxol, a microtubule stabilizing agent, blocks the replication of *Trypanosoma cruzi*. **Proceedings of the National Academy of Sciences, USA** 78:4571–4575.

Beeley, L. (1984). Drug-induced sexual disfunction and infertility. **Adverse Drug Reactions in Acute Poisoning, Reviews** 3:23–42.

Blanco, A., Aoki, A., Montamat, E.E. & Rovai, L.E. (1983). Effect of gossypol upon motility and ultrastructure of *Trypanosoma cruzi*. **The Journal of Protozoology** 30:648–651:

Boveris, A., Sies, H., Martino, E.E., Docampo, R., Turrens, J.F. & Stoppani, A.O.M. (1980). Deficient metabolic utilization of hydrogen peroxide in *Trypanosoma cruzi*. **Biochemical Journal** 188:643-–648.

Boyd, M.R., Catignani, G.L., Sasame, H.A., Mitchell, J.R. & Stiko, A.W. (1979). Acute pulmonary injury in rats by nitrofurantoin and modification by vitamin E, dietary fat, and oxygen. **American Reviews of Respiratory Diseases** 120:93–99.

Brajtburg, T., Medoff, G., Kobayashi, G.S., Elberg, S. & Finegold, C. (1980). Permeabilizing and hemolytic action of large and small concentrations of polyene antibiotics on human erythrocytes. **Antimicrobial Agents and Chemotherapy** 18:586–592.

Brener, Z. (1979). Present status of chemotherapy and chemoprophylaxis of human trypanosomiasis in the Western hemisphere. **Pharmacology and Therapeutics** 7:71–90.

Buhrley, L.E. & Legrande, C.E. (1973). Catalase activity in rat testicular preparations, distribution and changes in activity induced by hypophysectomy and X-irradiation. **Fertility and Sterility** 24:956–961.

Coburn, M., Sinsheimer, P., Segal, S., Burgos, M. & Troll, W. (1980). Oxygen free radical generation by gossypol: a possible mechanism of antifertility action in sea urchin sperm. **Biological Bulletin (Woods Hole, Massachussetts)** 159:468.

Corsi, G., Palazzo, G., Germani, C., Scorza Barcellona, P. & Silvestrini, B. (1976). 1-Halobenzyl-1-H-indazole-3-carboxylic acids. A new class of antispermatogenic agents. **Journal of Medicinal Chemistry** 19:778–783.

Couto, A.S., Zingales, B., De Lederkremer, R.M. & Colli, W. (1985). *Trypanosoma cruzi*: metabolic labeling of trypomastigotes sialoglycolipids. **Experientia** 41:736–738.

Crane, M.S.J. & Dvorak, J.A. (1982). Influence of monosaccharides on the infection of vertebrate cells by *Trypanosoma cruzi* and *Toxoplasma gondii*. **Molecular and Biochemical Parasitology** 5:333–341.

Cruz, F.S., Marr, J.J. & Berens, R.L. (1982). Prevention of transfusion-induced Chagas' disease during blood transfusion. **Transactions of the Royal Society of Tropical Medicine and Hygiene** 76:633–635.

De Peyster, A., Quintanilha, A., Packer, L. & Smith, M.T. (1984). Oxygen radical formation induced by gossypol in rat liver microsomes and human sperm. **Biochemical and Biophysical Research Communications** 118:573–579

De Souza, W. & Meyer, H. (1974). On the fine structure of the nucleus of *Trypanosoma cruzi* in tissue culture forms. Spindle fibers in the dividing nucleus. **The Journal of Protozoology** 21:48–52.

De Souza, W. (1984). Cell biology of *Trypanosoma cruzi*. **International Review of Cytology** 86:197–283.

Docampo, R., Cruz, F.S., Boveris, A., Muniz, R.P.A. & Esquivel, D.M.S. (1978). Lipid peroxidation and generation of free radicals, superoxide anion and hydrogen peroxide in β-lapachone-treated *Trypanosoma cruzi* epimastigotes. **Archives of Biochemistry and Biophysics** 186:292–297.

Docampo, R. & Moreno, S.N.J. (1984a). Free-radical intermediates in the antiparasitic action of drugs and phagocytic cells. In Pryor, W.A., ed. **Free Radicals in Biology**, vol. 6, New York, Academic Press, pp. 243–288.

Docampo, R. & Moreno, S.N.J. (1984b). Free radical metabolites in the mode of action of chemotherapeutic agents and phagocytic cells on *Trypanosoma cruzi*. **Reviews of Infectious Diseases** 6:223–238.

Docampo, R. & Moreno, S.N.J. (1985). Biochemical toxicology of antiparasitic compounds used in the chemotherapy and chemoprophylaxis of American trypanosomiasis (Chagas' disease). **Reviews in Biochemical Toxicology** 7:159–204.

Duncan, G.W. & Pharris, B.B. (1970). Effects of nonsteroidal compounds on fertility. **Federation Proceedings** 20:1232–1239.

Dvorak, J.A. & Hyde, T.P. (1973). *Trypanosoma cruzi* interaction with vertebrate cells *in vitro*.I. Individual interactions at the cellular and subcellular levels. **Experimental Parasitology** 34:268–283.

Fairlamb, A.H., Blackburn, P., Ulrich, P., Chait, B.T. & Cerami, A. (1985). Trypanothione: a novel bis(glutathionyl)spermidine cofactor for glutathione reductase in trypanosomatids. **Science** 227:1485–1487.

Floridi, A., De Martino, C., Marcante, M.L., Apollonj, C., Scorza-Bracellona, P. & Silvestrini, B. (1981). Morphological and biochemical modifications of rat germ cell mitochondria induced by new antispermatogenic compounds: studies *in vivo* and *in vitro*. **Experimental and Molecular Pathology** 35:314–331.

Floridi, A. & Lehninger, A. (1983). Action of the antitumor and antispermatogenic agent lonidamine on electron transport in Erlich ascites tumor mitochondria. **Archives of Biochemistry and Biophysics** 226:73–83.

Goncalves, A.M., Vasconcellos, M.E., Docampo, R., Cruz, F.S., De Souza, W. & Leon, W. (1980). Evaluation of the toxicity of 3-allyl-β-lapachone against *Trypanosoma cruzi* bloodstream forms. **Molecular and Biochemical Parasitology** 1:167–176.

Gorus, F., Finsy, R. & Pipeleers, D.G. (1986). Alloxan toxicity in human and canine spermatozoa. Possible biochemical basis for a species difference in sensitivity. **Biochemical Pharmacology** 35:1726–1729.

Gutteridge, W.E. & Coombs, G.H. (1977). **Biochemistry of Parasitic Protozoa**, London, Macmillan Press.

Hammond, D.J., Cover, B. & Gutteridge, W.E. (1984). A novel series of chemical structures active *in vitro* against the trypomastigote form of *Trypanosoma cruzi*. **Transactions of the Royal Society of Tropical Medicine and Hygiene** 78:91–95.

Hewlett, E.L., Pearson, R.D., Zilberstein, D. & Dwyer, D.M. (1985). Anti-protozoal activity of tricyclic compounds. **Science** 230:1063–1064.

Hinrichsen-Kohane, A.C., Hinrichsen, M.J. & Schill, W.B. (1984). Molecular events leading to fertilization – A review. **Andrologia** 16:321–341.

Holland, M.K. & Storey, B.T. (1981). Oxygen metabolism of mammalian spermatozoa. Generation of hydrogen peroxide by rabbit epididymal spermatozoa. **The Biochemical Journal** 198:273–280.

Holland, M.K. & Storey, B.T. (1982). Production of superoxide and activity of superoxide dismutase in rabbit epididymal spermatozoa. **Biology of Reproduction** 27:1109–1118.

Hoskins, D.D., Brandt, H. & Ascott, T.S. (1976). Initiation of sperm motility in the mammalian epididymis. **Federation Proceedings** 37:2534–2542.

Jones, R. & Mann, T. (1977). Damage to ram spermatozoa by peroxidation of endogenous phospholipids. **Journal of Reproduction and Fertility** 50:261–268.

Juliano, C., Martinotti, M.G. & Cappuccinelli, P. (1985). *In vitro* effect of microtubule inhibitors on *Trichomonas vaginalis*. **Microbiologica** 8:31–42.

Kappus, H. & Sies, H. (1981). Toxic drug effects associated with oxygen metabolism, redox cycling and lipid peroxidation. **Experientia** 37:1233–1241.

Kappus, H. (1986). Overview of enzyme systems involved in bioreduction of drugs and in redox cycling. **Biochemical Pharmacology** 35:1–6.

Lenz, R.W. & Cormier, M.J. (1982). Effects of calmodulin-binding drugs on the guinea pig spermatozoon acrosome reaction and the use of these drugs as vaginal contraceptive agents in rabbits. **Annals of the New York Academy of Sciences** 383:85–95.

Lopes, J.N., Cruz, F.S., Docampo, R., Vasconcellos, M.E., Sampaio, M.C.R., Pinto, A.V. & Gilbert, B. (1978). *In vitro* and *in vivo* evaluation of the toxicity of 1,4-naphthoquinone and 1,2-naphthoquinone derivatives against *Trypanosoma cruzi*. **Annals of Tropical Medicine and Parasitology** 72:523–531.

Lopo, A., Glabe, C.G., Lennarz, W.J. & Vacquier, V.D. (1982). Sperm-egg binding events during sea urchin fertilization. **Annals of the New York Academy of Sciences** 383:405–425.

Mancini, P.E. & Patton, C.I. (1981). Cyclic 3′,5′-adenosine monophosphate levels during the developmental cycle of *Trypanosoma brucei* in rat. **Molecular and Biochemical Parasitology** 3:19–31.

Marr, J.J. & Berens, R.L. (1983). Pyrazolpyrimidine metabolism in the pathogenic Trypanosomatidae. **Molecular and Biochemical Parasitology** 7:339–356.

Marr, J.J. & Docampo, R. (1986). Chemotherapy for Chagas' disease. A perspective of current therapy and considerations for future research. **Reviews of Infectious Diseases** 8:884–903.

Mason, R.P. & Holtzman, J.L. (1975). The role of catalytic superoxide formation in the O_2 inhibition of nitroreductase. **Biochemical and Biophysical Research Communications** 67:1267–1274.
Meistrich, M.L., Goldstein, L.S. & Wyrobek, A.J. (1985). Long-term infertility and dominant lethal mutations in male mice treated with adriamycin. **Mutation Research** 152:53–65.
Meyer, H. & De Souza, W. (1976). Electron microscopic study of *Trypanosoma cruzi* periplast in tissue culture. I. Number and arrangements of peripheral microtubules in the various forms of the parasite's life cycle. **The Journal of Protozoology** 23:285–290.
Montamat, E.E., Burgos, C., Gerez De Burgos, N.M., Rovai, L., Blanco, A. & Segura, E.L. (1982). Inhibitory action of gossypol on enzymes and growth of *Trypanosoma cruzi*. **Science** 218:288–289.
Moreno, S.N.J. & Docampo, R. (1985). Mechanism of toxicity of nitrocompounds used in the chemotherapy of trichomoniasis. **Environmental Health Perspectives** 64:199–208.
Nadavukaren, M.J., Sorensen, R.M. & Tone, J.N. (1979). Effect of gossypol on the ultrastructure of rat spermatozoa. **Cell and Tissue Research** 204:293–296.
O'Rand, M.G. (1982). Modification of the sperm membrane during capacitation. **Annals of the New York Academy of Sciences** 383:392–402.
Olson, G.E. & Orgebin-Crist, M.C. (1982). Sperm surface changes during epididymal maturation. **Annals of the New York Academy of Sciences** 383:372–390.
Ono, T. & Nabakayashi, T. (1980). Studies on the effect of bleomycin on *Trypanosoma gambiense*. **Biken Journal** 23:143–155.
Ouaissi, M.A. & Capron, A. (1986). Studies on the role of fibronectin in *Trypanosoma cruzi*-cell interaction. **Memorias do Instituto Oswaldo Cruz** 8l(Suppl):BQ-26.
Pearson, R.D., Manian, A.A., Hall, D., Marcus, J.L. & Hewlett, E.L. (1984). Antileishmanial activity of chlorpromazine. **Antimicrobial Agents and Chemotherapy** 25:571–574.
Pearson, R.D., Manian, A.A., Marcus, J.L., Hall, D. & Hewlett, E.L. (1982). Lethal effects of phenothiazine neuroleptics on the pathogenic protozoan *Leishmania donovani*. **Science** 217:369–370.
Penketh, P.G. & Klein, R.A. (1986). Hydrogen peroxide metabolism in *Trypanosoma brucei*. **Molecular and Biochemical Parasitology** 20:111–121.
Pereira, M.E.A., Loures, M.A., Villalta, F. & Andrade, A.F.B. (1980). Lectin receptors as markers for *Trypanosoma cruzi*. **Journal of Experimental Medicine** 152:1375–1392.
Peterson, D.S., Wrightsman, R.A. & Manning, J.E. (1986). Cloning of a major surface antigen gene of *Trypanosoma cruzi*. Identification of a nonapeptide repeat. **Nature** 322:566–568.
Peterson, F.J., Combs, G.F., Holtzman, J.L. & Mason, R.P. (1982). The effect of selenium and vitamin E deficiency in the toxicity of nitrofurantoin. **Journal of Nutrition** 112:1741–1746.
Piras, M.M., Henriquez, D. & Piras, R. (1985). The effect of proteolytic enzymes and protease inhibitors on the interaction *Trypanosoma cruzi*-fibroblasts. **Molecular and Biochemical Parasitology** 14:151–163.
Piras, M.M., Henriquez, D. & Piras, R. (1987). The effect of fetuin and other sialoglycoproteins on the *in vitro* penetration of *Trypanosoma cruzi* trypomastigotes into fibroblastic cells. **Molecular and Biochemical Parasitology** 22:135–143.
Piras, R., Boschetti, M.A. & Piras, M.M. (1986). Identification of surface proteins related to the adhesion process of *Trypanosoma cruzi* trypomastigotes to Vero cells. **Memorias do Instituto Oswaldo Cruz** 8l(Suppl):BQ-25.

Piras, R., Piras, M.M. & Henriquez, D. (1983). *Trypanosoma cruzi*–fibroblastic cell interaction necessary for cellular invasion. In: **Cytopathology of Parasitic Diseases**, London, Pitman, pp. 31–55 (Ciba Foundation Symposium 99).

Previato, J.O., Andrade, A.F.B., Pessolani, M.C.V. & Mendonca-Previato, L. (1985). Incorporation of sialic acid into *Trypanosoma cruzi* macromolecules. A proposal for a new metabolic route. **Molecular and Biochemical Parasitology** 16:85–96.

Pryor, W.A. (1976). The role of free radical reactions in biological systems. In Pryor, W.A., ed. **Free Radicals in Biology**, Vol. 1, New York, Academic Press, pp. 1–49.

Rangel-Aldao, R., Allende, O., Triana, F., Piras, R., Henriquez, D. & Piras, M. (1987). Possible role of cAMP in the differentiation of *Trypanosoma cruzi*. **Molecular and Biochemical Parasitology** 22:39–44:

Reyes, J., Aallen, J., Tanphaichitr, N., Bellvé, A.R. & Benos, D.J. (1984). Molecular mechanisms of gossypol action on lipid membranes. **The Journal of Biological Chemistry** 259:9607–9615.

Ruben, L., Egwuagu, C. & Patton, C. (1983). African trypanosomes contain calmodulin which is distinct from host calmodulin. **Biochemica et Biophysica Acta** 758:104–113.

Saha, A.K., Mukherjee, T. & Bhaduri, A. (1986). Mechanism of action of Amphotericin B on Leishmania donovani promastigotes. **Molecular and Biochemical Parasitology** 19:195–200.

Sampaio, S.A.P., Castro, R.M., Dillon, N.L. & Martins, J.E.C. (1971). Treatment of mucocutaneous leishmaniasis with amphotericin B: report of 70 cases. **International Journal of Dermatology** 10:179–181.

Schatten, G. (1982). Motility during fertilization. **International Review of Cytology** 79:59–163.

Schauer, R., Reuter, G., Muhlpfordt, H., Andrade, A.F.B. & Pereira, M.E.A. (1983). The occurrence of N-acetyl- and N-glycolylneuraminic acid in *Trypanosoma cruzi*. **Hoppe-Seyler's Zeitschrift für Physiologische Chemie** 364:1053–1057.

Schiff, P.B. & Horwitz, S.B. (1980). Taxol stabilizes microtubules in mouse fibroblast cells. **Proceedings of the National Academy of Sciences, USA** 77:1561–1565.

Seebeck, T. & Gehr, P. (1983). Trypanocidal action of neuroleptic phenothiazines in *Trypanosoma brucei*. **Molecular and Biochemical Parasitology** 9:197–208.

Sher, A. & Snary, D. (1983). Specific inhibition of the morphogenesis of *Trypanosoma cruzi* by a monoclonal antibody. **Nature** 300:39–40.

Sokol-Anderson, M.L., Brajtburg, J. & Medoff, G. (1986). Amphotericin B-induced oxidative damage and killing of Candida albicans. **The Journal of Infectious Diseases** 154:76–83.

Solari, A. J. & Duschak, V. (1985). Intermediate filaments induced by taxol in cultures of *Crithidia fasciculata*. **Microscopía electrónica y Biología celular** 9:163–175.

Souto-Padron, T. & De Souza, W. (1985). Sialoglycoproteins and sialoglycolipids contribute to the negative surface charge of epimastigote and trypomastigote forms of *Trypanosoma cruzi*. **Biochimica et Biophysica Acta** 814:163–169.

Tellez-Iñon, M.T., Ulloa, R.M., Torruella, M. & Torres, H.N. (1985). Calmodulin and Ca^{2+}-dependent cyclic AMP phosphodiesterase activity in *Trypanosoma cruzi*. **Molecular and Biochemical Parasitology** 17:143–153.

Tournade, A. (1913). Difference de motilité des spermatozoides prélèves dans les divers segments de l'epididyme. **Comptes Rendus Societe Biologie** 74:738.

Turrens, J.F. (1986a). Inhibitory action of the antitumor agent lonidamine on mitochondrial respiration of *Trypanosoma cruzi* and *Trypanosoma brucei*. **Molecular and Biochemical Parasitology** 20:237–241.

Turrens, J.F. (1986b). The potential of antispermatogenic drugs against trypanosomatids. **Parasitology Today** 2:351–352.
Voorheis, H.P. & Martin, B.P. (1981). Characteristics of the calcium-mediated mechanism activating adenylate cyclase in *Trypanosoma brucei*. **European Journal of Biochemistry** 116:471–477.
Voorheis, H.P., Bowles, D.J. & Smith, G.A. (1982). Characteristics of the release of the surface coat protein from bloodstream forms of *Trypanosoma brucei*. **Journal of Biological Chemistry** 257:2300–2304.
Weinbach, E.C., Costa, J.L., Nelson, B.D., Clagget, C.E., Hundal, T., Bradley, D. & Morris, S.J. (1986). Effects of tricyclic antidepressant drugs on energy-linked reactions in mitochondria. **Biochemical Pharmacology** 35:1445–1451.
Whorton, M.D. (1982). Male occupational reproductive hazard. **The Western Journal of Medicine** 137:521–524.
Zingales, B. & Colli, W. (1985). *Trypanosoma cruzi* interaction with host cells. **Current Topics in Microbiology and Immunology** 117:129–152.

New approaches to male fertility regulation: interference with spermiogenesis

MONTRI CHULAVATNATOL

Department of Biochemistry, Faculty of Science, Mahidol University, Bangkok 10400, Thailand.

Abstract

The use of pharmacological agents offers an important option in the development of a male antifertility method using spermatids as the targets. Compounds that can pass through the blood–testis barrier to reach the spermatids may have some chance of interfering with spermiogenesis. Possible compounds are those that are inhibitory to the key events in spermiogenesis, namely acrosomal glycoprotein synthesis, energy generation from lactate and axoneme formation. It may also be possible to identify a temperature-sensitive event in spermatids and to interfere with it using a chemical agent.

Key words

Spermatid, acrosomal glycoprotein synthesis, lactate metabolism, axoneme, temperature-sensitivity.

Introduction

Many attempts have been made to regulate male fertility by interfering with either spermatogenesis or sperm maturation using hormones or pharmacological agents. On the basis of the cell biology of spermatids and the physiology of spermiogenesis, two general considerations arise concerning approaches to intervene with spermiogenesis:

(1) Because of the blood–testis barrier and the location of the spermatids towards the luminal side of the seminiferous tubules, antibodies to any spermatid-specific antigens may not reach the spermatids.

(2) Spermatid differentiation requires the support of Sertoli cells. Although normal Sertoli cell functions depend on FSH, spermatid differentiation is not directly affected by hormones. Thus, any hormonal approach to alter spermiogenesis should be considered as indirect.

In this chapter, some approaches for interfering with spermiogenesis will be discussed that may lead to the development of male contraceptives.

Pharmacological agents and the blood–testis barrier

Pharmacological agents or their metabolites must be able to cross the blood–testis barrier to reach the spermatids and to interfere with spermatid survival or differentiation. With regard to the transport of compounds across the blood–testis barrier, it would be useful to evaluate those agents that have been found to reduce sperm count. Among these toxic compounds, many suppress spermatogenesis by unknown mechanisms (Bartke *et al.*, 1986) but some, such as gossypol, act at least partially by damaging round spermatids (Baccetti *et al.*, 1986).

Specific events and possible interventions

Many basic events in spermiogenesis have been investigated extensively in recent years. These advances have been made possible by the development of methods for spermatid isolation (Bellvé *et al.*, 1977; Grootegoed *et al.*, 1986; Nakamura *et al.*, 1984a). Some of these basic events in spermiogenesis will now be discussed and possible interventions by pharmacological agents will be proposed.

Acrosomal glycoproteins

Many studies have demonstrated that round spermatids actively synthesize several glycoproteins (Clermont & Tang, 1985; Gerton & Millette, 1986). Some of these glycoproteins are packaged into the acrosome. Since acrosomal enzymes such as acrosin and hyaluronidase have crucial roles in fertilization, it will be expected that spermatozoa without such acrosomal enzymes may be infertile. To produce spermatozoa without the essential acrosomal enzymes, one may interfere with glycoprotein synthesis and transport to the acrosomal vesicle in round spermatids. This should be possible since much is known about the mechanisms of glycosylation and transfer of glycosylated proteins to the lysosome in other cells (Dunphy & Rothman, 1985; Griffiths & Simons, 1986; Kornfeld & Kornfeld, 1985; Schachter, 1986; von Figura & Hasilik, 1986). Moreover, there are several pharmacological agents that can inhibit these processes. N-linked glycoprotein synthesis requires the participation of dolichol as carbohydrate carrier. Both dolichol and dolichol kinase are present in rat testis (Berkowitz & Nyquist, 1986; Nyquist & Holt, 1986). The dolichol cycle is sensitive to bacitracin, amphomycin and tunicamycin (Schwarz & Datema, 1980). The processing of the N-linked carbohydrate chains of the glycoproteins in rough endoplasmic reticulum and the Golgi can be inhibited by several compounds: nojirimycin, castanospermine, bromoconduritol and swainsonine (Fuhrmann *et al.*, 1985).

The transport of glycoproteins through the Golgi–endoplasmic reticulum–lysosome (GERL) complex involves specific recognition. The glycoproteins to be transported to the lysosome contain mannose-6-phosphate which binds specifically to mannose-6-phosphate receptor (MPR). The glycoprotein–MPR complex is brought to the lysosome where the acidic pH dissociates the complex. The glycoprotein is then left in the lysosome while the MPR recycles to the Golgi. The MPR recycling is sensitive to chloroquine inhibition (von Figura & Hasilik, 1986). Assuming that acrosomal enzymes are glycosylated, processed and transported to the acrosomal vesicle in a similar manner, it may be possible to produce such defective (or infertile) 'acrosin-less sperm' by administering the above-mentioned inhibitory agents. Without any experimental data, it will not be possible to predict which inhibitor will be effective. Neither can it be determined if the effective dose will not alter glycoprotein synthesis in other organs. Nevertheless, this approach should be evaluated experimentally.

Metabolism

Pachytene spermatocytes and round spermatids utilize lactate rather than glucose as an energy source (Nakamura *et al.*, 1984b, 1986; see chapter by Grootegoed & den Boer, this volume). FSH-stimulated Sertoli cells convert glucose to lactate which is then supplied to the spermatids (Jutte *et al.*, 1983; Mita *et al.*, 1982). In spermatids, lactate is metabolized *via* pyruvate, the Krebs' cycle and oxidative phosphorylation in the mitochondria to yield ATP (Grootegoed *et al.*, 1984, 1986; Nakamura *et al.*, 1984a). Inhibition of the Krebs' cycle by fluoroacetate (which is converted to fluorocitrate) impairs spermatogenesis *in vitro* (Sullivan *et al.*, 1979). Deprivation of ATP suppresses protein synthesis (Nakamura *et al.*, 1981) and threatens the survival of the spermatids (Jutte *et al.*, 1982). The strong dependence of spermatids on lactate is a unique feature that may be exploited. If the properties of the enzymes involved in lactate metabolism in the spermatids are studied, specific inhibitors may be found. Some of these inhibitors may have an antispermiogenic effect.

Axoneme formation

Proper formation of sperm tails is vital to sperm motility and hence to sperm transport to the site of fertilization. Axoneme, the force-generating structure of the sperm tails, consists of microtubules and dynein ATPase. Tubulin polymerization into microtubules which occurs in the round spermatid can be inhibited by demecolcine (Gerton & Millette, 1984). Centrioles are involved in the organization of the axoneme structure. Dissociation of the centrioles has been observed to cause spermatid malformation in man (Holstein *et al.*, 1986). Another human defect, Kartagener's syndrome, is caused by the absence of dynein ATPase (Baccetti *et al.*, 1981). It would therefore appear that axoneme formation in the spermatids may be sensitive to manipulation. More detailed studies on tubulin synthesis, polymerization and dynein ATPase synthesis may uncover new sites for pharmacological intervention of the axeneme formation.

Temperature effect

Body temperature is known to inhibit normal spermatogenesis in the mammalian testis. Round spermatids seem to be sensitive to the body temperature. Both protein synthesis (Nakamura & Hall, 1980) and hexose transport (Hall & Nakamura, 1981) by the spermatids of the rat are optimal at scrotal temperature (33-34° C) rather than at body temperature (37° C). However, it should be noted that body temperature only partially, but not completely, suppresses these processes. It may be worthwhile to identify biochemical events in spermatids that are highly sensitive to body temperature. Then one may ask if there is a pharmacological agent that can suppress such biochemical events instead of body temperature.

Conclusion

Using spermatids as the target, future antifertility agents may be developed from the inhibitors of acrosomal glycoprotein synthesis, lactate metabolism, axoneme formation or temperature-sensitive processes.

References

Baccetti, B., Burrini, A.G., Pallini, V. & Renieri, T. (1981). Human dynein and sperm pathology, **Journal of Cell Biology** 88:102–107.

Baccetti, B., Bigliardi, E., Burrini, A.G., Renieri, T. & Selmi, G. (1986). The action of gossypol on rat germinal cells. **Gamete Research** 13:1–17.

Bartke, A., Hahn, D.W., Foldesy, A.G. & McGuire, J.L. (1986). Experimental studies in the development of male contraceptives. In: Zatuchni, G.I., Goldsmith, A.A., Spieler, J.M. & Sciarra, J.J., eds. **Male contraception: advances and future prospects**. Philadelphia, Harper & Row, pp. 158–182.

Bellvé, A.R., Millette, C.F., Bhatnagar, Y.M. & O'Brien, D.A. (1977). Dissociation of mouse testis and characterization of isolated spermatogenic cells. **Journal of Histochemistry and Cytochemistry** 25:480–494.

Berkowitz, L. & Nyquist, S.E. (1986). Dolichol kinase activity in the developing rat testis. **Biology of Reproduction** 34:518–526.

Clermont, Y. & Tang, X.M. (1985). Glycoprotein synthesis in the Golgi apparatus of spermatids during spermiogenesis of the rat. **Anatomical Record** 213: 33–43.

Dunphy, W.G. & Rothman, J.E. (1985). Compartmental organization of the Golgi stack. **Cell** 42:13–21.

Fuhrmann, U., Bause, E. & Ploegh, H. (1985). Inhibitors of oligosaccharide processing. **Biochimica et Biophysica Acta** 825:95–110.

Gerton, G.L. & Millette, C.F. (1984). Generation of flagella by cultured mouse spermatids. **Journal of Cell Biology** 98:619–628.

Gerton, G.L. & Millette, C.F. (1986). Stage-specific synthesis and fucosylation of plasma membrane proteins by mouse pachytene spermatocytes and round spermatids in culture. **Biology of Reproduction** 35:1025–1035.

Griffiths, G. & Simons, K. (1986). The trans Golgi network: sorting at the exit site of the Golgi complex. **Science** 234:438–443.

Grootegoed, J.A., Jansen, R. & van der Molen, H.J. (1984). The role of glucose, pyruvate and lactate in ATP production by rat spermatocytes and spermatids. **Biochimica et Biophysica Acta** 767:248–256.

Grootegoed, J.A., Jansen, R. & van der Molen, H.J. (1986). Effect of glucose on ATP dephosphorylation in rat spermatids. **Journal of Reproduction and Fertility** 77:99–107.

Hall, P.F. & Nakamura, M. (1981). Influence of temperature on hexose transport by round spermatids of rats. **Journal of Reproduction and Fertility** 63:373–379.

Holstein, A.F., Schill, W.B. & Bruecker, H. (1986). Dissociated centriole development as a cause of spermatid malformation in man. **Journal of Reproduction and Fertility** 78:719–725.

Jutte, N.H.P.M., Jansen, R., Grootegoed, J.A., Rommerts, F.F.G., Clausen, O.P.F. & van der Molen, H.J. (1982). Regulation of survival of rat pachytene spermatocytes by lactate supply from Sertoli cells. **Journal of Reproduction and Fertility** 65:431–438.

Jutte, N.H.P.M., Hansen, R., Grootegoed, J.A., Rommerts, F.F.G. & van der Molen, H.J. (1983). FSH stimulation of the production of pyruvate and lactate by rat Sertoli cells may be involved in hormonal regulation of spermatogenesis. **Journal of Reproduction and Fertility** 68:219–226.

Kornfeld, R. & Kornfeld, S. (1985). Assembly of asparagine-linked oligosaccharides. **Annual Reviews of Biochemistry** 54:631–664.

Mita, M., Price, J.M. & Hall, P.F. (1982). Stimulation by follicle-stimulating hormone of synthesis of lactate by Sertoli cells from rat testis. **Endocrinology** 110:1535–1541.

Nakamura, M., Fujiwara, A., Yasumasu, I., Okinaga, S. & Arai, K. (1982). Regulation of glucose metabolism by adenine nucleotides in round spermatids from rat testes. **Journal of Biological Chemistry** 257:13945–13950.

Nakamura, M. & Hall, P.F. (1980). The mechanism by which body temperature inhibits protein biosynthesis in spermatids of rat testis. **Journal of Biological Chemistry** 255:2907–2913.

Nakamura, M., Hino, A., Yasumasu, I. & Kato, J. (1981). Stimulation of protein synthesis in round spermatids from rat testes by lactate. **Journal of Biochemistry** 89:1309–1315.

Nakamura, M., Okinaga, S. & Arai, K. (1984a). Metabolism of pachytene primary spermatocytes from rat testes pyruvate maintenance of adenosine triphosphate level. **Biology of Reproduction** 30:1187–1197.

Nakamura, M., Okinaga, S. & Arai, K. (1984b). Metabolism of round spermatids; evidence that lactate is preferred substrate. **American Journal of Physiology** 247:E234–E242.

Nakamura, M., Okinaga, S. & Ari, K. (1986). Studies of metabolism of round spermatids: glucose as unfavorable substrate. **Biology of Reproduction** 35:927–935.

Nyquist, S.E. & Holt, S. (1986). Rat testicular dolichols: distribution and accumulation with age. **Biology of Reproduction** 34:89–95.

Schachter, H. (1986). Biosynthetic controls that determine the branching microheterogeneity of protein- bound oligosaccharides. **Biochemistry and Cell Biology** 64:163–181.

Schwarz, R.T. & Datema, R. (1980). Inhibitors of protein glycosylation. **Trends in Biochemical Sciences** (March) 65–67.

Sullivan, J.L., Smith, F.A. & German, R.H. (1979). Effects of fluoroacetate on the testis of the rat. **Journal of Reproduction and Fertility** 56:210–217.
von Figura, K. & Hasilik, A. (1986). Lysosomal enzymes and their receptors. **Annual Reviews of Biochemistry** 55:167–193.

V Strategies for research

Strategies for research

At the end of the formal presentations, the participants were divided into three groups in order to identify specific areas of research that might yield new directions for the future development of fertility regulation.
The three major areas were:

(1) Golgi, lysosomes and acrosome formation.
(2) Centrioles and formation of cilia and flagellae.
(3) Molecular aspects of spermatid function.

Group 1: Golgi, lysosomes and acrosome formation (Dr A. Tartakoff)

This group discussed four major areas:

- acrosome composition and genesis,
- spermatid plasma membrane structure,
- use of ionophores in understanding spermatid function and possibly as agents for fertility regulation, and
- Sertoli cell-spermatid interactions.

The consensus of opinion was that, compared to general knowledge in the field, relatively little is known about this system in spermatids. Thus, questions concerning the mannose-6-phosphate receptor and its function, the spectrum of acid hydrolases in the acrosome, whether the acrosome contains proteoglycans, O-glycans, or sulfated glycans, whether the acrosomal H^+ pump is similar to other endosomal or lysosomal pumps, and the morphological pathways taken by molecules through the spermatid Golgi to the developing acrosome are all areas of great importance in our understanding of spermatid function about which little is known.

Equally, little is known about the details of the spermatid plasma membrane. Are there novel surface antigens? Are membrane proteins anchored *via* glycophospholipids, similar to Thy-1 and other membrane molecules? What is the relationship between the plasma membrane and developing structures such as the mitochondrial sheath or the fibrous sheath? These questions, among many raised, are basic to understanding spermatid interactions with Sertoli cells as well as their development into sperm. The interesting question was raised of whether Sertoli cell–spermatid interactions involve a cell adhesion molecule (CAM) related to the generalized adhesion molecules that have been described in the nervous system (NCAMs) and elswhere. No information is available on this.

Pharmacologically, the question of the effects of ionophores on reproductive function arises because it has been shown that the carboxylic ionophore monensin causes infertility in rats. Cattle apparently consume large amounts of monensin, but it is not known whether the drug affects their fertility.

An important suggestion from this group was to explore more carefully drugs such as monensin that are known to affect ion balance in cells for their potential antifertility effects.

Finally, this group raised questions about well-known aspects of cells in other systems to show the relative lack of information in spermatids. Thus, what is known about spermatid cytoskeleton (intermediate filaments), about nuclear lamins and their relationship to spermatid differentiation, about residual bodies, etc? It was felt that basic knowledge on these topics is absolutely necessary prior to the design of any contraceptive intervention.

Group 2: Centrioles and the formation of cilia and flagellae (Dr J. Rosenbaum)

This group discussed two general areas: motility and development of the flagellum and the relationships between this and the centriole from which it arises.

Motility. One of the main problems is that any attack on the flagellar axonemal motility mechanisms could interfere with ciliary movements in other tissues (i.e. respiratory epithelium). It was suggested that basic comparative biochemistry be done on sperm flagella and respiratory cilia *in the same mammal*, looking for differences in, e.g., dynein ATPases, radial links, etc., so that these differences could be exploited in the design of either drugs and/or antibodies against specific flagellar motility components. This is a 'basic' research project – it does not take into account that there may be difficulties in 'delivery' of the treatment of the axonemal components, even if a specific component could be identified.

It is clear that the flagellar membrane proteins should be biochemically defined, antibodies made to them, or drugs designed that interact with them, and these agents (drugs/antibodies) delivered to either the male or the female. These membrane protein antibodies/drugs could also be designed to interfere with sperm/egg interactions. Most participants felt that a good place for research emphasis would be on these membrane proteins – principally because of their *accessibility.*

Among the membrane protein systems that should be defined, in addition to those involved in sperm/egg interactions, are those membrane receptors involved in sperm cyclic AMP metabolism. Of importance here is the possibility that the activation of this system (deoxyadenosine) may be *unique to testes*. Since cAMP has been shown in a number of higher organisms to be important in sperm activation, this membrane-activated system requires more research.

The *activation* of the sperm upon ejaculation may involve a membrane Mg^{2+}-ATPase (calcium-calmodulin activated), since sperm appear to require the pumping out of Ca^{2+} in order to be activated to motility. The characterization and purification of this putative calcium pump is of importance to the future design of drugs/antibodies to inhibit, specifically, these membrane proteins.

One should also take advantage of pathological (including genetic) as well as environmental conditions which give rise to immotile sperm without otherwise affecting the individual.

(1) For example, men come to many clinics with immotile sperm, but with normal sperm numbers. In certain cases this may be an *immunological problem*. What are the antibodies against? The serum of these individuals could be screened with Western Blots against whole sperm and fractions thereof, including isolated flagellar membranes to give a lead as to what specific sperm proteins we should be directing our efforts towards. In another area of science, this same approach has opened up the field of centrosomal/kinetochore proteins (i.e. the screening of patients with *schleroderma*) in which human monoclonal antibodies have been identified and effectively used in dissecting the structure of these organelles.

(2) Occasionally, there are man-caused infertilities which may fortuitously give insight into mechanisms of sperm immotility, e.g. in a village in India, pesticide use (deldrin) gave immotile sperm with normal sperm numbers, ostensibly without otherwise affecting the individuals.

Development. Interruption of the development of the sperm tail will ultimately lead to an infertile sperm.

There was discussion of the fact that the post-meiotic, haploid genome is transcriptionally active and that this might be exploited in terms of interceding in sperm development. For example, sperm protamine (for DNA packaging) is one of the proteins synthesized at this time from mRNAs that have been stored for several days. Other abundant proteins whose synthesis should be studied in detail are the outer dense fiber proteins.

The manchette and its role in shaping the nucleus were discussed at some length. Considerable information is known about ultrastructural changes that occur in this organelle during spermatid differentiation, but almost nothing is known about its biochemistry. How do its microtubules function (if they do at all) in nuclear shaping? Are the tubulins of these unique? According to published reports, the post-meiotically synthesized sperm tubulin may be as much as 20% different in amino acid composition when compared to other (e.g., brain) tubulins. Thus, they lend themselves to selective attention as potential points of attack in fertility regulation. Similar approaches should be made on the outer dense fibers and the fibrous sheath.

The next topic discussed by this group was the centriole and its movement during spermatid differentiation to take up a place at the opposite pole of the cell from the developing acrosome. The flagellum will not form if the centriole does not move into position. How does this movement occur? Microtubules? Actomyosin? Kinesin? Can one use currently existing centriolar/centrosomal antibodies to interfere with this movement, or drugs that interfere with cytoskeleton? There was a general discussion on *delivery* to the testis of antibodies against newly defined membrane/axonemal/centriolar proteins. Is it feasible? Fertility clinic work indicated that injections should *not* be discounted. What is difficult is the *subsequent monitoring of antibody titres* – when are booster injections necessary, etc?

Lastly, the group turned to consider the neck of the spermatozoon. What makes it unique? Clearly, the answer to this is tied into the questions about the centriole, but there was general agreement that it was an important topic to delve into further.

The general feeling of this discussion group was similar to the previous group. That is, there are significant gaps in knowledge at the basic level on sperm tail formation, but that it could be a fruitful approach in developing antifertility agents.

Group 3: Aspects of cellular and molecular biology of spermatids (Dr I.B. Fritz)

The following aspects were discussed:

(1) Requirements for practical considerations of potential male contraceptive agents.
- Agents must effectively block one or more stages of germinal cell development, and thereby eliminate the capacity for fertilization.
- The effects must be reversible.
- The usage of the agent must be acceptable to society, and permit ready compliance by individuals receiving the agent.

(2) Caveats concerning above.
- No 'quick fix' is likely, and assignments of priority to areas of research to be proposed must therefore be considered arbitrary.
- More fundamental knowledge concerning processes of germinal cell development and factors governing germinal cell development is needed in nearly every area. Only with this information can truly effective means be found to satisfy all criteria listed above.
- However, it is appropriate to recommend utilization of existing knowledge and technology, intelligently applied, while searching for deeper knowledge. It is in this sense that potential target areas are recommended below.

Potential areas for exploration are deemed to be of biological interest at a fundamental level and to offer potential short-term gains.

(1) Sertoli cells and intact cytoarchitecture of the seminiferous tubule are required for germinal cell development.
   - Target specific inhibitors to Sertoli cells by utilizing covalent binding to an FSH analogue which binds specifically and uniquely to the FSH-receptor.

(2) Other techniques can be used to disturb function of spermatogonia in the basal compartment, although there is a problem of reversibility if there is damage to the seminiferous tubule barrier (irreversible orchitis?) as well as the problem of stem cell deficiency if spermatogonia are destroyed.
   - Temperature sensitivy. What is (are) the most vulnerable stage(s) of germ cell development?
   - Determine reversibility: primates.

(3) Progression of spermatozoa from testis through rete testis to epididymis.
   - Basic processes of spermiation: premature release?
   - Transport through the rete: role of ions, specific proteins? Fundamental functions of rete testis epithelial cells.
   - Maturation during epididymal transit. Inhibition of epididymal epithelial cells: local anti-androgens – similar to disruption of the androgen receptor locally or to occupancy of the receptor by a 'toxic', competitive analogue. Identify which of the proteins synthesized by epididymal epithelial cells and become bound to sperm are crucial for 'competence'.

(4) Ejaculation
   - Sperm-bound poisons, targeted specifically (e.g. alpha-chlorhydrin type compound never allowed to reach brain, etc.)

Although some favoured a conservative and non-interventional approach such as temperature elevation in the scrotum (if this should prove reversible), the group was divided on this. Interventions at the Sertoli cell–FSH level were generally regarded with considerable wariness, in contrast to the views of Group 2. However, the group was unanimous in the feeling that there is no substitute for gaining a thorough and fundamental understanding of processes involved. 'Nature is never embarassed by the difficulties of the analysis.'

List of invited participants

*Dr R.J. Aitken	MRC Reproductive Biology Unit, Centre for Reproductive Biology, Edinburgh, Scotland, UK.
Dr A.R. Bellvé	College of Physicians and Surgeons, Columbia University, New York, USA.
*Dr G. Bialy	National Institute of Child Health and Human Development, Contraceptive Development Branch, Bethesda, Maryland, USA.
*Dr J.A. Blaquier	Instituto de Biologia y Medicina Experimental, Laboratorio de Esteroides, Buenos Aires, Argentina.
Dr A.P. Boulton	Courtauld Institute of Biochemistry, The Middlesex Hospital Medical School, London, UK.
Dr J.A. Bermudez	Instituto Nacional de la Nutricion, Salvador Zubiran, Mexico D.F., Mexico.
*Dr W.J. Bremner	Veterans Administration Medical Center, Endocrinology/Metabolism, Seattle, Washington, USA. (unable to attend)
Dr L. Burzio	Instituto de Bioquímica, Facultad de Ciencias, Universidad Austral de Chile, Valdivia, Chile. (unable to attend)
Dr M.A. Cerbon	Instituto Nacional de la Nutricion, Salvador Zubiran, Mexico D.F., Mexico.
Dr Y. Clermont	McGill University, Department of Anatomy, Montreal, Quebec, Canada.
Dr M. Chulavatnatol	Mahidol University, Department of Biochemistry, Bangkok, Thailand.
*Dr P. Crabbé	International Organization for Chemical Sciences in Development, Coyoacan, Mexico D.F., Mexico.
Dr V. Diaz-Sanchez	Instituto Nacional de la Nutricion, Salvador Zubiran, Mexico D.F., Mexico.
Dr R. Docampo	Instituto de Microbiologia, Universidade Federal de Rio de Janeiro, Rio de Janeiro, Brazil.
Dr J.M. Edwardson	University of Cambridge, Department of Pharmacology, Cambridge, UK.
Dr J. Garza Flores	Instituto Nacional de la Nutricion, Salvador Zubiran, Mexico D.F., Mexico.
Dr I.B. Fritz	University of Toronto, Banting and Best Department of Medical Research, Toronto, Ontario, Canada.
*Dr J.A. Grootegoed	Erasmus Universiteit Rotterdam, Department of Biochemistry, Rotterdam, Netherlands.
Dr D.W. Hamilton	University of Minnesota, Department of Cell Biology and Neuroanatomy, Minneapolis, Minnesota, USA. (Convenor)
Dr A. Hasilik	Physiologisches-Chemisches Institut der Universität Münster, Münster, Federal Republic of Germany.

Dr N.B. Hecht	Tufts University, Department of Biology, Medford, Massachusetts, USA.
*Dr M. Kapur	All India Institute of Medical Sciences, Department of Surgery, New Delhi, India. (unable to attend)
*Dr A.G. Khomassuridze	Human Reproduction Institute, Public Health Ministry of Georgian SSR, Tbilisi, USSR.
Dr R. Kuriyama	University of Minnesota, Department of Cell Biology and Neuroanatomy, Minneapolis, Minnesota, USA.
Dr F. Larrea	Instituto Nacional de la Nutricion, Salvador Zubiran, Mexico D.F., Mexico.
Dr R.W. Linck	University of Minnesota, Department of Cell Biology and Neuroanatomy, Minneapolis, Minnesota, USA.
Dr S. Majumdar	Postgraduate Institute of Medical Education and Research, Department of Experimental Medicine, Chandigarh, India.
*Dr S.A. Matlin	The City University, Department of Chemistry, London, UK.
*Dr E. Nieschlag	Max Planck Clinical Research Unit for Reproductive Medicine, Münster, Federal Republic of Germany.
Dr M. Parvinen	University of Turku, Institute of Biomedicine, Department of Anatomy, Turku, Finland.
Dr G. Perez-Palacios	Instituto Nacional de la Nutricion, Salvador Zubiran, Mexico D.F., Mexico.
Dr. D.M. Phillips	The Population Council, New York, USA.
*Dr Qian Shao-zhen	Jiangsu Family Planning Institute, Department of Basic Research, Nanjing, People's Republic of China.
Dr E. Raff	Indiana University, Biology Department, Bloomington, Indiana, USA.
Dr J. Rosenbaum	Yale University, Department of Biology, New Haven, Connecticut, USA.
*Dr R.S. Swerdloff	UCLA School of Medicine, Harbor-UCLA Medical Center, Torrance, California, USA.
Dr A. Tartakoff	Case Western Reserve University, School of Medicine, Cleveland, Ohio, USA.
Dr A. Ulloa-Aguirre	Instituto Nacional de la Nutricion, Salvador Zubiran, Mexico D.F., Mexico.
*Dr C. Wang	University of Hong Kong, Department of Medicine, Queen Mary Hospital, Hong Kong.
Dr Wang Yi-fei	Shanghai No. 2 Medical College, Shanghai, People's Republic of China.
Dr L. Wilson	University of California, Department of Biological Sciences, Santa Barbara, California, USA.

FHI Representative

Dr R. Foldesy	Family Health International, Research Triangle Park, North Carolina, USA.

WHO Representatives

Dr E. Diczfalusy	Karolinska Institute, Reproductive Endocrinology Research Unit, Stockholm, Sweden.
Dr G.M.H. Waites	World Health Organization, Special Programme of Research, Development and Research Training in Human Reproduction, Geneva, Switzerland. (Convenor)

WHO Research Trainees

Dr O. Antunez-Flores	Family Planning Clinic and Department of Reproductive Biology, Mexico City General Hospital, Mexico.
Dr S. Berrios	University of Chile, Faculty of Medicine, Santiago, Chile.
Dr D. de la Cruz Aguilera	Research and Training Centre in Human Reproductive Biology, University Hospital, Torreon, Mexico.
Dr J. Recio Vega	Research and Training Centre in Human Reproductive Biology, University Hospital, Torreon, Mexico.
Dr L. Hernandez-Richards	St Thomas' Hospital, Ministry of Health, Panama.
Dr Wang Gengxin	Hebei Medical College, Department of Histology and Embryology, Shijiazhuang, People's Republic of China.
Dr Xu Wenhao	Shanghai No. 2 Medical College, Shanghai, People's Republic of China.

*Denotes members of the Steering Committee of the WHO Task Force on Methods for the Regulation of Male Fertility.

Index